AF736137

HIGH PERFORMANCE FASTENERS & PLUMBING

A Guide to Nuts, Bolts, Fuel, Brake, Oil and Coolant Lines, Hoses, Clamps, Racing Hardware and Plumbing Techniques

Mike Mavrigian

HPBOOKS

HPBooks
Published by the Penguin Group
Penguin Group (USA) Inc.
375 Hudson Street, New York, New York 10014, USA
Penguin Group (Canada), 90 Eglinton Avenue East, Suite 700, Toronto, Ontario M4P 2Y3, Canada
(a division of Pearson Penguin Canada Inc.)
Penguin Books Ltd., 80 Strand, London WC2R 0RL, England
Penguin Group Ireland, 25 St. Stephen's Green, Dublin 2, Ireland (a division of Penguin Books Ltd.)
Penguin Group (Australia), 250 Camberwell Road, Camberwell, Victoria 3124, Australia
(a division of Pearson Australia Group Pty. Ltd.)
Penguin Books India Pvt. Ltd., 11 Community Centre, Panchsheel Park, New Delhi—110 017, India
Penguin Group (NZ), 67 Apollo Drive, Rosedale, North Shore 0632, New Zealand
(a division of Pearson New Zealand Ltd.)
Penguin Books (South Africa) (Pty.) Ltd., 24 Sturdee Avenue, Rosebank, Johannesburg 2196, South Africa

Penguin Books Ltd., Registered Offices: 80 Strand, London WC2R 0RL, England

While the author has made every effort to provide accurate telephone numbers and Internet addresses at the time of publication, neither the publisher nor the author assumes any responsibility for errors, or for changes that occur after publication. Further, publisher does not have any control over and does not assume any responsibility for author or third-party websites or their content.

HIGH PERFORMANCE FASTENERS & PLUMBING

Copyright © 2008 by Mike Mavrigian
Cover design by Tresa Rowe/Bird Studios
Cover photos by Mike Mavrigian
Editing, design and production by Michael Lutfy

All rights reserved.
No part of this book may be reproduced, scanned, or distributed in any printed or electronic form without permission. Please do not participate in or encourage piracy of copyrighted materials in violation of the author's rights. Purchase only authorized editions.
HPBooks is a trademark of Penguin Group (USA) Inc.

First edition: January 2008

ISBN: 978-1-55788-523-4

PRINTED IN THE UNITED STATES OF AMERICA

10 9 8 7 6 5 4 3 2 1

NOTICE: The information in this book is true and complete to the best of our knowledge. All recommendations on parts and procedures are made without any guarantees on the part of the author or the publisher. Tampering with, altering, modifying, or removing any emissions-control device is a violation of federal law. Author and publisher disclaim all liability incurred in connection with the use of this information. We recognize that some words, engine names, model names and designations mentioned in this book are the property of the trademark holder and are used for identification purposes only.

CONTENTS

ACKNOWLEDGMENTS

Thanks to the following for their assistance in compiling some of the information included in this book.

Aeroquip, ARP, Aurora Bearings, Aviaid Oil Systems, Barnes Dry Sump Oil Systems, Breeze Industrial Products, Clampco, Earl's Performance Plumbing, Emhart Teknologies, Evans Cooling, Bob Fall, Fall Automotive Machine, Fine Lines, Gearhead Tools, Goodridge, Goodson Shop Supplies, Hanson Rivets, Holley Performance Products, Jiffy-Tite, Koul Tools, Lock-N-Stitch, Marson, Maryland Metrics, McMaster-Carr, Meziere Enterprises, Moroso Performance Products, Peterson Fluid Systems, Seals-It, Speedway Engineering, Russell, Snap-On Tools, Textron Fastening Systems and XRP.

INTRODUCTION

This book addresses both fasteners and plumbing as they relate to automotive performance applications, in an easy-to-understand format. Subjects include threaded fasteners (bolts, nuts, etc.), understanding different styles of bolts/screws, nuts and washers, how to measure different styles of fasteners, and more. Also included are discussions on fastener torque, thread repair, a special chapter that focuses on wheel fasteners, spherical bearing rod ends, explanations of pipe thread, the use of studs, how to use thread taps and dies, and more. The plumbing section covers the popular –AN stainless braided hose assemblies (understanding –AN sizes and how to assemble your own hoses), as well as tips on oil, coolant, fuel and brake fluid plumbing, fluid tubing and tube assembly tips, and much, much more. We've also included helpful source guides for both fastener and plumbing products. We think you'll find this book helpful and useful for all of your street and race projects.

—Mike Mavrigian

Chapter 1
THREADED FASTENERS

THREADED FASTENERS

In terms of automotive applications, we encounter two basic types of threaded fasteners: straight thread (straight thread bolts, nuts and studs in both fractional and metric formats) and sheet metal screws.

In automotive applications, we commonly deal with fractional-inch or metric straight-thread bolts, nuts and studs, which usually range in size from 1/4" diameter to 1/2" diameter (fractional), and 6mm through 14mm (metric). In addition, we often deal with fasteners that are referred to with the generic term "machine screws." Machine screws feature straight threads, and are smaller in diameter than 1/4". Common sizes include #4, #6, #8, #10 and #12. Machine screws are commonly used for street rod or race car fabrication jobs, for mounting a variety of components that require smaller-diameter fasteners. Don't let the term confuse you. Machine screws feature straight threads and are designed to be mated to either nuts or tapped holes (unlike sheet metal screws which bite into a parent material). Don't let the term "screw" confuse you. All threaded bolts can also be referred to as screws.

I've included more detailed information regarding machine screws later in this chapter.

Most inch-format fasteners are identified directly by their diameter measured in fractional inches. For instance, a 1/4" bolt features approximately 1/4" diameter thread (measured at the bolt shank's major diameter...at the thread peaks). Actual diameters (if you were to measure with a dial caliper or a mic) will usually be a few thousandths of an inch less, depending on the tolerance of the particular bolt. For instance, a 1/4" bolt might actually feature threads that measure .245" instead of .250".

Fractional-format fasteners are always referred to in inch fractional numbers (1/4", 5/16", 3/8", 7/16", 1/2", 9/16", 5/8", 11/16", 3/4", 7/8", 1", etc.). Metric thread fasteners are referred to by their diameter in millimeters (4mm, 5mm, 6mm, 7mm, 8mm, 9mm, 10mm, etc.).

Instead of blindly scavenging through coffee cans of miscellaneous bolts and nuts in an attempt to match a bolt or nut that you're holding in your hand, take the time to measure and identify with accuracy. Once you understand how to properly measure a threaded fastener, you'll quickly find that doing it the right way is easier and faster, and definitely less frustrating. After all, if you're building an engine, you wouldn't sort through a box of rod bearings and decide to use a bearing shell size that "appears" to fit. The same holds true for threaded fasteners. Being able to identify and understand fastener sizing is critical.

First of all, let's clarify bolt or nut diameter. When we refer to a threaded fastener's diameter, we're talking about the *thread diameter*. This has absolutely nothing to do with the size of the drive head. I've seen so many mistakes in various

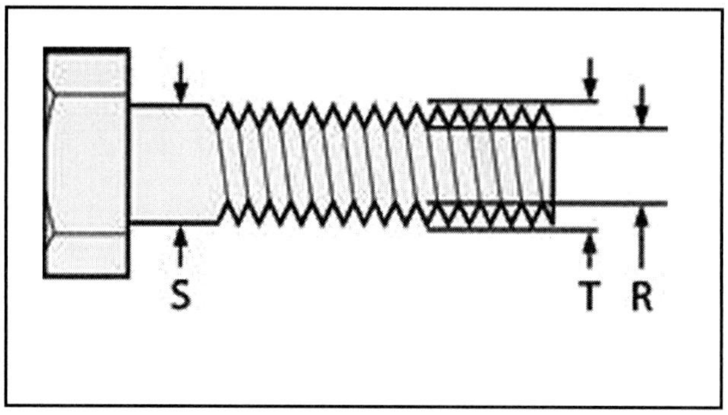

S: Shank diameter
T: Major diameter
R: Root, or minor diameter

publications over the years that identify bolts based on the size of the wrench required for service. Just because a certain hex-headed bolt requires a 9/16" wrench, this does not mean that it's a 9/16" bolt! In most cases, as an example, if the bolt requires a 9/16" wrench, we know from experience that the bolt's thread diameter is probably 3/8". Again...fastener size refers to the thread diameter and has absolutely nothing to do with the size of the wrench needed. If you accept this premise, you'll avoid gobs of frustration.

FASTENER SIZING

All bolts, screws, studs and nuts are identified for size based on thread diameter and thread pitch. In the case of bolts, studs or screws, length is also part of the sizing label.

The first number always represents thread diameter. The second number always represents thread pitch. In the case of male threaded fasteners such as bolts, screws and studs, the third number represents length. Following are three examples of male threaded fastener identification.

Fractional Inch Screw

Example: 1/4 x 20 x 1 1/2. This refers to a bolt or screw that features a 1/4" diameter thread, a thread pitch of 20 and a total shank length of 1 1/2".

Fractional Stud

Example: 1/4 x 20 x 2. This refers to a threaded stud that features a 1/4" thread diameter, a thread pitch of 20 and an overall length of 2".

Metric Screw

Example: 10 x 1.5 x 30. This refers to a bolt or screw that features a 10mm-diameter thread, a thread pitch spacing of 1.5mm, and a total shank length of 30mm.

Diameter

Bolt, screw and stud diameters refer to just that...the outer diameter (also called the major diameter) of the thread. With inch fractional thread, the labels can seem somewhat misleading, since if you actually measure thread diameter with a caliper, you'll generally find that the measured diameter is a few thousandths under what you'd expect. For instance, measuring a 1/4" bolt (where we know that 1/4 = 0.250), you'll usually find that the diameter actually measures 0.243—0.246". A 3/8" bolt may actually measure 0.370" instead of 0.375". A small fudge factor of tolerance is designed in to assure fit.

Thread diameter (the diameter by which you will identify the fastener's diameter size) is also called the major diameter. This represents the distance from thread peak to opposing thread peak (the highest surface of the thread).

The minor diameter (also called the root diameter) is measured from the thread valleys (the deepest area of the threads.

Thread Pitch

Pitch is often misunderstood, but it's easy to grasp once you learn the basics. Thread pitch refers to the thread count or thread spacing (in layman's terms, whether the thread is "fine" or "coarse"). It's important to also understand that inch-fractional thread pitch and metric thread pitch are measured differently.

Inch-fractional thread pitch represents the number of threads that would exist along a one-inch length of thread; while metric thread pitch represents the distance between adjacent threads, measured in millimeters.

Fractional Thread Pitch—Let's say we have a bolt that's labeled as 5/16 x 18 x 3/4. This means that we have a bolt that features a 5/16" thread diameter, a thread pitch of 18 and a shank length of 3/4".

The thread pitch, in this case, 18, indicates that the threaded section of the bolt features a thread spacing that would allow 18 threads within a one-inch length. It doesn't matter how long the shank really is....the point is that the threads are spaced so that a one-inch length would include 18 threads.

In fractional fasteners, the more threads that can exist along a one inch length, the "finer" the thread pitch is. As a result, the greater the pitch number, the finer the threads. Here are a few examples:

Thread Dia.	Pitch
1/4"	20 (coarse)
1/4"	28 (fine)
5/16"	18 (coarse)
5/16"	24 (fine)
3/8"	16 (coarse)
3/8"	24 (fine)

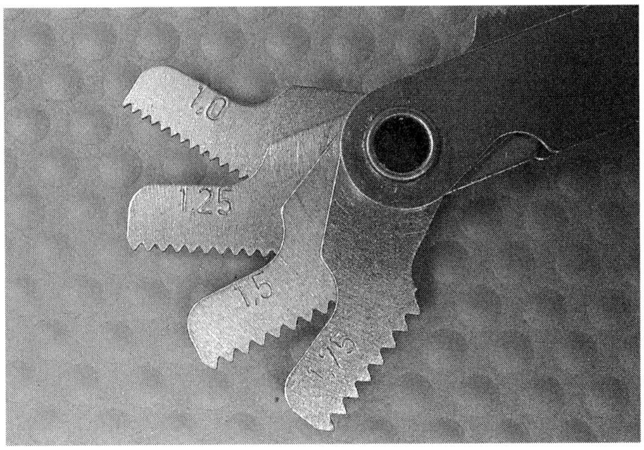

Shown here is a metric pitch gauge. Each blade is clearly marked for thread pitch. A selection of pitch blades will be included, covering every metric thread pitch that you'll ever encounter.

As you can see, the coarse pitch numbers in each case are lower than the same-diameter's fine thread count.

Inch lower pitch number = coarse thread
Inch higher pitch number = fine thread

In the inch world, we have two distinct thread pitch groups (since there are usually two pitches offered for each given thread diameter size). In inch fractional terms, we have coarse thread and we have fine thread. **Note**: "fine" thread may also be referred to as "SAE" thread, and coarse thread may be referred to as "USS" thread.

Metric Thread Pitch

Let's say we have a bolt that's labeled 8 x 1.25 x 25. This means that we have a bolt that features an 8mm thread diameter, a thread pitch of 1.25 and a shank length of 25mm.

The thread pitch, in this case, 1.25, indicates that two adjacent threads are spaced 1.25mm apart (the distance from one thread to the next thread). Since we're measuring the distance between threads on metric fasteners, the less the distance, the more threads exist along the length. The greater the distance between threads, the fewer threads exist along the length. As a result, metric pitch numbers graduate opposite those of inch fractional pitch numbers. Here are some examples:

Thread Dia.	Pitch
10mm	1.50
10mm	1.25

Since the 10mm bolt that features a pitch of 1.50 has a thread-to-thread spacing of 1.50mm; and the 10mm bolt with a 1.25 pitch has a thread-to-thread spacing of only 1.25mm, the 1.25 pitch is considered "finer" than the 1.50 pitch.

Shown here are two pitch gauge sets. The top set is for measuring fractional thread pitch, and the lower set is for metric thread pitch.

Here we check thread pitch on a bolt using a pitch gauge. This is a 1/2"-diameter bolt. Using our set of fractional pitch gauges, we find a perfect pitch match with the 20-pitch blade, confirming that this bolt features a 20 pitch, which, when dealing with a 1/2" fastener, is commonly referred to as a "fine" pitch (a 1/2" coarse thread would have a pitch of 13). Remember: when dealing with fractional sizes, the pitch number refers to the number of threads found along a 1" length of threaded area. With metric fasteners, the pitch number refers to the distance between threads. For example, a metric bolt with a pitch of 1.25 means that the distance from one thread peak to the next adjacent peak is 1.25mm.

Metric lower pitch number: fine thread
Metric higher pitch number: coarse thread

In the metric world, the distinction between coarse and fine isn't as clear-cut. Depending on thread diameter, certain pitches are considered as standard, while other pitches may be considered coarse, fine or extra-fine.

My suggestion is to stop using the terms coarse and fine altogether, because these are vague terms at best, whether you're dealing with inch or metric stuff. Instead, get into the habit of using the actual dimension nomenclature to accurately identify any threaded fastener. After all, isn't the label of 3/8 x 24 x 1 1/4 much more precise as opposed to asking for a 3/8" course bolt that's a bit longer than an inch long? If you use precise dimensional terms, you'll eliminate confusion, second-guessing and wasted trips to the parts store.

Measuring Diameter

It's important to note that there is a difference between theoretical thread diameter and actual thread diameter. Thread diameter will always be slightly less than labeled, or theoretical, diameter, due to manufacturing tolerances and because the thread peaks on a machine screw, bolt or stud must be slightly smaller than the theoretical diameter in order to allow a fitment tolerance. Listed on page 4 are examples of measured thread diameters of bolts

METRIC THREAD PITCH TABLE

Shows thread pitch available in various thread diameters, and indicates the general use terms of standard, fine and extra-fine thread pitch. Thread pitch indicates distance, in mm, between threads.

Bolt Dia.	Standard	Fine	Superfine
2	0.4	—	—
2.5	0.45	—	—
3	0.5	—	—
4	0.7	—	—
5	0.8	—	—
6	1.0	—	—
7	1.0	—	—
8	1.25	1.0	—
10	1.5	1.25	1.0
12	1.75	1.5	1.25
14	2.0	1.5	—
16	2.0	—	—
18	2.5	—	—
20	2.5	—	—

FRACTIONAL/INCH THREADS PITCH TABLE

Shows thread pitch and common coarse/fine definition. Thread pitch indicates number of threads along a 1-inch length.

Bolt Dia..	Coarse UNC	Fine UNF
#2	56	—
#3	48	—
#4	40	—
#5	40	—
#6	32	—
#8	32	—
#10	24	32
#12	24	—
1/4	20	28
5/16	18	24
3/8	16	24
7/16	14	20
1/2	13	20
9/16	12	18
5/8	11	18
3/4	10	16
7/8	9	14
1	8	12 or 14
1 1/8	7	12
1 1/4	7	12
1 1/2	6	12

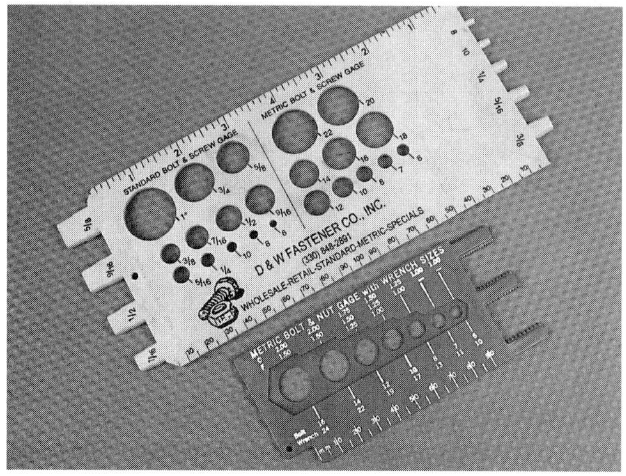

Bolt sizing cards are readily available. Shown here is a yellow card that handles both fractional and metric sizes, and a green card dedicated to metric sizes.

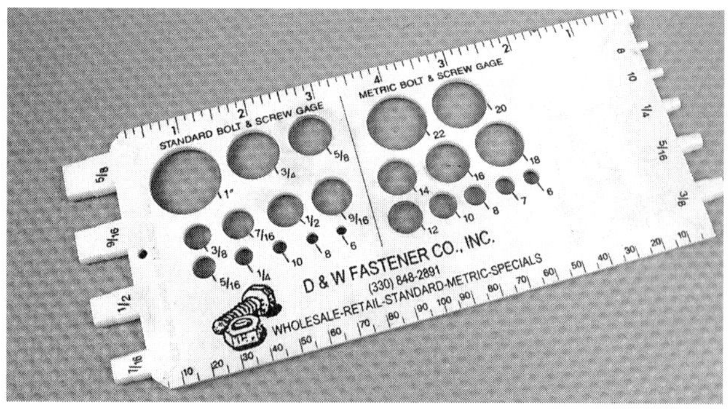

This sizing card handles both fractional and metric sizes. Simply insert a bolt into the hole that most closely fits in order to identify bolt diameter. This card also features incremented scales (rulers) to allow measuring bolt shank length, as well as sizing studs to determine nut diameters.

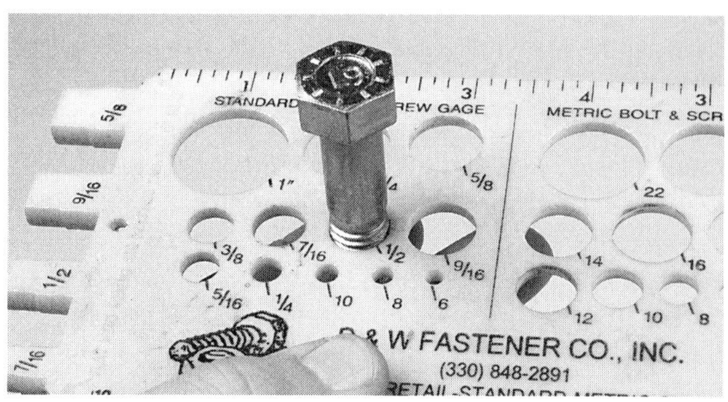

Identifying a 1/2" diameter bolt using the sizing card.

(actual measurements taken using a calibrated micrometer, measuring the bolt at the thread peaks).

Bear in mind that these actual diameters shown here represent an average measurement, since variances of a few thousandths of an inch will be seen depending on the fastener manufacturer.

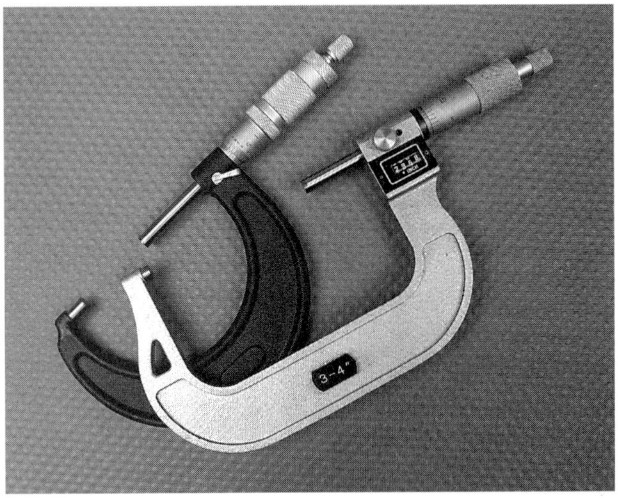

Micrometers may also be used to measure thread diameter as well as shank length, if you don't have a sizing card handy.

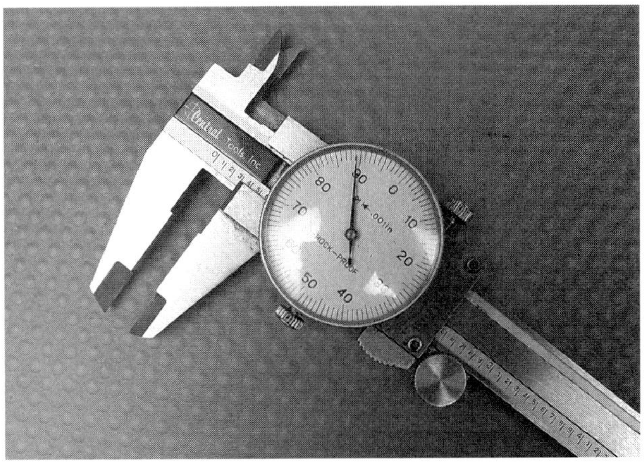

A caliper (vernier, digital or dial as shown here) may also be used to measure bolt diameter and shank length. While you may or may not have these precision measuring instruments in your tool inventory, picking up an inexpensive sizing card makes the job easier and faster.

LENGTH

The majority of bolts and screws are always measured for length by measuring the distance between the underhead and the tip of the threads (in other words, total length minus the head). This applies to any hex headed cap screw, socket head cap screw, button head, pan head, carriage bolt, truss head, etc. (any screw or bolt that features a flat underhead).

However, if the bolt or screw is the flat-top design (where the top is flat and the underhead is chamfered for a countersunk fit), length always denotes overall length, from top of the head to the tip of the threads. Stud length refers to overall length, from tip to tip.

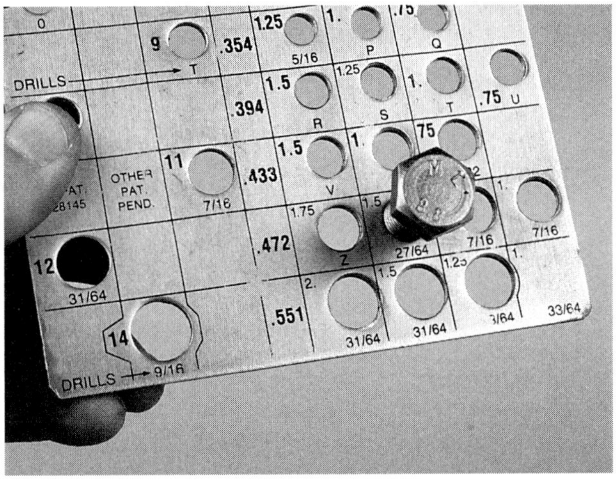

This aluminum sizing card features smooth holes for diameter measuring, along with handy threaded holes to check thread pitch. Shown here is a metric thread sizing card. This is the easy way, eliminating the need for a pitch gauge. You still need pitch gauges in your tool chest, but this type of card can prove handy. We grabbed a metric bolt to test. We found that it was a perfect fit in the 12mm hole (see the left side of the card). We then tried to thread it into several of the 12mm holes, to find that it threaded smoothly into the 1.5 pitch hole. So, we know that this bolt is 12mm x 1.5. After you deal with thread fasteners long enough, you'll be able to visually identify diameter and pitch without the need for tools. But it's always nice to have a few cards lying around to verify, or to help you out on those days when you're a little bleary-eyed.

FRACTIONAL BOLTS/STUDS

Labeled Diameter	Measured Diameter	Theoretical Diameter
1/4"	0.245"	0.2500"
5/16"	0.310"	0.3125"
3/8"	0.370"	0.3750"
7/16"	0.430"	0.4375"
1/2"	0.495"	0.5000"

METRIC BOLTS/STUDS

Labeled Diameter (inches)	Measured Diameter
5mm	0.1969"
6mm	0.2362"
8mm	0.3150"
10mm	0.3937"
12mm	0.4724"
14mm	0.5512"

We decided to show measured diameter in inches, because it's easier for most people in the U.S. to measure in this manner, especially given that most automotive enthusiasts/builders own calipers and micrometers that are formatted for inches.

METHODS OF MEASURING BOLTS

Shank diameter can be checked in one of two ways: to actually measure the shank diameter with a caliper or micrometer, or to use a bolt sizing card. A bolt sizing card features a series of holes, individually labeled for diameter. Simply insert the bolt shank through the holes until you achieve a close-tolerance fit and read the size marking for that hole. That's the quick "cheater" way of identifying diameter.

Thread pitch is measured using a pitch gauge. Two pitch gauges should be obtained (one for metric pitch and one for inch pitch). Simply match

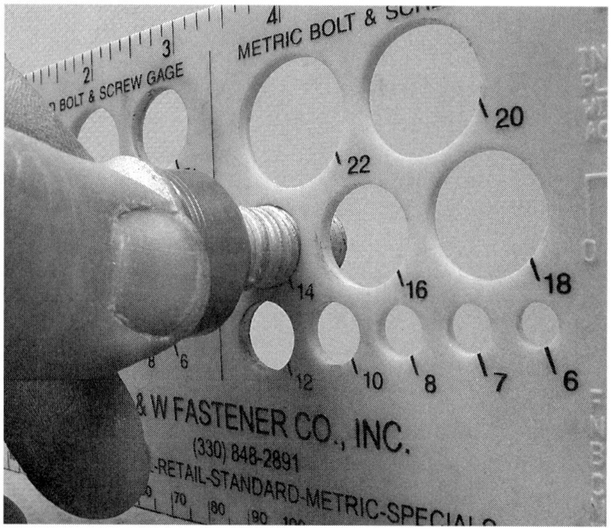

Here, we're checking thread diameter of a Porsche wheel bolt. Simply insert the bolt into holes until you locate the one that most exactly fits. As you can see here, the 14mm hole provides an exact fit for our wheel bolt. Sizing cards are easy to use and rapidly clear up any confusion that you might have had regarding thread diameter.

the bolt's threads to the toothed gauge blades until an exact profile match is found. Each pitch gauge is marked with its pitch number. Or, you can compare a questioned bolt to a bolt of a known pitch by attempting to mesh the threads together. Whether you're using a pitch gauge or a known-size bolt, the thread mesh must be a perfect fit. "Close" isn't close enough. If the threads don't match perfectly, it isn't a match.

Shank length is measured from the underside of the bolt head to the shank tip. A caliper or straight ruler can be used for this. Most bolt sizing cards feature two edge rulers (one in inch format and one in metric format).

Hex Head Bolt Length: Measured from the underside of the head to the tip of the threaded shank.

12-Point Bolt Length: Measured from the underside of the head to the tip of the threaded shank.

Button Head Bolt Length: Measured from the underside of the head to the tip of the threaded shank.

Sizing cards often include nut-sizing posts. This card offers nut posts, clearly labeled, for both inch and metric nuts.

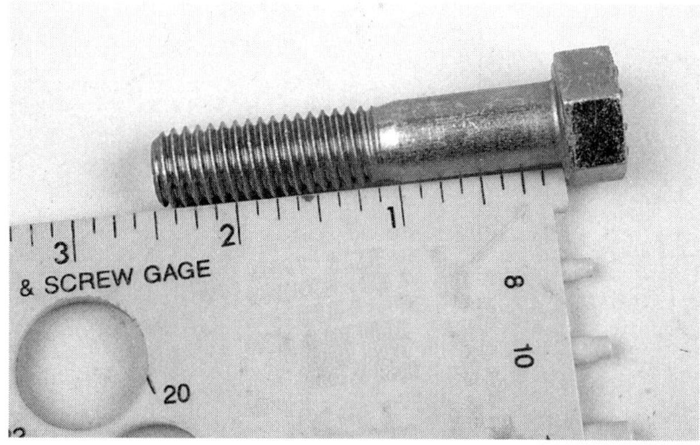

Using the sizing card, this hex head bolt measures 2 1/2" in length. Hex head bolt length is always measured from the underside of the bolt head to the threaded tip of the shank.

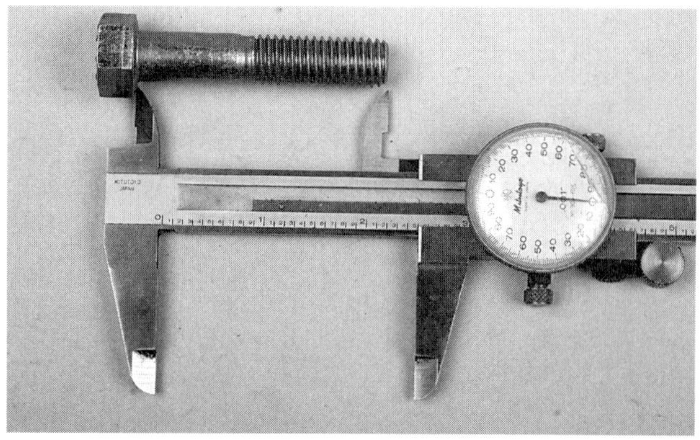

Here we measure shank length of a hex headed bolt using a dial caliper. Never include the head height when identifying a hex head bolt for length.

Cap Screw Length: Measure from the underside of the head to the tip of the threaded shank.

Flat-Top Screw Length: Measure overall length, from the top of the head to the tip of the threaded shank.

Stud Length: Measure overall length.

MACHINE SCREWS

Machine screw size designations indicate both thread diameter and thread pitch. The diameter number represents the diameter of the thread. In the case of machine screws, a code designation is used that translates into thread diameter (machine screw diameters are referred to by single or double-digit numbers...#4, #6, #8, #12, etc.). The pitch number is based on the fractional format. For instance, a 32 pitch features 32 threads per inch of shank length.

Machine screws are available in a variety of materials and head styles. Materials may include nylon, brass, steel and stainless steel. Stainless is a popular choice, especially for street rod applications

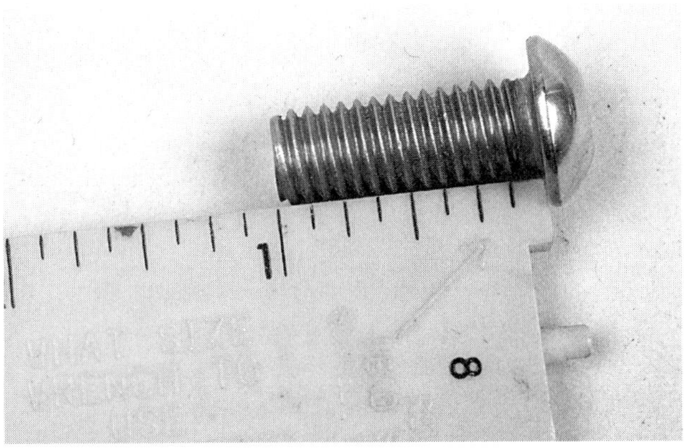

Button head screws are also measured for length from the underside of the head. Here we identify a 1" long button head screw.

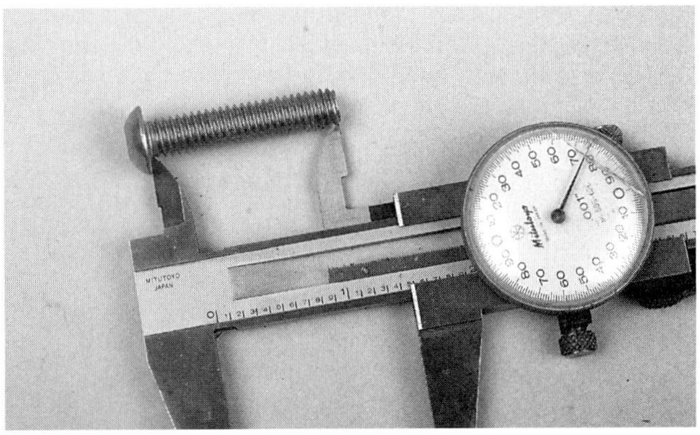

Here we measure the length of a button head screw using a dial caliper. This length, while actually measuring 1.475", is boxed/labeled as 1 1/2" in length. Due to manufacturer tolerances, length won't always be exactly as advertised. This screw length is 0.025" shy of the advertised length, which isn't uncommon.

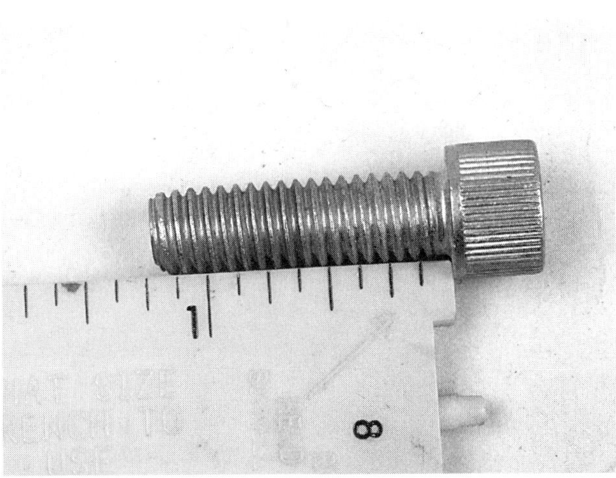

A cap screw length checked on the sizing card. Sizing cards offer ruled edges (one in inches and one in millimeters) in addition to providing holes to check diameter. This cap screw is advertised as being 1 1/4" long. We can see here that it's actually a few thousandths shorter, but that's no big deal.

MACHINE SCREW SIZES

(These are common sizes/pitches. Smaller sizes exist, but generally have no automotive application)

Diameter Designation	Thread Pitch Designation	Actual Thread Diameter Standard (Inches)
#4	40	0.112"
#6	32	0.138"
#8	32	0.164"
#10	24, 32	0.190"
#12	24	0.216"
Metric (in mm)		
M1	0.25	1.0
M1.2	0.25	1.2
M1.4	0.30	1.4
M1.6	0.35	1.6
M2	0.40	2.0
M2.5	0.45	2.5
M2.6	0.45	2.6
M3	0.50	3.0
M3.5	0.60	3.5
M4	0.70	4.0
M5	0.80	5.0
M6	1.0	6.0

(since stainless heads are easily polished and will resist corrosion).

Common head styles may include button head, flat-top, Phillips, slotted, pan head, socket head cap screw. These small-diameter machine screws are ideal choices for mounting dash plates, small adel or line clamps, etc., due to their conveniently small size and appearance.

Instead of referring to theoretical thread diameter, the "actual thread diameters" listed in the chart are measurements that you would note if you actually measured the diameter, at the thread peaks, of the specific size machine screw. This will help you to identify machine screw diameter if you had one in your hand but were unsure about its size.

As far as the number size designation of machine screws is concerned, to be honest, I'm not sure where these numbers were originated. However, the diameter identification numbers (#4, #6, #8, #10) are also used in sheet metal screw sizes.

I've included information regarding metric machine screws for those who might be performing a restoration on vintage import vehicles. As far as common availability is concerned (at your local hardware stores or fastener supply houses), it's likely that you'll only have access to inch-format machine

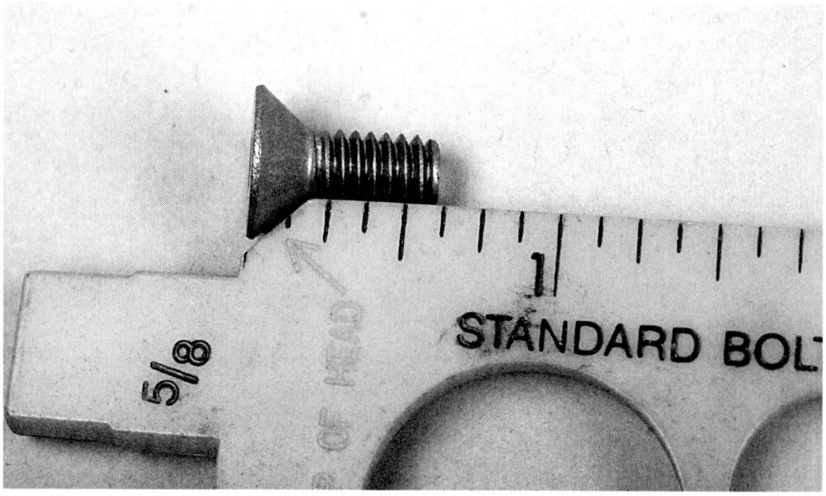

Flat-top screws are measured for length by measuring overall length, from the top of the flat to the tip of the threaded shank. Unlike hex head, cap screw or button head bolts, we do include head height when determining length whenever measuring or ordering any flat-top screw. This flat-top screw is advertised as 5/8" long. Remember...it's not the just the thread that we measure when identifying a bolt or screw for length....we're measuring the total "depth" length that will insert into the material or component. Since a flat-top's chamfered underhead inserts past the surface of a parent object, we must consider the overall length of a flat-top screw. Note that a good sizing card (like the one we're using here) will feature a chamfered end of the rule. This angle-cut end is designed for measuring flat-top screws. Nestle the screw head's chamfer against the angle on the card to measure screw length.

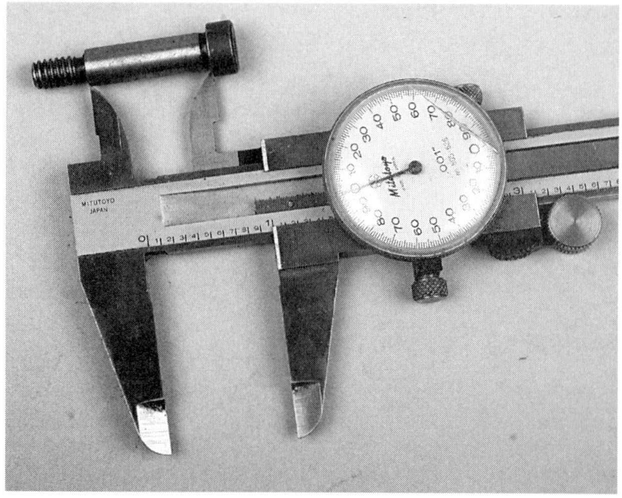

Shoulder screws feature a bare shank of a specified diameter, terminating with a threaded shank of a specific diameter and length. This example features a 1" shoulder shank length with 5/16" diameter, accompanied by a short 1/4" diameter threaded tip.

screws in sizes #4, 6, 8, 10 and 12. Thread pitch is measured the same as with larger diameter fasteners. Inch-format pitch refers to the number of threads along a 1-inch length; while metric pitch numbers refer to the distance (in millimeters) between threads.

BOLT/SCREW HEAD STYLES

A wide variety of head styles exist (the array is amazingly broad, with styles you've probably never seen nor will ever encounter). Here we'll discuss those head styles most routinely used in performance automotive applications.

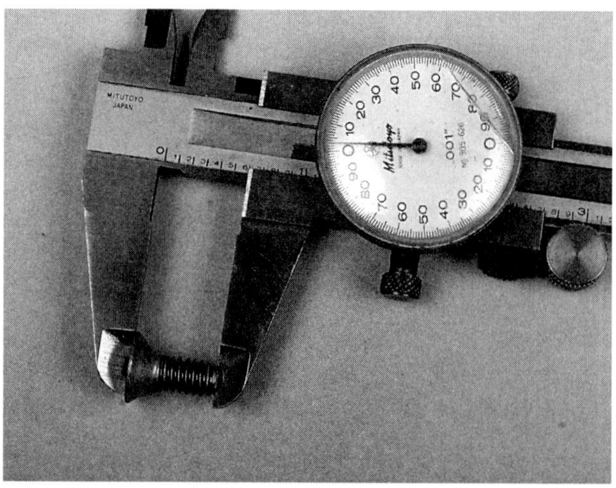

Here we measure a flat-top screw using a dial caliper. Again, always measure any flat-top screw for overall length.

Hex head bolts feature a drive head that features six flat sides. This is the most common style used throughout the world, and the style you're most familiar with. This requires the use of a hex wrench (although 12-point wrenches will fit as well).

12-point bolts feature a dozen engagement points around the circumference of the head. This style offers a more precise wrench fit, and is often preferred for applications that require precision tightening and repeated on-off use. A 12-point wrench is required.

Socket-head cap screws are screws/bolts (pick the term you like) that feature a drive head that has a round outer diameter, a flat top and bottom surface, and a female hex socket in the top male hex wrench (commonly referred to as an Allen wrench, even though Allen is a brand name) is required. Cap screws are suitable for tight spaces (where a socket, closed or open-end wrench would be

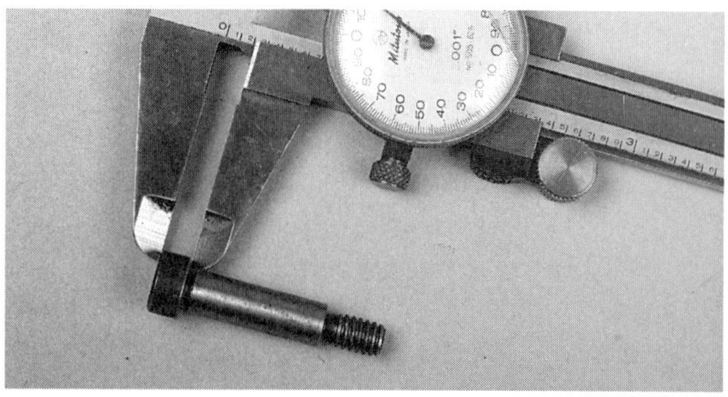

The head of a shoulder screw is typically round, featuring a female hex drive. The head of a shoulder screw usually offers a fairly low profile. This 5/16"-shoulder x 1/4-20 shoulder screw head measures 0.220" in height.

A hex head bolt features six flat sides.

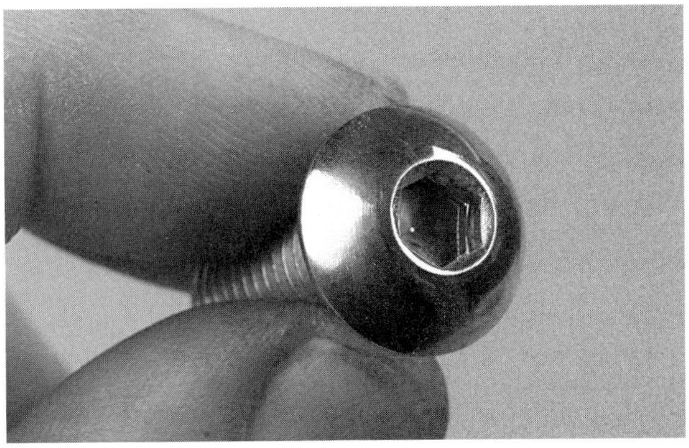

A button-head screw features a head with a flat underside and a domed, or "button" shape head. This button-head screw features a hex socket drive (female hole with six flat sides). Button heads are also available in phillips and torx drives

difficult or impossible to use). Quite often, custom builders simply prefer the appearance of cap screws, even if access isn't an issue.

Torx drive screws feature a 6-point extended-tip star drive pattern. Most Torx screws that you'll encounter feature a female Torx recess, requiring the use of a male Torx bit. However, male Torx screws are also out there (though not as common), requiring a Torx style socket wrench. The Torx drive style provides a no-slip engagement for the tool, and greater surface area for drive contact. Lots of people whine about dealing with Torx, but the reality is that it's a great design. Just buy a set of Torx bits and be done with it. Torx fasteners are sized according to an alpha-numeric format. Sizes (for the range that you may encounter in automotive applications) range from T9 (0.093" major diameter) to T60 (.512" major diameter).

Button-head screws feature a flat head underside and a domed top surface, with a female hex recess. Use of a male hex wrench is required. Button heads are often suitable for applications where a lower head profile is needed (for clearance issues), or for

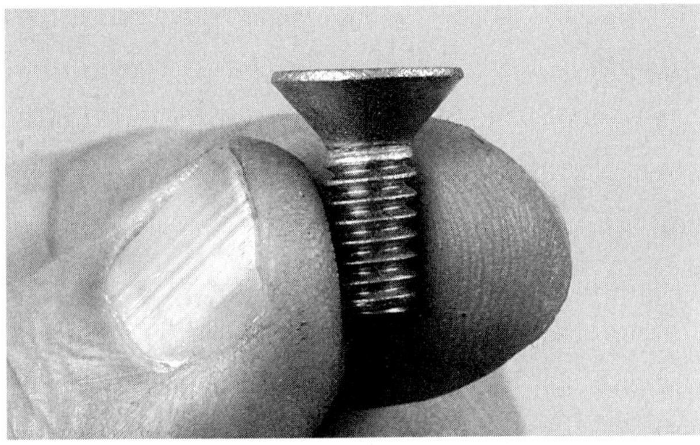

Flat-top screws feature a...you guessed it....flat-top head design. The head underside is chamfered to allow the top of the head to be flush-installed to a mating surface, providing the hole features a chamfered hole entrance. Flat-top screws are offered with hex sockets, as well as slotted and phillips drives.

areas where the driver's body might make contact, since the domed head doesn't feature any sharp raised edges to snag clothing, etc. As with cap screws, button heads are often chosen for customs, street rods or race cars simply based on appearance.

Shoulder screws feature a cylindrical bare shoulder area under the head, that serves as a bearing area or as a radial spacer. The threaded area is smaller in diameter than that of the shoulder. Shoulder screws can be used for a number of

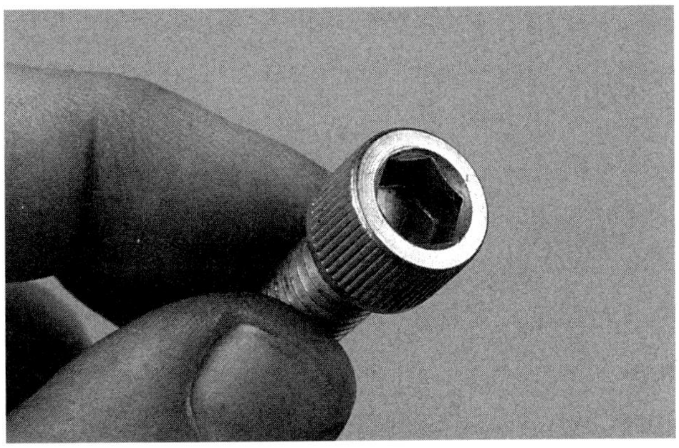

Socket-head cap screws feature a round outer head surface, relatively tall head, and a hex socket drive. Some cap screws feature a smooth-surfaced head perimeter, while others (like the example shown here) features a series of flutes, or grooves, which provide a grip area in case the hex socket is damaged. The grooves/ridges help to accommodate the use of pliers or vise grips. Socket-head cap screws are so-named because the head is shaped like a cap, and the drive style features a female six-sided hex"socket."

applications, including precision-aligning mating holes, to serve as pivots, etc. A good example is use in steering column shaft connection, where a cross-drilled steering shaft tube fits over a cross-drilled steering shaft. The bare shoulder area serves as a load bearing spacer, when the bare shoulder area alone contacts all hole edges. The exposed thread shank then accepts a retaining nut. Shoulder screws can be ordered in a wide range of closely-incremented lengths, so you'll be able to find the correct length for your needs. Head drive styles include hex socket (the most popular), Phillips and slotted. Shoulder screws are offered in both "standard" and "precision" versions, with the precision versions offering tighter diameter tolerance control. Of course, you'll pay more for the precision versions.

Flat-top screws feature a tapered head underside and a flat top surface. The drive style may feature

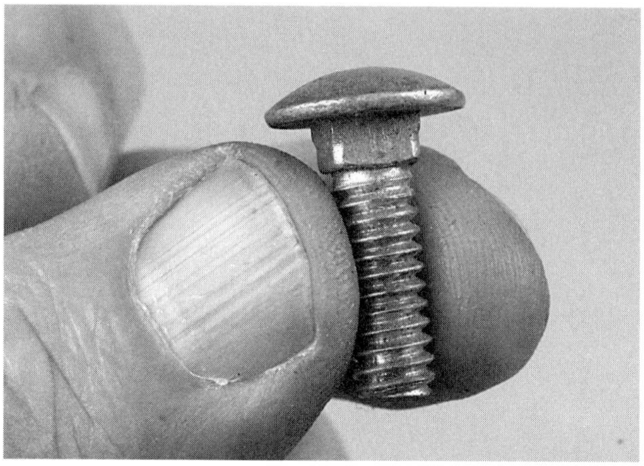

Carriage-head bolts feature a domed, round head and an oversized square area under the head. The square is designed to fit into a square hole, which prevents the bolt from rotating while attaching or removing a nut. Since carriage bolts feature no drive provision on the head, this square locating nub is needed. Carriage head bolts are used wherever a smooth domed surface is required or desired, either for function or appearance. Personally, I hate these things, but lots of folks prefer them for applications such as truck beds, running boards, fender brace fasteners on older cars and trucks, etc. I'd rather use button heads, but if you're performing a loyal restoration, sometimes you don't have a choice.

either a female hex, a straight slot or a cross-slot (Phillips). While these different drive styles exist among flat-tops ranging in size from #4 through about 3/8" or 10mm, in general, larger sizes are only commonly found in the female hex drive style. Flat-tops are ideal where the fastener must be installed flush with the adjacent mating surface. Installation requires a chamfered hole, to allow the top flat surface of the screw to sit flush with the material surface.

TORX

The Torx drive style features a 6-point star pattern. Although many folks hate Torx fasteners (only because it's something different and because they may not own a set of Torx bits), in fact, the Torx drive style is a great design. Because of the distinct extended points, this is considered a no-slip design...as long as you use the correct size wrench bit, and the drive socket is clean, your chances of ever stripping out the drive socket are extremely remote. Torx design is similar to a spline engagement. Multiple "fingers" engage around the perimeter, making it very, very tough to slip or break tool engagement. While this style of fastener drive may have once been considered a "tamper proof" design (when the style was new and the required wrenches weren't common), in reality it's simply a better design that prevents "rounding out" during service. A set of Torx wrench bits are not that expensive. Just buy a set so you'll have 'em when you need 'em. By the way, Torx drive isn't used by the car makers only. We're seeing more Torx employed in high-end aftermarket components as well since engineers appreciate the value of this fastener drive (on ATI Super Damper crank balancers and Jesel timing systems to name only two). In cases where a female drive socket is required or preferred, the Torx pattern provides much more engagement reliability than a female hex socket (female hex sockets are commonly

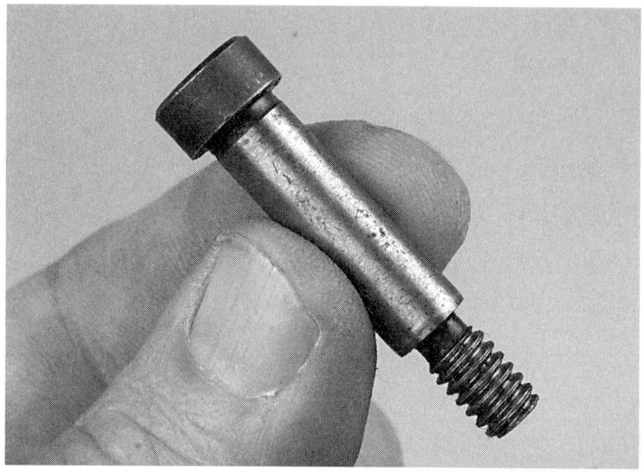

Shoulder screws serve as a centering device, load-bearing or pivot surface, depending on the application. The trick is to select a shoulder length that exactly matches (or a couple of thousandths less than) the total attachment material thickness, so that only the threaded tip protrudes. In that way, this type of screw serves as a nice, fat locating pin, that just happens to offer the convenience of a drive head and a threaded tip for secure placement. The example shown here features a 5/16" diameter shoulder and a 1/4" diameter thread. Drop this through a 5/16" hole and secure the exposed tip with a 1/4" x 20 nut, and you're in business. Just remember that shoulder screws do not offer the same tensile strength as other fasteners of the same material in the same thread size. Even though the shoulder on this screw is 5/16" in diameter, it's only going to offer the tensile strength provided by the smaller 1/4"-diameter threaded stem. Shoulder bolts are employed not in high tensile strength/clamping load applications, but rather in applications that demand exact hole centering/alignment, and for applications where the shoulder can serve as a pivot pin.

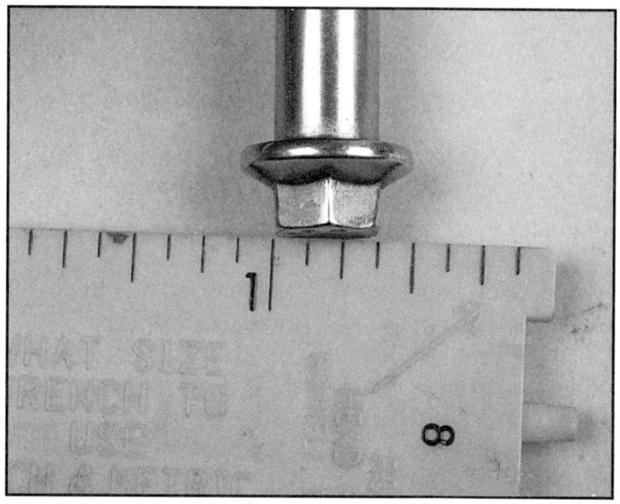

You can clearly see here that the hex head on this 3/8" bolt measures a mere 3/8" from flat-to-flat. "Common" or "traditional" hex head bolts that feature a 3/8" thread diameter would normally feature a 9/16" head.

Speaking of reduced-size bolt heads for restrictive spaces, a common application involves the mounting of exhaust header flanges to the cylinder head. Depending on the engine and the style of header, space can be at a premium due to bolt proximity to adjacent header tubes. Header mounting bolts have been designed that specifically address this common problem. From left: a reduced-diameter 12-point bolt head. Next, a common-sized socket-head cap screw (even though not a dedicated style, the round head doesn't require as much space as a full-size hex head, and a slender male hex bit wrench requires less working room than a female socket). Next, a 12-point hardened-steel bolt with a thicker base flange. Finally, a hex-head bolt with base flange where the hex is smaller than normal. All four examples shown here feature 3/8" thread diameter. However, the three male-head bolts shown here all require a 3/8" wrench, as opposed to a 9/16" wrench that would be required for a "common" 3/8" bolt. The heads require less installed space, and the wrench (due to its smaller size) requires less working room.

TORX SIZES

Torx Size	Major Diameter (peak to peak)	Max Torque (peak to peak)
T1	0.031"/0.81mm	0.02–0.03 Nm
T2	0.036"/0.93mm	0.07–0.09 Nm
T3	0.046"/1.10mm	0.14–0.18 Nm
T4	0.050"/1.28mm	0.22–0.28 Nm
T5	0.055"/1.42mm	0.43–0.51 Nm
T6	0.066"/1.70mm	0.75–0.90 Nm
T7	0.078"/1.99mm	1.4–1.7 Nm
T8	0.090"/2.31mm	2.2–2.6 Nm
T9	0.098"/2.50mm	2.8–3.4 Nm
T10	0.107"/2.74mm	3.7–4.5 Nm
T15	0.128"/3.27mm	6.4–7.7 Nm
T20	0.151"/3.86mm	10.5–12.7 Nm
T25	0.173"/4.43mm	15.9–19 Nm
T27	0.195"/4.99mm	22.5–26.9 Nm
T30	0.216"/5.52mm	31.1–37.4 Nm
T40	0.260"/6.65mm	54.1–65.1 Nm
T45	0.306"/7.82mm	86–103.2 Nm
T47	0.312"/.97mm	110–127 Nm
T50	0.346"/8.83mm	132–158 Nm
T55	0.440"/11.22mm	218–256 Nm
T60	0.519"/13.25mm	379–445 Nm
T70	0.610"/15.51mm	630–700 Nm
T80	0.690"/17.54mm	943–1048 Nm
T90	0.784"/19.92mm	1334–1483 Nm
T100	0.871"/22.13mm	1843–2048 Nm

referred to as "Allen").

I'm aware of thirteen Torx sizes (at least for the range that may apply to automotive applications). These include T9, T10, T15, T20, T25, T27, T30, T40, T45, T47, T50, T55 and T60 (**Note:** Torx sizes T1, T2, T3, T4, T5, T6, T7 and T8 also exist, but I've never seen these used in automotive applications. Larger sizes also exist but are uncommon in automotive applications, including T70, T80, T90 and T100).

Torx wrench bits (and sockets, for those rare instances when you encounter male-drive-head Torx fasteners) are identified for size using the T-alpha-numeric sizing system.

If you ever want to identify Torx wrench bit size by measuring (maybe the marking has worn off), I measured my sets and listed the results here. The diameter shown here was measured across the major diameter (peak spline to opposing peak spline).

A Torx socket-head (female Torx) screw. Note the six spline fingers for wrench engagement.

The wrench-to-fastener engagement provides a much more positive engagement than would be found in a socket-head, Phillips or slotted drive.

Here's a low-head Torx screw. Similar to socket head cap screws, Torx screws are available with various head heights.

Flywheel bolts are another example of head designs that accommodate a specific need. Primarily because of space restrictions, engine flywheel mounting bolts must feature a fairly low profile. However, because of the need for secure flywheel clamping to the crankshaft, a large bolt head footprint is also desired. Here are two typical flywheel bolts. The bolt on the left features an ultra-low profile in a 12-point design, while the bolt on the right features a low profile hex head design. Bolt head diameter, head thickness and drive style (obviously as well as thread diameter and shank length) is dictated by the specific flywheel/crankshaft/clutch application. If your crankshaft features 7/16" x 20 threaded holes in the crank base flange, don't grab just any flywheel bolt. Always be sure to use only the specific flywheel bolt recommended for your application.

CROSS-DRILLED AND HOLLOW BOLTS

Special-duty bolts may feature cross-drilled holes (a hole through the threaded shank or through the bolt head) for locking purposes. If a hole is drilled through near the tip of the shank, this allows the addition of a cotter pin or safety wire for added security against loss. Race car applications, may, at the discretion of the builder, employ a series of cross-drilled bolts on such items as oil pans, allowing stainless steel safety wire to be routed from one bolt head to another, in series. This prevents all of the bolts from loosening, providing the safety wire is oriented so that wire tension pulls at the bolt in the direction of tightening (preventing the bolt from rotating in the opposite loosening direction.

Cross-drilled bolt heads can come in mighty handy to remedy problem leaks. For instance, when my race team campaigned 2.0L Neons in 24-hour races, we found that the oil drain plug loved to loosen from its aluminum pan. We cross-drilled the head of the plug, ran safety wire from the plug head to a hole we drilled through one of the cast pan's cooling fins, and no more loose plug.

Banjo bolts are designed for routing liquids. The bolt shank is hollow, intersected by a cross-drilled hole at a specific distance from the underside of the head. Typically used in applications such as brake hose-to-brake caliper connections, or for fuel connection to carburetors or fuel logs. The use of a banjo bolt requires a banjo fitting (straight hollow shank that connects to a hose or line, with a round eyelet-type head that accepts pass-through of the banjo bolt. The inside of the fitting hole features an open feed from the hose or line. When the banjo bolt is installed, fluid fed from the hose or line enters the banjo fitting, and fluid routes from the feed hole inside the fitting to the cross-drilled hole in the banjo bolt. Fluid then routes through the

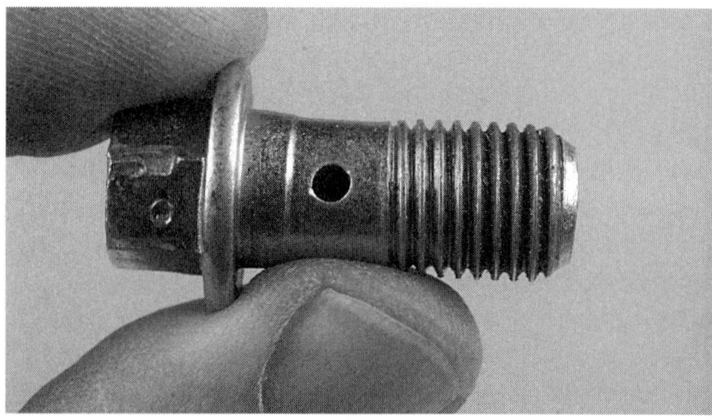

A banjo bolt features a side orifice on the shank, positioned to align with the banjo fitting. Fluid passes from the fitting housing I.D. to the orifice, then through the hollow bolt shank.

A banjo bolt features a hollow shank for fluid delivery. This is a brake caliper banjo bolt. If a copper or aluminum crush washer won't seal (copper seen here), try Stat-O-Seals.

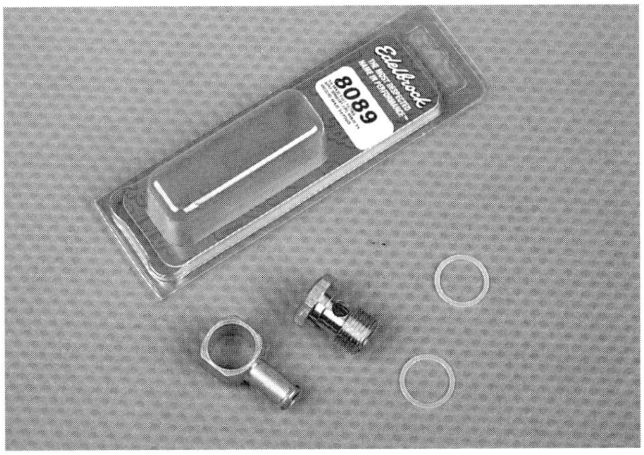

Carburetor inlet fittings are available in both slip-on hose (nipple) and threaded –AN versions.

hollow bolt shank. Any application that requires a banjo setup will require typically two sealing washers (one on each side of the fitting eyelet). This seals the bolt head to the fitting, and the fitting to the parent material.

NUTS

Nuts feature female threaded holes, allowing attachment to an exposed bolt or stud thread body. Nuts are available in all thread sizes in inch and metric formats. The outside face of the nut permits tool/wrench engagement. Nut driving styles typically include square (four-sided), hex (six-sided) and 12-point. Specialty nut driving styles (round bodies with dimple holes, etc.) exist but are rarely applicable for automotive use, so we'll ignore those.

Most nuts that you'll encounter are open (the threaded hole passes all the way through the nut body). However, when using an enclosed, or capped nut (such as acorn nuts that are sometimes preferred by rodders for their neat appearance on items such as air cleaner lids or valve covers), always make sure that the stud will not bottom-out inside the nut. If the nut bottoms-out, it will not be able to deliver any clamping force at all, which will lead to rattles or vibrating parts, leakage, or nut loss due to vibration.

Square nuts are only used in some vintage automotive applications, where the builder is intent on maintaining originality (for instance, running board nuts on older truck bed rails or running boards). Square nuts have their place, when wrench access to the nut is difficult, and when the nut drops into a channel that will prevent it from rotating during installation or removal. Otherwise, avoid them.

Hex nuts (obviously the most common) require the use of a hex or 12-point box, open or socket

All externally-mounted banjo bolt assemblies require two sealing washers: one between the bolt head and fitting, and another between the fitting and component.

Installing a banjo fuel fitting onto a carb. This allows positioning of the fuel inlet hose 90-degrees to the fuel bowl, providing an option in terms of routing for clearance.

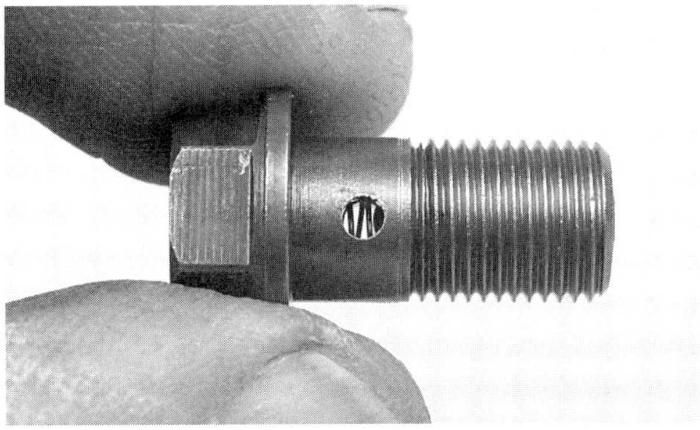

This is a banjo bolt from a Honda B18C1 engine. A series of these bolts secure special oiling nozzles inside the block that spray oil to the underside of the pistons for improved lube and piston cooling. The bolt shank is hollow. A side hole intersects the main hole to allow oil to travel from the main oil galley (with oil running through the main hole), and exiting the side hole, which allows oil to exit through the nozzle fitting. Notice the internal spring, which provides a degree of oil pressure control.

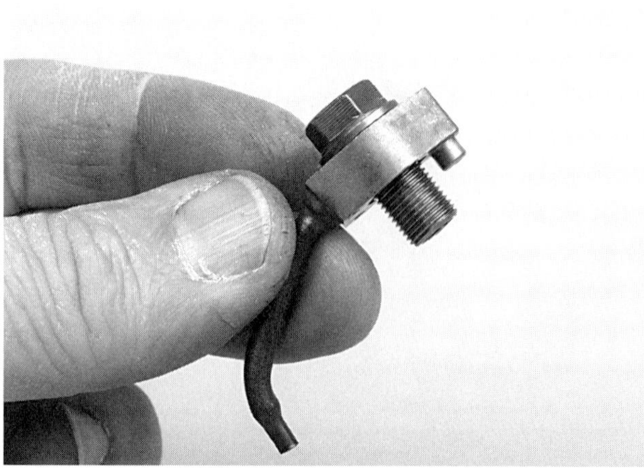

Here's an oil nozzle assembled to a banjo bolt. The base of the nozzle head mounts flush against the block wall. Banjo bolts allow fluid routing in a 90-degree angle from the fluid source to the intended output.

Hex nuts feature six flats as the drive profile, allowing the use of a 6-point or 12-point wrench.

wrench, while 12-point nuts can be serviced only with 12-point wrenches.

Nut materials are offered in a wide array, including plastic, steel, aluminum, titanium and other alloys. Typically you'll be using steel, stainless steel or titanium.

Nut heights vary as well, from full-height nuts to very thin, low-profile nuts that are used in constricted spaces or as "jam" nuts to prevent unwanted rotation of an assembly.

Nut styles also vary in terms of the base design. Nuts may feature a larger flange base (often called a flange nut) that provides a larger "footprint" at the mating surface, for wider, more even dispersion of the clamping load (especially useful if a separate flat washer isn't being used). A fixed-flange nut features a larger-diameter base as part of the nut...not a separate component. A spinning-flange nut will feature a "washer" that's captive-installed onto the nut, and is allowed to rotate independently of the nut. The free-spinning type flange is a good choice where you don't want the washer to rotate against the component surface.

Commonly, nuts used in suspension and steering systems may feature a series of notches, or cutouts. Once tightened to required torque value, a security pin (cotter pin, or cotter key) may be passed through one of the notches, through a cross-drilled hole in a joint stud, and through the opposing nut notch.

Nut Retention

In order to "lock" a nut in place to minimize its chances of loosening under vibration or through heat cycles, various methods of nut retention may be used. These include the use of loc washers, the application of anerobic chemical compounds, or the use of self-locking nuts.

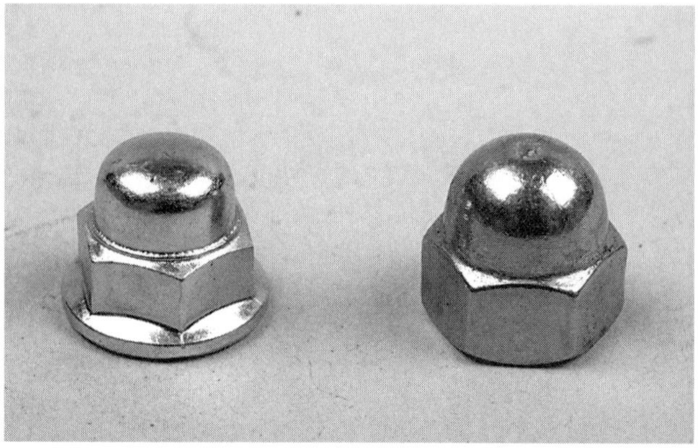

Acorn or "capped" nuts. These are available in flanged style (left) and non-flanged. Acorn nuts are available, as are most nut styles, in zinc-plated steel, stainless steel and chrome-plated steel.

Lock Nuts—Self-locking nuts commonly used in automotive applications feature one of two approaches. The most common is the nylon-insert, nylon-locking or "nyloc" nut (all interchangeable terms, with the latter actually referring to a brand name). This type of nut features the use of a soft plastic ring (crimped captive in the nut exit) that provides an interference fit that deforms over the bolt threads, providing a mechanical grip.

Another type of self-locking nut features no separate plastic grip ring, but rather features a slightly undersized exit hole or a slightly deformed exit hole that seizes the bolt threads. These are often referred to as "stover" or "prevailing torque" nuts.

While locking nuts may be reusable in some cases, on a critical assembly (suspension components, etc.) it's always best to replace a locking nut instead of reusing the nut. A used nylon-insert nut's grip ring may have relaxed over time (especially if exposed to high heat) and may no longer provide the needed locking grip. A stover-type nut's exit hole can easily become round after repeated applications (removing/installing), and may no longer offer mechanical retention. When it comes to locking nuts, the best course is to always install fresh nuts whenever possible or practical to ensure that you'll obtain the full locking benefit.

Locking Compounds—The use of thread locking compounds is an excellent method of securing a threaded connection, but the compound must be chosen carefully, since a variety of locking strengths are available. These locking compounds are "anerobic" in nature, which means that they harden and cure only in the absence of air. Once a coated

Nyloc nut, or nylon self-locking nut. A hard nylon insert ring is held captive in the exit end of this nut via a swagged collar. This nylon ring, featuring an I.D. smaller than the thread diameter, allows the bolt thread to bite into the nylon, creating an interference fit. This prevents the nut from backing off. Nylon insert nuts should only be used once, but in reality can be reused several times. However, since it's impossible to predict how many times the nylon ring will provide sufficient grip against the bolt threads, you're taking a chance anytime you reuse nylon locking nuts. Especially if the application is on a race car, it's best to always install a fresh nylon locking nut.

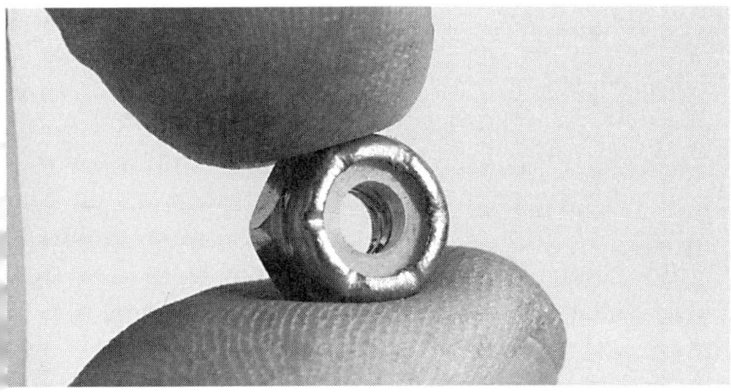

Here's a stainless steel nylon locking nut in size 8–32. Nylon locking nuts are good choices for small sizes where you can't or shouldn't apply lots of tightening torque. Nyloc nuts also allow you to avoid the use of locking washers, especially handy for small fasteners in tight spaces where you'd rather nut deal with two pieces (nuts and loc washer) as opposed to one (the nyloc nut). However, keep in mind that you can't finger-spin a nyloc nut as far as a non-locking nut due to the nylon insert's grip on the screw threads.

nut or bolt is installed, the compound only then begins to cure. Locking compounds vary from "low strength" to "high strength." Depending on the maker, the specific compound may be labeled as low strength, medium strength or high strength; or it may be identified by a number (for example, Loctite's 241 thread locker is considered by most as a medium-strength material, while their 271 version is a high-strength material.

In general terms, a low- or medium-strength thread locking compound is suitable for fasteners that may require future disassembly, since the

chemical bond can usually be broken mechanically with a wrench. The high-strength locking compounds are reserved for more permanent applications, such as locking a bushing or sleeve in a hole, etc. If you were to apply a high-strength locking compound on, say, a 1/4"-diameter bolt, chances are very high that you'll probably break the bolt head off when trying to remove it.

Always read the application instructions before choosing a locking compound to make sure that it's suitable for your application.

Note: All of these locking compounds are, in simple terms, liquid plastic. I use the term plastic only because they break down, or plasticise, when exposed to sufficient heat. If you need to remove a stubbornly locked fastener, instead of trying to force the locking compound to break its bond, heat may be applied, which will soften the compound. The specific level of heat required to release the chemical lock will vary depending on the manufacturer's model (this info may be printed on the package or may be obtained from the maker). More information on thread locking and sealing chemicals appears in Chapter 2.

Castle Nuts—Castle nuts feature notches to allow pass-through of a security pin (cotter pin) as an

This used nylon locking nut reveals the threads that were cut into the nylon ring insert by the mating bolt. Once the threads have been cut, the nylon insert won't provide as much interference grip to the bolt threads as a new nylon locking nut will. Also, prolonged exposure to radiant heat will cause the nylon insert to soften and conform to the bolt threads, reducing the insert's grip. It's best to always replace with new nyloc nuts instead of reusing old ones.

Nylon locking low-profile (incorrectly and generically called jam nuts only due to their short profile) nuts allow the use of a locking nut in tight spaces where a full-height nut might not clear, and for applications where the exposed bolt threads might not be long enough to take advantage of the nylon ring engagement of a full-height nyloc nut.

An all-metal self-locking nut. This style goes by several names, the most common of which is the "stover" nut. The thread exit is deformed out-of-round. This non-round exit area provides the locking grip at the bolt threads, providing a more secure interference fit at the bolt threads. Also, a stover-type nut won't be affected by heat, since there's no nylon insert to soften and platicise.

Whenever using a nyloc or stover nut, make sure that the locking area is fully engaged by the bolt threads, not just "touching" the grip area. Otherwise, you're wasting your time and money. This photo shows a nyloc nut being used at the minimum of its locking feature. The bolt tip must at least pass through the insert. If the thread engagement to the nylon insert doesn't meet this minimum engagement depth, use a longer bolt.

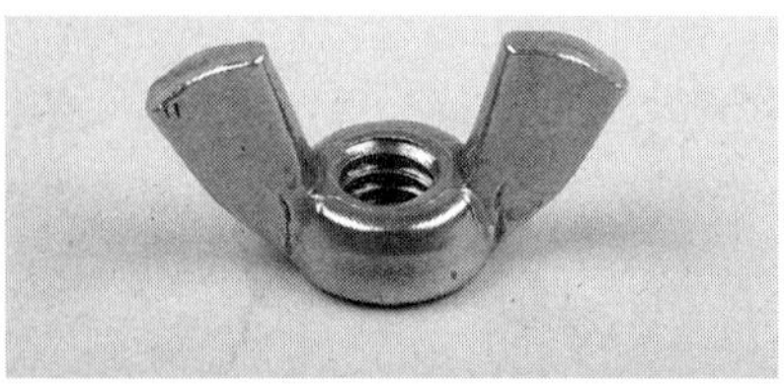

Wing nuts are so named because, well, they look like they have a pair of wings. This style of nut allows servicing by hand (use your fingers instead of a tool). Wing nuts are readily available in 1/4" and 5/16" thread diameters, but can be tricky to find in other sizes (mostly because local hardware stores usually carry only the most common fasteners; and because most towns don't have a decent specialty fastener supply house).

This big boy (1" stover nut) features a distinct raised shoulder. This shoulder area at the nut exit is the portion of the nut that is deformed into an oval for thread grip. In order to take advantage of the locking feature, the bolt threads must fully engage this deformed exit area.

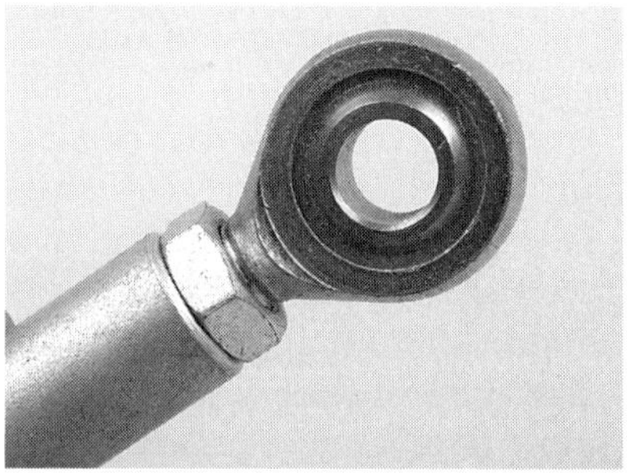

Low profile jam nuts are commonly used to secure radius rod ends. Once the rod end is adjusted (for total link assembly length), tighten the jam nut against the link. This prevents the rod end's threads from rotating.

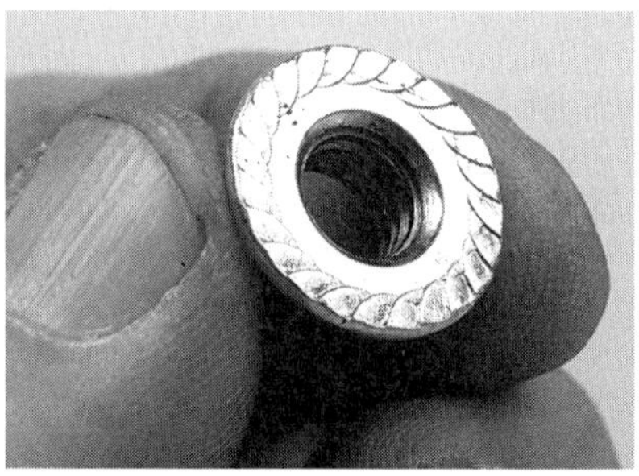

The underside of this flange nut shows the locking feature. The serrations are cut on an angle that allows tightening (the serrations don't fight you as the nut rotates clockwise), but restricts loosening (counterclockwise rotation) because the raised edges of the serrations bite into the parent metal. This is fine where you don't mind score marks on the parent metal (once the nut is removed, the metal surface that was hidden under the flange will be gouged).

The most common use for wing nuts in the automotive field is for securing air cleaner housings. They're common, and they don't require tools, so many people view these as convenient. Note: whenever you plan to install any nut against a chromed, polished or painted air cleaner lid, always use a smooth flat washer (polished stainless or nylon) to prevent scratching/digging into the finish. Actually, a rubber or fabric washer is often a good choice, since it will provide a bit of frictional grip between the wing nut and the lid, minimizing the chance for nut loosening. A very common mistake is to tighten the boogers out of a wing nut. Just snug it down fingertight. If you overtighten, you'll distort the air cleaner lid, you'll make it difficult to remove later, and you can damage the carburetor (if you tighten the nut excessively, you can create stresses in the carb casting than can lead to vacuum leaks or even cracks).

added safeguard against unwanted rotation.

WASHERS

Washers of various types are used for, well, a variety of reasons. Flat washers are generally used to either help spread the clamping load of the bolt or nut over a larger area of material surface, or to achieve a more accurate torque value, or to prevent the underside of the bolt head or nut from digging into a soft, painted or plated material surface, for the sake of appearance, or any combination of these reasons. Essentially, the use of a flat washer provides a bearing surface between the underside of the bolt head and the parent surface. Locking washers are, quite obviously, designed to prevent unwanted loosening of a bolt or nut during vehicle operation.

Flat washers are offered in a range of inner and outer diameters, thickness and hardness.

Specialty flat washers used on many race engines may also feature a slight chamfer on one side of the washer's inner diameter. For critical applications such as cylinder head clamping, where cylinder head bolts are being used, this chamfer is placed facing up to compensate for a slight radius at the head underside. If cylinder head studs are being used, the hardened washers will likely be flat with no chamfer. Pay strict attention to the manufacturer of the fasteners for advice on washer orientation.

For custom assembly of areas other than the engine internals, bear in mind that flat washers are available in various thicknesses. This is something that you can take advantage of if installed bolt length is an issue, where you're trying to fine tune the bolt's installed height. You don't need to be

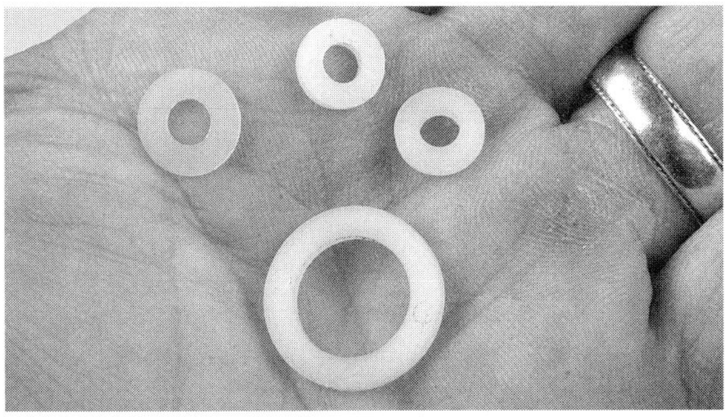

Nylon and delrin flat washers allow insulating between the screw head and parent surface and are useful where high-compression loads are not required. I use them frequently for attaching light materials together (for example, a light fiberglass or carbon fiber scoop to a fiberglass or composite hood, where I simply want a bearing surface between the nut and parent material). Nylon and delrin washers are available in a range of thicknesses, from a few thousandths to chunky thick.

In addition to the hardware store USS and SAE washer families, super-small-O.D. & precision-fit washers are also available, generically called "mil-spec" (military specification). Pictured here is a 3/8" SAE washer (left) and a 3/8" mil-spec washer. Mil-spec washers are great when a smaller washer footprint is desired, or where space is limited.

Here's an example of a mil-spec washer on a 3/8" socket-head cap screw. The washer O.D. is only slightly larger than the screw's head diameter. The I.D. fit is a bit tighter too to reduce off-center slop. If the smaller footprint is acceptable for the application, this also makes a neater presentation.

stuck with whatever thickness washer is available at your local hardware store. Check with a local fastener supply store or go on-line and check one of a national fastener suppliers (McMaster-Carr, Grainger, etc.).

Flat washers are also available in a variety of outer diameters, again to accommodate specific requirements. Bear in mind that all washers, regardless of type, are sized according to the bolt or stud being used. A 1/2" washer will accommodate a 1/2"-diameter bolt, etc. The inside diameter of the washer must match the bolt diameter.

Both unhardened (grade 2) and hardened flat washers are available. Don't automatically assume that you need a hardened washer in every instance. A softer (grade 2) washer can prevent galling against a surface in some cases. However, where hardened washers are recommended (or included) with a component, use them.

While flat washers are available in a variety of outer diameters (from specialty fastener suppliers), in general terms an SAE washer will feature a closer-tolerance screw fit, are generally thinner and provide a smaller O.D. USS flat washers on the other hand, are larger in O.D. and thicker. When you need a flat washer of a specific thickness, hardness grade and outer diameter, check availability from the specialty suppliers. This will enable you to select exactly the thickness and O.D. that you want.

Also keep in mind that flat washers are available in a wide range of materials, depending on the application. Materials include steel (in various hardness grades), stainless steel, aluminum, copper, brass, bronze, titanium, nylon, delrin, fiberglass, PVC, hard-fiber, phenolic, and more.

Sealing Washers

Sealing washers are flat washers designed to not only provide a frictional-contact surface between the bolt head and parent material, but to provide liquid sealing as well. Two basic types of sealing washers include all-metal crush washers and gasket-assisted washers.

Examples of sealing crush washers can be found in banjo bolt applications on disc brake calipers (at the brake hose to caliper connection). Crush washers are offered in various styles and materials, including copper and aluminum flat sealing washers, as well as special soft aluminum washers that feature an integrated "rubber" O-ring. Soft metals such as copper and aluminum seal by crushing under clamping load when the steel bolt tightens against a steel or cast item such as a brake caliper. For problem sealing (for example, if you have a brake caliper that stubbornly leaks after several attempts using new copper or aluminum washers), or for sealing against an aluminum parent material, an aluminum/O-ring washer is the ticket. These are sometimes referred to as Stat-O-Seals (a brand name) or Dowty seals (a British version). This type of O-ring–equipped washer often comes in handy for other applications. For instance, on water pump bolts that may enter an open water jacket. In addition to applying a liquid-resistant thread sealer on the threads, adding an O-ring type flat washer under the bolt head can provide added insurance against leaks. Another type of gasket-assisted sealing washer is the bonded sealing washer, which features a metal (usually aluminum or stainless steel) top, bonded to a neoprene or silicone

The Stat-O-Seal washer features a soft aluminum housing, crimped over a synthetic rubber O-ring–type insert. These work great on brake caliper-to-hose-fitting applications. This type offers a twofold benefit, since it acts as both a crush washer (aluminum housing) and an O-ring seal.

The bonded sealing washer features an alloy top and a bonded gasket underside. Bonded sealing washer gaskets are available in neoprene (temp range –45 to +250 degrees F); and silicone rubber (-160 to +500 degrees F).

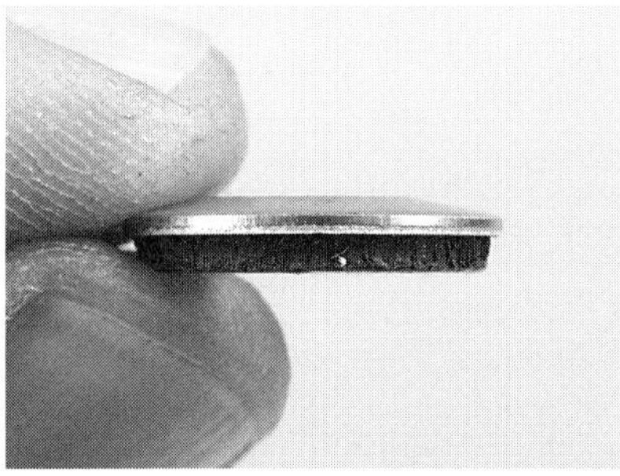

Bonded sealing washers feature a slightly convex top washer, which flattens out as the screw is tightened. This "spring" action also helps to provide a degree of washer locking ability.

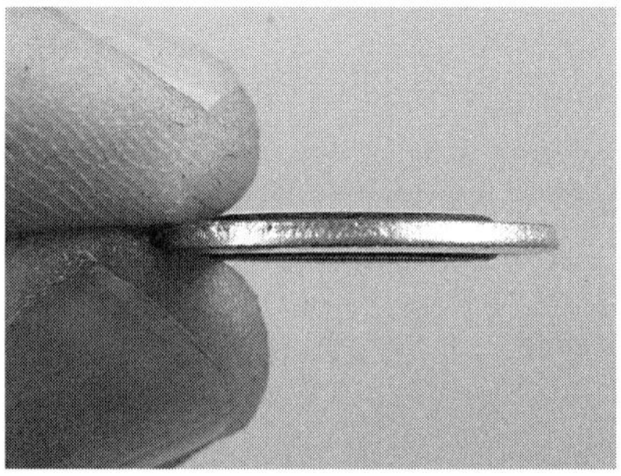

Side view of a 3/8" Stat-O-Seal. The O-ring insert protrudes slightly relative to the metal housing. This provides the additional O-ring crush to insure sealing.

rubber round gasket. These can be used at various locations to assist in sealing water or oil.

Locking washers (often referred to as loc washers) encompass several styles, including external star, internal star, serrated, and the most common, the split loc. The purpose of a lock washer is obvious: to prevent the screw from loosening during vehicle operation. Locking washers can't be used everywhere, so don't toss them into the mix at whim. Cylinder head bolts or studs/nuts, main cap bolts or studs/nuts are two examples where locking washers are not needed and should not be used.

The most common type of locking washer is the split loc, which is slightly spiraled, with a gap. These are mini springs with sharp edges at the gaps. The spring action places a load against the material and underhead of the screw, and the gap edges "bite" into the parent material and screw underhead, in an attempt to prevent unwanted loosening rotation. Star loc washers feature a series of twisted fingers that bite into the material (the fingers are on the O.D. (external star) or in the I.D. (internal star). Serrated washers feature a series of shallow cuts, or serrations, which bite into the material.

AIRCRAFT FALLACY

Especially within the street performance and race industry, a product may be labeled as "aircraft quality," "aircraft grade" or "aircraft type." Quite often these terms are used to enhance the product's image. However, with regard to fasteners, using "aircraft" grade isn't necessarily as good an idea as you might think in all situations.

While we normally associate a higher level of quality to anything identified as being good enough for use in an airplane or helicopter, "aircraft" grade fasteners are not necessarily stronger than an "automotive" counterpart fastener. The reason: aircraft assemblies commonly employ multiple fasteners for various component installations, using an approach of redundancy (the more fasteners the better to spread clamping loads over a wider area ... if a few fail, the odds are good that the component will remain installed because there are many remaining fasteners to get the job done). So, if we compare one "automotive" grade bolt with a similar sized "aircraft" bolt, we cannot automatically assume that the aircraft version will be stronger. Since automotive-related components usually don't use as many fasteners as an aircraft system might use, we must use the strongest fastener that suits the specific application, especially for hard-pounding/high-stress racing applications. In other words, we can't afford to spare any threaded fasteners in a "what if" scenario. We must make good use of every fastener we have.

THREADED FASTENER TIPS

Galling

A note of caution when using any stainless steel bolt or nut: always apply a thin coat of a thread lubricant before assembly, since dry stainless fasteners can be prone to galling (when the threads begin to smear and "weld" to their mating threads). Galling of stainless steel threads is most common when installing/tightening a stainless steel fastener quickly. This galling phenomenon is often referred to as "cold welding," where the softer alloy fastener's protective surface oxides are wiped off, allowing surprisingly fast thread-clogging (where the alloy threads smear and transfer to the female threads...or in the case of a stainless steel bolt in an

US SHEET METAL SCREW DIAMETER TABLE

Size	Thread Diameter (Decimal)	Thread Diameter (Fractional)
4	0.11	7/64
6	0.14	9/64
8	0.17	11/64
10	0.19	3/16
12	0.22	7/32
14	0.25	1/4

(Sheet metal screw thread diameter is measured at the thread peaks...the major diameter.)

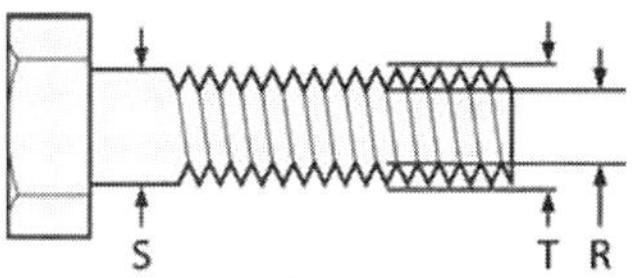

S: Shank diameter
T: Thread diameter (major diameter)
R: Root, or minor diameter

US MACHINE SCREW DIAMETER TABLE

Size	Thread Diameter (Decimal)	Thread Diameter (Fractional)
4	0.11	7/64
6	0.14	9/64
8	0.16	5/32
10	0.19	3/16
12	0.21	7/32

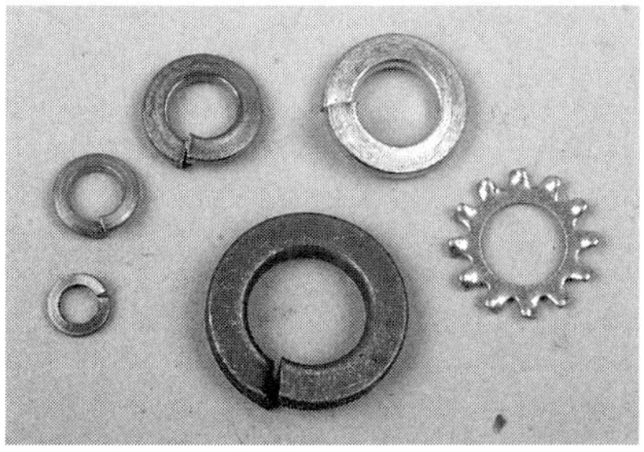

Pictured here are split loc washers (clockwise from left: #6, 1/4", 5/16" and 7/16"). An external star washer is seen at the far right.

An external star washer. The small fingers feature a slight twist, providing edges that bite into the parent surface.

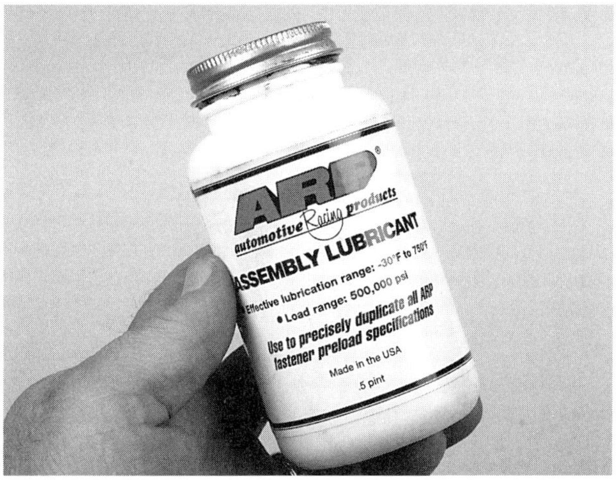

Always follow the fastener maker's or the component maker's instructions for fastener torque. Pay close attention to recommended thread lubricant, because in order to achieve the desired clamping load, the torque value will differ for the use of engine oil versus moly lube! Be sure to follow the torque-value spec for the type of lube you are using. In order to prevent thread galling, moly is an excellent product. Be especially aware of the dangers in galling stainless steel thread. Whenever using stainless steel threaded fasteners, a light application of moly lube is highly recommended.

luminum hole, the aluminum female threads may ransfer...either way, the threads lock together).

You'd be surprised how much frictional heat can ouild up at the thread mating surfaces, and alloy naterials just don't like that. When a stainless bolt or nut galls and seizes, quite often your only ecourse is to cut the fastener off (sometimes esulting in drilling out the remaining shank naterial). To avoid galling, apply a bit of anti-seize ubricant to the threads before assembly. If a ocking anerobic compound is to be used, this may rovide the needed lubricity as well. Just avoid nstalling stainless steel or aluminum threaded asteners dry.

Thread galling can be a big problem. As I noted arlier, this is most common with alloy fasteners, ut it can occur anytime if conditions aren't right. Vhen a bolt seizes in place, the installer is quick to blame the bolt, when the female thread is usually at fault. The reason: most bolt threads are rolled instead of cut during manufacturing. Thread rolling produces a smoother thread flank. Female threads, on the other hand, are usually cut (using a tap), and therefore have rougher threads. This can occur in production runs where nuts are made in bulk (or where female threads are cut in mounting holes in a component), due to wear and contamination of the taps being used, or because the taps may have been run at too high a speed. When a bolt seizes, you're still "stuck" with the problem of removing it, but don't be too quick to blame the bolt itself.

Whenever you're performing assembly, try to take the time to inspect both male and female threads for condition and cleanliness. If you're using an alloy fastener (stainless steel, etc.), apply the appropriate lube to the threads before insertion. A moly-based anti-seize always works well, as long as you know the correct torque value for moly lubrication. Refer to the chapter on torque values for more information. Just be aware that torque values will be different for dry, oiled or moly-lubed fasteners!

Here's a tightening tip: when a bolt and nut system is featured, whenever possible (where access allows), always tighten or loosen the nut, while holding the bolt stationary. This draws the bolt, allowing you to achieve proper clamping force, while avoiding spinning the bolt. While you may not encounter any problems by rotating the bolt, in some cases this might result in galling the bolt shank against the parent material. It's just "kinder" to the bolt to keep it stationary if the nut can be accessed and turned.

Thread Engagement

Thread engagement length is an issue that should never be ignored. Obtaining the proper shear strength for the treaded connection is as important as selecting the proper grade. As a general rule of thumb, the thread engagement (the length of male thread that connects to the mated female thread) must at least equal the diameter of the thread. For instance, if you're using a 1/2"-diameter bolt, the threads on that bolt must physically engage the mating nut or female threads in the parent material by at least 1/2". This is a general rule of thumb for steel and iron applications. For aluminum threads, since the shear strength of aluminum is less than it is for steel, an engagement distance of more than the bolt diameter is preferred. In other words, while you may be able to get away with 1/2" of thread engagement if a 1/2"-diameter bolt is screwed into a steel component's threaded hole, if the same bolt is screwed into an aluminum component, you should have more than the minimum 1/2" of thread engagement distance. In any situation, the more practical thread engagement distance you have, the better. This increases the connection's shear strength.

Right Hand/Left Hand Threads

You're probably aware that threads can be designed in both right-hand or left-hand directions. Most threaded fasteners feature a right-hand thread (where the bolt or nut is rotated clockwise to tighten and counterclockwise to loosen). Left-hand thread is also available (where the bolt or nut is turned counterclockwise to tighten and clockwise to loosen). Left-hand threads have their place. For instance, some car makers (older Mopars and some newer Porsches, as examples) choose to employ left-hand thread wheel fasteners on one side of the car, in an effort to counteract centrifugal force as the car is operated at high speed (it's simply an effort to prevent accidental loosening).

However, the most popular use of left-hand thread is found in mounting or locating rod assemblies that provide ease of length adjustment. A common example is a steering tie rod adjuster. If the adjuster features right-hand thread at one end, and left-hand thread at the opposite end, the adjuster may be rotated in place (without the need to disconnect the tie rod) in order to adjust length to achieve toe-in adjustment.

Another very popular application for left-hand thread is found on adjusting rods for street rod/custom component installation, such as A/C compressors, power-steering compressors, alternators, etc. Instead of using a stationary bracket that features a long slot (requiring the use of a wrench or pry bar to set belt tension; and relying on the tightness of the set bolt to hold it in place), a threaded adjuster rod can be adjusted for length (and resulting belt tension) by loosening the jam nuts located at each end, and rotating the center-located adjuster nut or rod. A popular setup would include a spherical bearing rod end (commonly called a "heim joint) at each end, with one featuring right-hand thread and the other with left-hand thread. Again, this allows precise and reliable length adjustment without the need to disconnect the adjuster or to use a pry bar to hold belt tension.

Left-hand taps and dies are readily available, although not at most auto parts or hardware stores. If you need a left-hand tape or die (if you're making your own adjuster or if you need to clean up existing left-hand threads), these tools are available from any decent machine-tool supply house (McMaster-Carr is but one example. This is a company that offers a huge selection of industrial tools and materials, and ships nationwide. Their phone number is 330-995-5500. Their online catalog can be viewed at www.mcmaster.com).

If you can't find what you need at a local machine tool or fastener supplier, you can always get it within a couple of days by ordering from any of a number of "mail order" suppliers.

WHITWORTH THREADS

Whitworth threads (called BSW for British Standard Whitworth) represented the first standardized thread in Great Britain in 1841. You may encounter these unique threads in some early British cars such as old Jaguars. This fractional thread form is now obsolete, being replaced in the early 1960s for Unified and metric threads. Whitworth features a 55-degree thread angle, and both the thread peaks and roots are radiused. In addition, Whitworth bolt heads and nuts require dedicated Whitworth wrenches (odd hex sizing).

The Whitworth family is considered a "coarse" thread group. While some fractional fasteners may appear to interchange (you may be tempted to use 1/4 x 20 USS nut on a 1/4 x 20 BSW bolt), do no do this! The thread angles, peaks and roots won't

BRITISH STANDARD WHITWORTH (BSW)

Nominal Diameter	Threads Per Inch	Pitch Depth	Thread Diameter	Major Diameter	Effective Diameter	Minor Diameter
1/8	40	0.02500	0.0160	0.1250	0.1090	0.0930
3/16	24	0.04167	0.0267	0.1875	0.1608	0.1341
1/4	20	0.05000	0.0320	0.2500	0.2180	0.1860
5/16	18	0.05556	0.0356	0.3125	0.2769	0.2413
3/8	16	0.06250	0.0400	0.3750	0.3350	0.2950
7/16	14	0.07143	0.0457	0.4375	0.3918	0.3461
1/2	12	0.08333	0.0534	0.5000	0.4466	0.3932
9/16	12	0.08333	0.0534	0.5625	0.5091	0.4557
5/8	11	0.09091	0.0582	0.6250	0.5668	0.5086
11/16	11	0.09091	0.0582	0.6875	0.6293	0.5711
3/4	10	0.10000	0.0640	0.7500	0.6860	0.6220
7/8	9	0.11111	0.0711	0.8750	0.8039	0.7328
1	8	0.12500	0.0800	1.0000	0.9200	0.8400

match, and you'll just end up damaging the fasteners, or you won't have proper engagement which will lead to loosening down the road. As a case in point, I once purchased an old Jaguar differential center section for a Cobra kit car that I was building. Once I began to disassemble the case, it became apparent that I was dealing with something weird. After learning that this assembly was nailed together with Whitworth thread, I junked it in favor of a later-model Jag rear. Trying to find dedicated Whitworth wrenches and replacement Whitworth bolts and nuts was simply too much of a hassle. If you have an old British component, and you can't seem to find a fractional or metric wrench to fit the hexes, that's a tip-off that it's probably Whitworth.

THREADED FASTENER TERMINOLOGY

Elastic Limit

The maximum stress to which a fastener may be subjected and still return to its original length upon release of the load. When the part is stressed beyond its elastic limit, it enters a "plastic" range where it will not return to its original length.

Yield Strength

The maximum stress that can be applied without permanent deformation of the fastener.

Tensile Strength

This is by far the most important aspect of a fastener's performance. High tensile strength allows use of a higher pre-load without exceeding that pre-load, thus avoiding fatigue failure. High pre-load fasteners are necessary for high-load applications such as cylinder head, connecting rod and main cap installations. However, if the tensile strength is too high (in excess of 200,000 psi), this can result in a fastener that is too brittle, and can reduce the fastener's ability to stretch and deform (ductility), which can lead to failure.

Fatigue Strength

A fastener's total tensile strength is also affected by the style of threads. Rolled threads (where the threads are created under pressure, rolling the thread material into the thread area after heat treating) creates a thread that is dense and radiused/rounded. Compared to threads that are cut (where sharp edges exist in the thread peaks and valleys), rolled threads create a stronger fastener because of the reduction in potential stress risers.

Another method of increasing a bolt or stud's fatigue strength is to distribute the stretch (as any fastener is tightened near or at its yield point, it stretches as it creates clamping force) along the shank and/or away from the threaded area. Using Pioneer's head studs as an example, they reduce the shank body diameter between the threaded ends in order to "move" the stretch away from the threaded areas. This provides increased fatigue strength, since the stretch is taking place along a smooth, radiused area instead of at the "notched" threaded sections (less chance of failure due to any potential stress riser points. Even though that maker already uses

Grade 5 steel bolt. Three radial line marks. 120,000 psi minimum tensile strength. This is the lowest grade bolt to use in automotive applications.

Grade 8 steel bolt. Eight radial markings identify a grade 8 bolt. Minimum 150,000 psi tensile strength.

L9 (grade 9) steel bolt. Rated at 180,000 psi tensile strength. These bolts feature a slightly taller-than normal head thickness. A series of nine radial marks accompanies the "L9" mark.

rolled threads, the reduced shank body diameter contributes to fatigue failure resistance).

THREADED FASTENER GRADES

The grade label specifies the minimum strength properties the fastener is supposed to meet. Industrial fasteners are also marked with a label that identifies the maker of the fastener. Grade markings on domestic fasteners meet ASTM and SAE specs, while metric fasteners meet ISO and SAE specs.

Bolts whose heads are unmarked can be assumed to carry a grade of either 1, 2 or 4. These bolts will be made from low or medium carbon steel. At best, they may be cold-drawn, but certainly have not been tempered.

Contrary to popular opinion, there are more than only three grades of bolts (the most popular being grades 3, 5 and 8). There are about 17 grades (other "grades" may exist in aerospace or other niche industries).

A threaded fastener's grade indicates its tensile strength and hardness. The higher the clamping load required, the higher the grade you'll need. In automotive applications, while it's difficult to make broad generalizations, you'll need at least Grade 8 (or 12.9 in metric) for high clamping load applications such as connecting rods, cylinder heads and main caps, flywheel bolts, clutch cover bolts, etc. Never use any threaded fastener below a Grade 5 (or 8.8 metric) for any automotive application. Always follow the vehicle or aftermarket component maker's recommendation for fastener grade.

Grade 5 fasteners are made from medium-carbon steel that has been quenched and tempered. Minimum tensile strength is 120,000 psi for bolts up to 1" diameter (tensile strength drops for diameters of larger than 1"). Grade 5 bolt heads are marked with three radial lines. Rockwell hardness for bolts up to 1" diameter is usually C25-34.

Grade 8 fasteners generally feature a minimum tensile strength of 150,000 psi and a Rockwell hardness of C38-39. A head marking of six radial lines indicate Grade 8.

Grade 9 fasteners (also called by trade names L9, PFC9, Tru-Torq and Bow Malloy) feature taller heads and 180,000 psi tensile strength and Rockwell hardness C38-42.

Inch Nut Grades

As far as nuts are concerned, there are basically three grades: 2, 5 and 8. In general terms, always use a nut of the same or higher grade as the bolt in use. Grade 2 nuts are not required to be marked. All grade 5 and grade 8 nuts (in the 1/4"–1 1/2" range) do require markings. The identification marks may be handled in one of three ways: grade 5 nuts may be marked with a single dot on the face of the nut and a radial or circumferential line at 120-degrees counterclockwise from the dot; or a dot at one corner and a radial line at 120-degrees clockwise from the nut; or one notch at each of the six corners of the nut. Grade 8 nuts may be marked with a dot on the nut face, with a radial or circumferential line at 60-degrees counterclockwise from the dot; or a dot at one corner and a radial line at 60-degrees clockwise; or two notches at each of the nut's six corners. Why can't this be simpler, with one type of mark for each grade? It doesn't really make much sense. It sounds as though there were too many cooks in the kitchen when this whole scenario was devised.

METRIC GRADES

As with most things metric, the numbers actually mean something. When a metric bolt head is marked for grade identification, you'll see two or three digits with a decimal point. The first number

Grade 8.8 metric steel bolt. This features a minimum tensile strength of 116,030 psi. In terms of strength, a metric 8.8 is roughly equivalent to a grade 5 fractional bolt.

refers to one tenth of the minimum tensile strength (in kgf/mm^2...it's not important for you to understand the increments of measure used here; simply know that the numbers actually represent the bolt's strength properties). The second figure, following the decimal point, represents one tenth of the ratio between the minimum yield stress and the minimum tensile strength, expressed as a percentage. As an example, let's consider a bolt head marked with an identification 8.8 label. The first number represents 1/10 the bolt's minimum tensile strength (in this case, 80 kgf/mm^2). The second number represents 1/10 of the ratio between tensile and yield. The higher the second number, the longer it takes to bring the bolt to yield.

Wanna know more? The first one or two numbers (the number before the decimal point) indicates minimum tensile strength in Mpa. (**Note**: Mpa is the symbol for megapascals, which is a metric unit of pressure or stress equal to one Newton per square millimeter, which equals about 145.038 pounds per square inch.) The final number (after the decimal point) indicates one-tenth of the ratio between minimum yield stress and minimum tensile stress. For example: if the bolt head is marked 8.8, this indicates a minimum tensile strength of 800 Mpa (about 116,030 psi) and a yield stress of 0.8 x 800, or 640 Mpa (about 92,824 psi). Metric screws are manufactured in grades 3.8 to 12.9, but for automotive applications, common grades include 8.8, 10.9 or 12.9 (tensile strength increases as the numbers grow). If we want to approximately compare metric grades to US grades, a metric 8.8 is roughly equivalent to a Grade 5. Grade 10.9 is roughly equal to a Grade 8; and 12.9 is roughly equal to a Grade 9. Metric nuts are marked with a single or double numerical symbol (8, 10 or 12).

METRIC BOLT GRADES

Grade Mark	Tensile Strength Min.	Yield Stress Min.
4.6	40	24
4.8	40	32
5.6	50	30
5.8	50	40
6.6	60	36
6.8	60	48
8.8	80	64
10.9	100	90
12.9	120	108
14.9	140	126

(all stress and strength values are in kgf/mm^2 units)

The higher the first number, the stronger the bolt, in terms of tensile strength. The higher the second number, the longer it will take to enter the yield point.

Note: Metric steel bolts are only required to show a grade mark if they are 6mm and larger and/or are grade 8.8 or higher. Nuts must be grade labeled if they are grade 8 or higher.

METRIC NUT GRADES

The grade system for nuts is a number that represents 1/10 of the specified "proof load" stress (in kgf/mm^2). The proof load stress corresponds to the minimum tensile strength of the highest grade bolt that can be used with that nut.

Grade Mark	Proof Load Stress (kgf/mm^2)
4	40
5	50
6	60
8	80
12	120
14	140

RECOMMENDED METRIC BOLT/NUT COMBINATIONS

Grade of Bolt	Grade of Nut
4.6	4
4.8	4
5.6	5
5.8	5
6.6	6
6.8	6
8.8	8
10.9	12
12.9	12
14.9	14

Grade 10.9 metric steel bolt. In terms of strength, this is roughly equivalent to a fractional bolt in grade 8.

Always match bolts and nuts of comparable grades (use a Grade 8 nut with a Grade 8 bolt; use a metric grade 10 nut with a 10.9 grade bolt, etc.). When dealing with metric fasteners, the 8.8 bolts are similar to a Grade 5. If you need higher tensile strength in metric, stick to 10.9 or 12.9.

Grade Caution

While it is common for grade 5 bolts to feature 3 slash marks, and for grade 8 bolts to feature 6 slash marks for identification purposes, it's important to note that not all grade 5 or grade 8 bolts will feature these marks! Quite often, specialty bolts (made for an automaker's specific requirements) may not feature any marks at all, or may feature a unique symbol. If you're performing a restoration, it's best to stick with factory-original bolts where grade may be in question. Otherwise, if the bolt head isn't marked (blank head), and if you don't know where the bolt came from, you should assume that it features a hardness grade of less than 5, and shouldn't be used. If the bolt was produced by a known quality maker, such as ARP, you can be sure that the bolt is correct for the application recommended by the maker. I'm simply saying that if you don't know where the bolt came from, and it's not marked, throw it away and buy known-quality bolts instead.

Grades of Steel

Threaded fasteners are "graded" or "classed" to identify their hardness and tensile strength. When choosing fasteners for a performance car or race car build, never blindly assume that harder is always better. Bolts, screws, studs and nuts should be chosen for the application at hand. Lower-grade or lower-class fasteners will tend to stretch more easily than higher-grade or higher-class fasteners, and higher grades/classes will be more resistant to stretching but may provide either overkill or damage to components that need to move more easily under thermal conditions. By the same token, don't always assume that you can use chromed or stainless fasteners anywhere you like, since they may not be up to the task. When bolts are chrome plated, they may become more brittle and less "elastic" as a result of hydrogen embrittlement during the electroplating process.

Stainless fasteners, depending on their grade, may not be able to withstand forces encountered in some engine and chassis applications. Stainless steel is an alloy of low carbon steel and chromium, offering enhanced corrosion characteristics.

A common misconception among hot rodders is that stainless steel is stronger than "regular" steel, which isn't the case. Because of the low carbon content in stainless steel, stainless steel cannot be hardened. When compared to steel, stainless steel is slightly stronger than an un-hardened Grade 2 steel fastener, but is significantly weaker than hardened Grade 5 or Grade 8 fasteners.

Grade 2 is a standard hardware grade steel, generally unacceptable for any automotive applications.

Grade 5 bolts are hardened for increased strength and are the most common bolts found in automotive applications. Grade 5 bolt heads feature three radial lines as a marking.

Grade 8 bolts have been hardened more than Grade 5 bolts, and feature six evenly spaced radial lines on the head as a marking. Grade 8 bolts are stronger than Grade 5 bolts, and are typically used for demanding applications such as suspension components.

Alloy steel bolts are made from a high-strength steel alloy (not to be confused with stainless steel), and are further heat treated for added strength. Alloy steel bolts are typically not plated, resulting in a dull black finish.

In other words, don't pick fasteners based only on appearance, especially for applications such as engine assembly, suspension, steering and brake systems. Do your homework and use the specific metals and grades recommended for specific applications.

Note: Grades are used to identify fractional fasteners, while Class is used for metric fasteners.

Note: Performance aftermarket fastener manufacturers offer much higher tensile-strength fasteners that are specifically designed for racing applications. ARP, for example, offers threaded fasteners that boast 170,000 psi, 190,000 psi and 220,000 psi tensile-strength ratings. In other words, instead of hitting the local hardware store or Internet source for imported who-knows-what-or-where-they-came-from fasteners, stick with reputable U.S. firms that are dedicated to supplying pro race engine and car builders. Be aware that a lot of overseas junk is floating around out there, some of it counterfeit (low grade junk marked as Grade 8, etc.). If you want to screw on a dash plaque, feel free to grab whatever works. However, when you're building an engine or are about to assemble a chassis, don't take chances. Know what you're buying!

Grade 2—Made from C1006 to C1022, C1215, 12L13 or 12L14 steel. For use in only low-strength applications. Minimum Rockwell hardness B49. Minimum tensile strength is only 44,225 psi.

Grade 5—Made from steels such as C1008–C1026. For use in medium-strength applications, with a Rockwell hardness of C25. Minimum tensile strength is 120,000 psi.

Grade 8—Made from stronger alloy steel such as C1021-C1045, for high-strength applications. Minimum Rockwell hardness C24. Tensile strength is minimum of 150,000 psi.

Grade 9—Commonly called L9. Rockwell hardness C38–42. Minimum tensile strength 180,000 psi.

Class 4/4.8/5.8—Comparable to Grade 2. Minimum Rockwell hardness is B71. Tensile strength is 60,900 psi minimum.

Class 6—Comparable to Grade 2. Minimum Rockwell hardness is B89, and minimum tensile strength is 87,000 psi.

Class 8—Comparable to Grade 5. Minimum Rockwell hardness is C20. Minimum tensile strength is 116,000 psi.

Class 10—Comparable to Grade 8. Minimum Rockwell hardness is C32. Minimum tensile strength is 150,800 psi.

Class 12—Exceeds Grade 8. Made from alloy steel C1035 and C1045. Minimum Rockwell hardness is C39. Minimum tensile strength is 176,900 psi.

Grades of Stainless Steel

18-8—Excellent corrosion resistance. Rockwell hardness B70. Minimum tensile strength 70,000 psi. Just a step above Grade 2 steel in terms of strength.

Type 316—Better corrosion resistance (alloy contains moly). Minimum Rockwell hardness B70. Minimum tensile strength 70,000 psi.

Type 400—Lower corrosion resistance but slightly stronger. Contains more carbon than 18-8 or 316. Minimum Rockwell hardness C28. Minimum tensile strength 75,000 psi.

ASTM F594 18-8 and Type 316—Same as 18-8 and 316 but meets strict ASTM standards.

ASTM A194 Grade 8M strain-hardened Type 316—Minimum Rockwell hardness C34. Minimum tensile strength 90,000 psi. Slightly greater strength than Grade 2 steel.

Alloy 20—Also called carpenter 20. Most resistant to stress corrosion and high-temperature sulfuric acid. Rockwell hardness B89. Minimum tensile strength 85,000 psi.

GALVANIC CORROSION

Galvanic corrosion occurs when dissimilar metals are in contact with one another. Especially in salt water conditions, a small electrical current flows from one metal to another. One metal will begin corroding faster than normal, acting as the anode, and the other metal will corrode more slowly than normal (acting as the cathode). The result is that the anode material will corrode much faster than the cathode material. Galvanic corrosion can be avoided by using only similar metals (steel bolt in steel components, or aluminum bolts in aluminum components, etc.). Also, by using stainless steel nuts on stainless steel bolts, steel nuts on steel bolts, etc. However, we all know that in some cases, dissimilar metals will be assembled (installing stainless steel nuts on steel bolts, etc.). In order to avoid galvanic corrosion (sometimes called electrolysis), apply a small amount of moly anti-seize compound or thread-locking compound to the threads before assembly, to act as an insulator between the two metals.

HEAD MARKING	GRADE/CLASS MATERIAL	NOMINAL SIZE RANGE (inches)	Mechanical Properties: PROOF LOAD (psi)	Mechanical Properties: MINIMUM Y IELD STRENGTH (psi)	Mechanical Properties: MINIMUM TENSILE STRENGTH (psi)
US BOLTS					
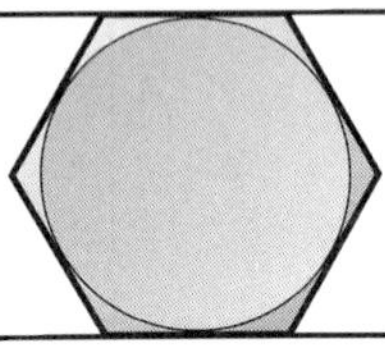	Grade 2 Low or Medium Carbon Steel	1/4 to 3/4	55,000	57,000	74,000
		Over 3/4 to 1-1/2	33,000	36,000	60,000
	Grade 5 Medium Carbon Steel, Quenched and Tempered	1/4 to 1	85,000	92,000	120,000
		Over 1 to 1-1/2	74,000	81,000	105,000
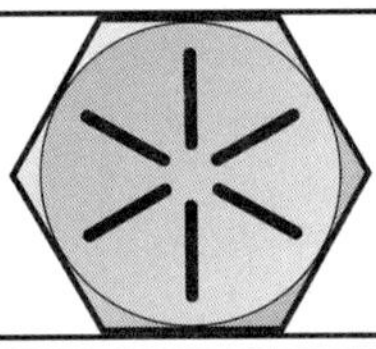	Grade 8 Medium Carbon Alloy Steel, Quenched and Tempered	1/4 to 1-1/2	120,000	130,000	150,000
Stainless markings vary. Most stainless is nonmagnetic	18-8 Steel Alloy with 17%–19% Chromium and 8%–13% Nickel	1/4 to 5/8		40, 000 min. 80,000–90,000 Typical	100,000–125,000 Typical
		3/4 to 1			100,000 Typical
		Above 1		40, 000 min. 45,000–70,000 Typical	80,000–90,000 Typical
METRIC BOLTS					
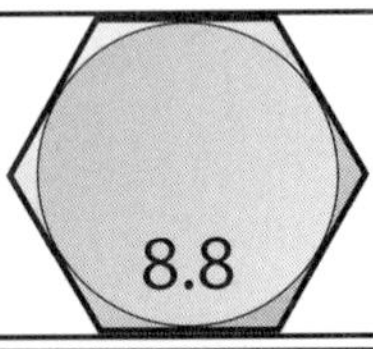	Class 8.8 Medium Carbon Steel, Quenched and Tempered	All Sizes Through 1 1/2	85,000	92,000	120,000
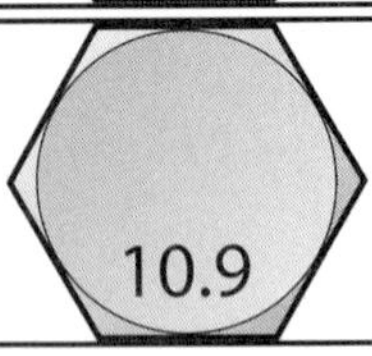	Class 10.9 Alloy Steel, Quenched and Tempered	All Sizes Through 1 1/2	120,000	130,000	150,000
Stainless markings vary. Most stainless is nonmagnetic	A-2 Stainless Steel Alloy with 17%–19% Chromium and 8%–13% Nickel	1/4 to 5/8		40, 000 min. 80,000–90,000 Typical	100,000 Typical
		3/4 to 1			80,000–90,000 Typical
		Above 1		40, 000 min. 45,000–70,000 Typical	100,000–125,000 Typical

Chapter 2

CHEMICAL THREAD LOCKERS, LUBRICANTS AND SEALANTS

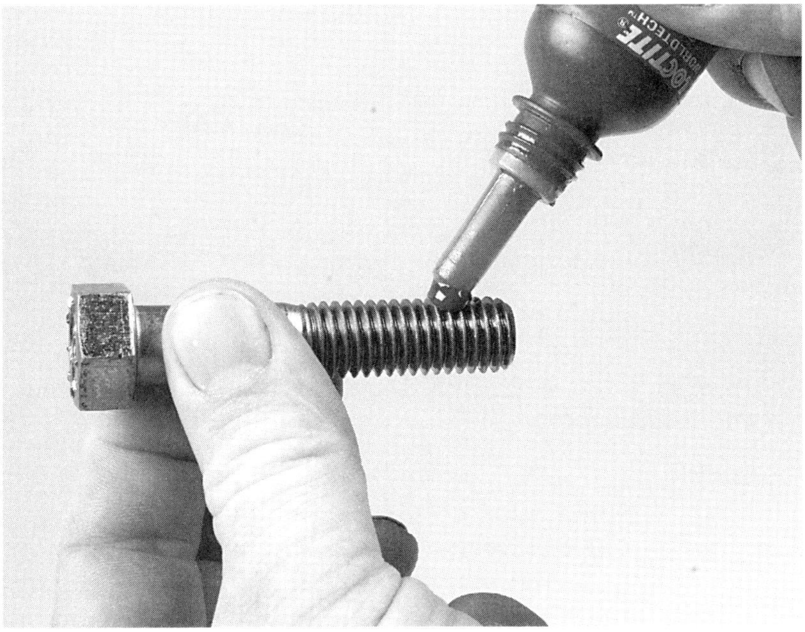

Shake the thread locker container well before applying. Apply to the threads. Avoid applying a bunch to the very tip, to avoid creating a hydraulic lock in a blind hole application.

Even the highest quality threaded fastener sometimes needs a bit of help to do its job properly. In terms of engine fasteners, certain bolts, studs and nuts can benefit from an applied chemical coating in order to achieve the perfect seal or position lock, to achieve correct torque value, or simply to prevent unwanted seizure caused by corrosion.

Application (or lack thereof) of chemical helpers can often create the difference between a reliable and an unreliable assembly.

THREAD LOCKERS

A chemical that "locks" a nut or bolt in place is generically referred to as a "thread lock." This is an anaerobic liquid or gel that, once cured, secures the fastener in place, preventing unwanted loosening. The term "anaerobic" means that the chemical is designed to cure in the absence of oxygen. Even though it may be applied to the threads of a bolt and set aside, curing won't occur until the bolt is installed.

When the bolt is installed, air/oxygen is squeezed out as the threads engage. Once the bolt is seated and all oxygen is eliminated in the thread area, the chemical then begins to cure and fully harden, cementing the bolt threads to the female threads. Basically, an anaerobic sealer fills the void that exists between threads, creating a single unified assembly.

An anaerobic sealer (when wet) also serves as a thread lubricant, aiding bolt installation and torque setting.

Thread locking anaerobics are available in a variety of adhesion strengths, and you need to be aware of the differences prior to application. Low-strength to medium-strength lockers are available for general use on just about any bolt application. This level of strength prevents bolts from loosening, while allowing future disassembly. However, some high-strength lockers are designed only for securing items in a semipermanent or permanent fashion, such as anchoring rocker studs, or adhering a press-fit sleeve.

The following list includes a number of popular and commonly available thread coatings offered by Fel-Pro, Loctite/Permatex and Silver Seal. This provides an overview of what's available and the characteristics of each.

Fel-Pro Gray Bolt Prep—A multi-purpose product that provides a number of services in one convenient package. It's a low-strength threadlocker, a bolt assembly lubricant, a thread sealer and, due to a graphite content, provides anti-seize protection during disassembly. Fel-Pro recommends this compound for all engine fastener locations.

Fel-Pro Tight #604—A multi-purpose anaerobic thread locker gel designed to lock and seal. This blue-colored, no-drip gel can be used to lock bolts and nuts, anchor studs, seal pipes and retain bearings and sleeves. It acts as a lubricant during assembly and cures to lock and seal against oil, water, coolant, gasoline, alcohol, air and hydraulic fluids. Specified

PERMATEX THREAD LOCKERS

Note: Permatex now offers thread lockers under version names instead of the older numbering system

Product	Color	Fasteners	Torque-Hold Break/Prevail	Cure Time Fixture/Full	Temp (°F)
Low Strength	Purple	up to 1/4" (6mm)	62/27	20 min/24 hrs	– 65–300
Apply before assembly for throttle body assembly screws, choke assembly screws, idle control screws, relay lock screws, body panel mounting fasteners, etc.					
Med. Strength	Blue	1/4"–3/4"(6–20mm)	115/53	20 min/24 hrs	– 65–300
Apply before assembly for oil pan bolts, alternator mounts, intake manifold, valve cover, disc brake caliper bolts, case cover bolts, shift lever nuts, etc.					
Surface Insensitive	Blue	1/4"–3/4"(6–20mm)	180/62	20 min/24 hrs	–65–300
Apply before assembly for oil pan bolts, valve and cam cover bolts, intake bolts, disc brake caliper bolts, rocker studs, alternator mounting, etc.					
Permanent Strength	Red	3/8"–1" (10–25mm)	168/150	20 min/24 hrs	–65–300
Apply before assembly for air tools, ring gear bolts, transmission shaft bolts, etc.					
High Strength	Red	3/8"–1" (10–25mm)	230/225	20 min/24 hrs	–65–300
Apply before assembly for threaded and slip fit joints on cylinder block and rocker arm studs, shock absorber bolts, ring gear bolts, frame bolts, etc.					
High Temperature	Red	3/8"–1" (10–25mm)	180/270	20 min/24 hrs	–65–300
Apply before assembly for camshaft sprocket bolts, crankshaft snout/flywheel bolts, idler bearings, press fit filler tubes, transmission input and output shaft threads, shock bolts, ring gear bolts, etc.					
Large Diameter	Red	1" and over	250/200	20 min/24 hrs	–65–300
Apply before assembly for frame brackets, suspension fasteners, shock absorber mount, wheel studs, etc.					
Penetrating Grade	Green	#2 to 1/2" (2.2-12mm)	90/30	20 min/24 hrs	–65—300
Apply after assembly for carb adjustment screws, distributor bolts, preassembled nuts and bolts, set screws					

for use on all metals, including copper, bronze, brass, stainless steel, plated steel and anodized aluminum.

Loctite 222—Low-strength thread locker for small fasteners that require occasional adjustment. For fasteners up to 1/4" (6mm) diameter. Safe on aluminum fasteners. Applications include carb/throttle body assembly screws, etc. Lavender/light purple in color.

Loctite 242—Medium-strength thread locker. The chemical is blue in color. An excellent general-purpose medium-strength thread locker.

Loctite 243—An enhanced version of 242. A medium-strength thread locker that offers superior locking for use on stainless steel and plated fasteners. It's also "import specified" by some foreign automakers. Specified as an approved OE "Dri-Loc" replacement. Offers good oil resistance. For fasteners 1/4" to 1" in diameter (6mm to 25mm). Applications include oil pan bolts, valve cover and cam cover bolts, intake manifold bolts, rocker studs, etc. Its primary interest to engine builders is its ability to provide a higher breaking strength for plated and stainless fasteners, making it

Even if locking washers are used, it's still a good idea to apply threadlocker to camshaft sprocket bolts.

Another good application for a threadlocker. Timing belt tensioner bolts should be locked in place.

ideal for custom and race engine applications. Blue in color.

Loctite 262—A refined version of a high-strength thread locker. Offers a "controlled lubricity" for reliable torque predictions, making it ideally suited for high volume production work. The viscosity allows it to flow around threads, but not drip off. Red in color.

Loctite 271—High-strength thread locker. An excellent general-purpose high-strength locker. Typical applications include rocker arm studs, differential bolts, connecting rod bolts, re-anchoring loose studs, etc. Recommended for use on 3/8" to 1" diameter bolts. #271 fills gaps up to .007". Operating range is -65 degrees F to +300 degrees F. Hardened 271 must be heated to 450 degrees F to disassemble. Red in color.

Loctite 272—A high-strength thread locker for high-heat and hostile engine environments (over 300 degrees F, up to 450 degrees F). Import specified. Heat or special tools are required for future disassembly. For fasteners 3/8" to 1" in diameter (10mm to 25mm). Applications include camshaft sprocket bolts, crankshaft attaching bolts, idler bearings, press-fit filler tubes, etc. Red in color.

Silver Seal Hylomar—A medium-strength thread locker (initially used only by Rolls Royce, and referred to as a "jointing compound"). Good for general-purpose use on all bolt sizes. This compound is solvent-dried, requiring that the parts air-dry for a few minutes before assembly. However, continued exposure to air won't hurt, as it's usable for an indefinite period of time after being applied to a bolt. It works well as a sealing agent, especially suited for open-deck cylinder head bolts. Dark blue in color.

THREAD LUBRICATION

Many engine fasteners require lubrication prior to use in order to achieve proper clamping torque, and to prevent seizing due to corrosion or galling. Lubrication of the threads (with the appropriate material for the task at hand) insures that torque readings will be accurate. Dry threads may tend to gall (metal of the hole transferring to the metal of the bolt), creating unnecessary friction, which leads to an artificially high torque reading.

For example, if the bolt threads experience undue friction, a torque wrench reading of 50 ft. lbs. may actually result in a torque setting of only 30 ft. lbs. In such a case, while the assembler thinks the torque is correct, the clamping force may be inadequate. And as we all know, improper and insufficient clamping force is often the culprit when a gasket fails to seal under stress.

While some bolts may require a seal to prevent pressurized liquids from escaping (where a bolt passes into a water jacket for instance), and others may require a thread locking material, virtually any thread coating that is chosen will help in achieving a more accurate torque reading.

Especially for critical bolts (cylinder head, main cap, connecting rod), remember to apply a small smear of lubricant to the underside of a bolt head or nut as well, to reduce friction between the bolt head (or nut) and the parent material during tightening and torquing. Also, if hardened washers are used, be sure to coat the washer faces and the I.D. of the washer, to reduce friction between the bolt under-head and washer and between the bolt shank and washer. This will further aid in achieving a more accurate torque setting. In other words, don't limit your lubrication to the threads only.

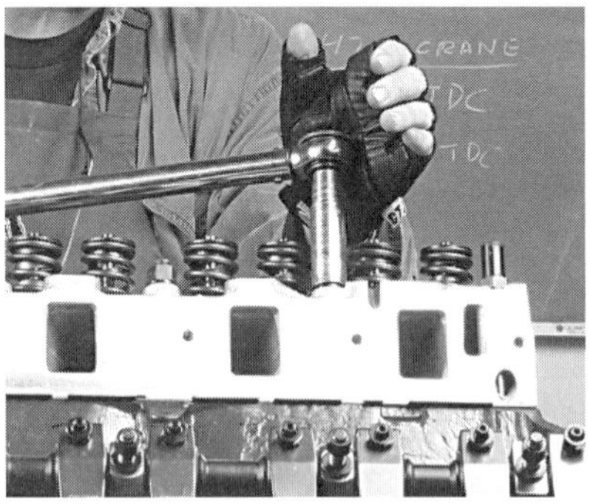

Always read the instructions beforehand. Be sure to apply the appropriate thread lube for the application at hand before installing and tightening any threaded fastener. Blindly installing and tightening fasteners without the proper lube, sealant or locker can lead to incorrect torque values, fluid or vacuum leaks or fasteners eventually loosened by heat or vibration.

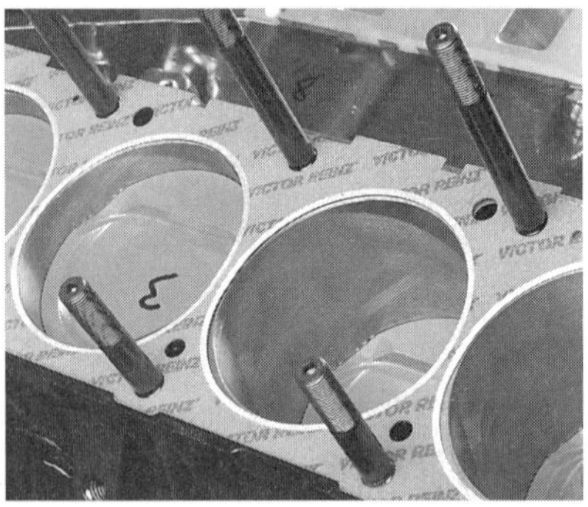

Be careful not to contaminate the head gasket or block deck with oil or moly. This can lead to inadequate sealing around water and oil ports.

Here an already moly-coated washer is being positioned into a head bolt hole on a Honda head. In this case, the stud top will be hidden from view, so the washer and nut are fully lubed prior to insertion.

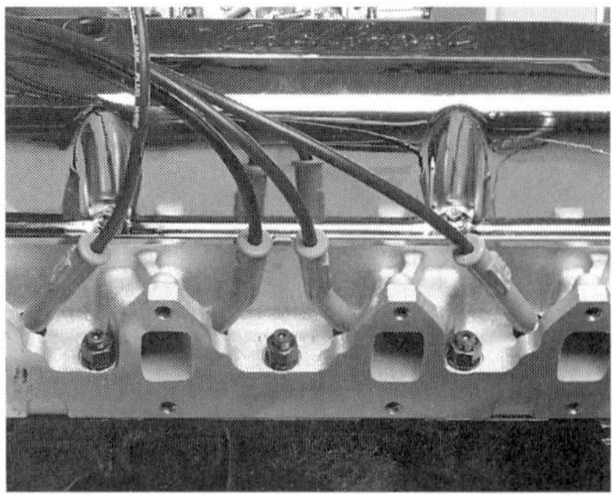

Whenever using washers and nuts on cylinder head studs, lube the exposed stud threads, the base of the nuts, and the faces and I.D. of the washers before installing. Follow the fastener maker's torque value specs, based on the type of lube you're using (oil or moly).

If you're using head studs, it's best to first install the studs (lubricate the bottom threads as required by the stud maker first), then install the head gasket, then install the head, then carefully lube the exposed stud threads. If the upper stud threads are too difficult to access with the head in place, be sure to apply the appropriate lube to the I.D. of each nut.

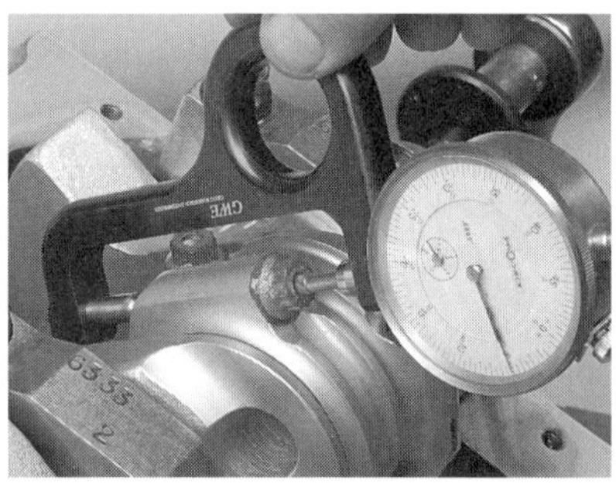

Whether you're tightening connecting rod bolts using torque value or stretch, be sure to follow the thread lube recommendations of the rod bolt maker. Remember: values will differ depending on which type of lube is used (oil or moly).

When performing AN hose end assembly, a light coat of lithium lube or penetrating lube will suffice.

It's always a good idea to lube AN hose end threads before assembly to ease threading and prevent galling.

Consider all areas of contact/friction.

As anyone who has used anti-seize compounds can tell you, this stuff can be very messy. It tends to migrate everywhere. Once it touches your fingers, it seems to show up everywhere in the shop, and it's nearly impossible to clean it from clothing. My advice is to wear latex gloves whenever handling the container and when using the stuff. Then, after you've stored the container, toss the gloves into the trash.

Some fastener manufacturers, like ARP for example, provide (or recommend) their own moly paste. In that case, use that specific product. ARP, for example, has their own formulation that they feel works best for specific head, main or rod bolt applications. Hey, maybe it's the same as another brand of lube, and maybe not. But when you're building a mega-horsepower monster and investing your kid's college fund in the engine, why take chances? If a fastener maker recommends a specific thread lube for their stuff, always follow their advice.

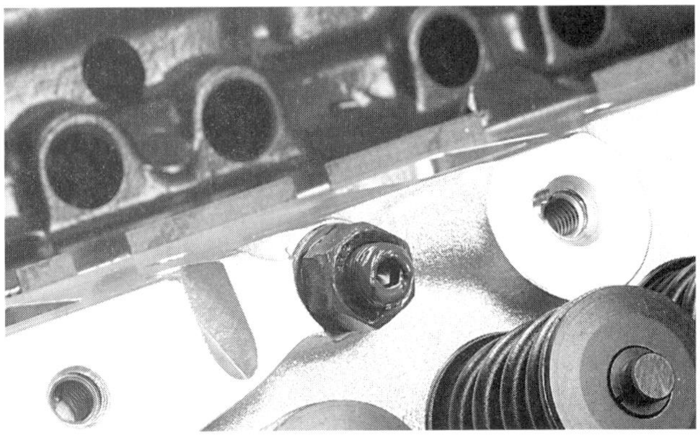

We applied ARP moly to the exposed stud threads on this cylinder head stud, as well as to the washer and nut. Consider all of the friction areas, not just the threads.

If the application calls for moly lube, apply a dab around the circumference of the threads. A little goes a long way, so there's no need to pile it on. Here we apply ARP moly to a head bolt's threads.

EXAMPLES OF SPECIFIC THREAD LUBES (PERMATEX)

Permatex Anti-Seize Lubricant—is a blend of aluminum, copper and graphite lubricants. Use during assembly to prevent galling, corrosion and seizing and to assure easier disassembly. Temperature range is –60 to 1600 degrees F. Salt, corrosion and moisture resistant. These are available in paste form (with brush applicator), tube form and in a non-aerosol spray version.

Permatex Copper Anti-Seize Lubricant—is a premium quality high-copper content blend anti-seize and thread lubricant designed to prevent galling, seizing and corrosion in high temperature conditions, such as exhaust header bolts, etc. Contains a high percentage of micro-fine copper flakes in a semi-synthetic grease carrier. Temperature range is –30–1800 degrees F. Also provides good electrical conductivity. Good for spark plug threads in aluminum heads, exhaust manifold bolts, oxygen sensors, knock sensors, thermostat housing bolts, fuel filter fittings and battery cable connectors.

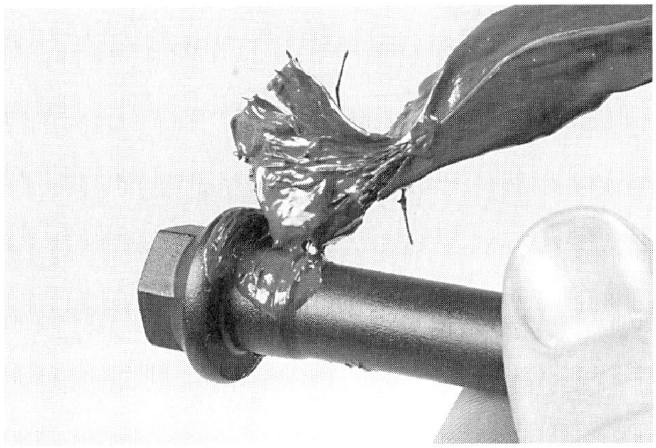

Here we apply moly lube to the underside of a cylinder head bolt. I can't stress this enough. Don't limit your lube to the threads only. By also lubricating the underside of the bolt head, you're reducing friction at all of the bolt's contact areas, enabling you to obtain a much more precise and even torque value during bolt tightening.

Permatex Nickel Anti-Seize Lubricant—protects metal parts from seizing and galling at temperatures up to 2400 degrees F. Recommended where copper contamination must be avoided, under conditions of extreme pressure and temperature, and with stainless steel and titanium and nickel alloys. Good for exhaust manifold bolts, exhaust system bolts, muffler clamps and tailpipe assemblies.

THREAD SEALANTS

A number of situations require the sealing of threads to prevent passage of fluids or pressure (fuel, oil, coolant, vacuum). A cylinder head bolt may pass into a water jacket or through an open head deck (exposing the threads to coolant or pressurized oil) or a pipe plug may be used to seal the crankcase or an oiling gallery, or a threaded fitting may enter a vacuum port area in an intake manifold. In any case, the threads must be sealant-coated in order to prevent leakage.

A number of sealing approaches are available. Some, like Teflon sealant, are ready to use immediately and cure over a 24-hour period. Some are anaerobic and will cure only when the bolt is installed. Others are solvent-dried, and must be allowed to breathe (allowing the solvents to evaporate) before assembly.

A choice of thread sealant can be a matter of personal preference, as a number of options exist.

Teflon thread sealant has been a long-time favorite used in sealing threaded fittings on fluid, hydraulic and pneumatic connections. Teflon paste lubricates threads and provides an immediate seal for threaded fittings on an engine such as sensors, plugs and plumbing connections for oil, fuel, vacuum, coolant or oil locations. Teflon paste also provides about a 24-hour window of opportunity for final adjustments. Citing Loctite's PST Teflon sealant as an example, the operating range is -65 degrees F to +400 degrees F.

Loctite's 565, citing another example, is an

If a plumbing fitting requires the use of a crush washer, no chemical sealants are needed.

Here's a JPP/BSPT plug for a Honda block. Apply a thread sealer to any water jacket or oil jacket plug or fitting that doesn't use a crush washer or O-ring.

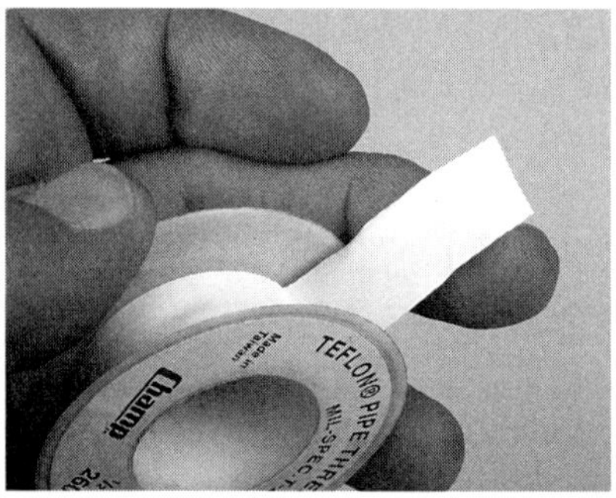

When using Teflon tape, cut the tape to length using a sharp blade or snips. Don't stretch and tear the tape, as this will create loose strands that can cause contamination.

These waterjacket plugs feature aluminum crush washers. Again, if the fitting uses crush washers or O-rings, no additional sealer should be applied.

Two common materials for sealing NPT threads. Thread sealant paste with Teflon is always a good choice, although some prefer to use Teflon tape, based on personal preference. If you do use tape though, make SURE that no excess tape falls into the fluid port. Small pieces of Teflon tape can easily clog fuel systems.

When using Teflon tape, wrap the tape in the opposite direction from the thread, to prevent unwinding the tape during fitting installation.

Once Teflon tape has been applied, gently squish it down into the threads to secure it.

anaerobic high-temperature thread sealant, designed to seal threaded connections at fluid, pressure and vacuum areas. It's import-specified, and approved as a Vibra-seal replacement. It carries a temperature rating of 375 degrees F, and withstands 10,000 psi. Applications include head bolt through-holes, vacuum valves, oil and coolant fittings, intake manifold switches, etc.

Hylomar, as we mentioned earlier, provides an excellent cylinder head bolt seal, and is also great for any threaded connection that needs extra fluid or pressure sealing capabilities.

Fel-Pro's Gray Bolt Prep provides a thread seal, while also offering thread lubrication during assembly.

A FEW TIPS

There are a few no-nos to avoid concerning chemical application to threaded fasteners. Consider how the bolt is going to be used. If it's a small bolt, or something that requires periodic adjustment or disassembly, don't use the strongest anaerobic locking compound available. Some of this stuff is very nasty, and will make bolt removal difficult if not impossible. Always read the application instructions on the chemical before use!

Most anaerobics contain dimethacrylate esters, which can irritate the skin. Avoid prolonged exposure to skin and avoid breathing vapors. In other words, use common sense.

The threads to be coated must be absolutely clean and dry before applying an anaerobic coating. To ensure the best results, it's always a good idea to use a primer first (some of the newer chemical lockers specifically don't require a primer, so always read the instructions before using!). These special primers are available from the major thread locker makers.

Choose the correct level of locking strength for the job at hand. When an anaerobic thread locker is labeled "high strength," believe it! High-strength formulas are nasty buggers that can create seemingly welded-in-place fastener situations when handled by inexperienced users. Always read the instructions with any locking compound before using it.

Can the compound be used immediately, or is there a waiting period? Some are solvent-dried, and require a few minutes' wait until the bolt can be installed. Some are ready for use at once. Again, read the instructions before using the material.

Never apply a locking compound to spark plug threads. There's no benefit to be gained, and locking the threads will only serve to make spark plug removal horrendously difficult.

Don't apply a silicone RTV to threaded fasteners

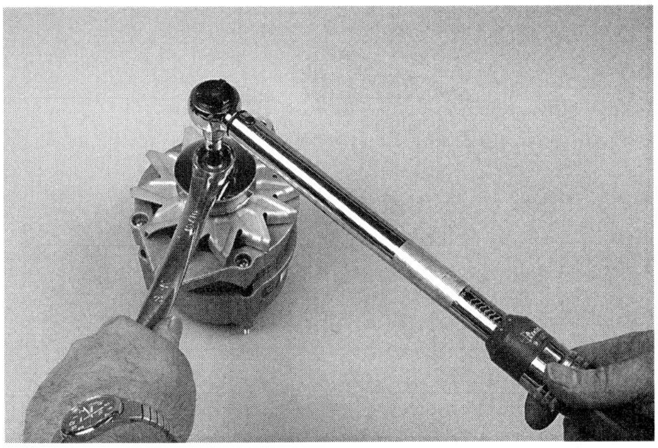

As long as you have access, you can tighten any threaded fastener to a specified torque value. Let's say you bought a replacement alternator, but you need to install the fan and pulley. Instead of using belt tension to hold the pulley in place, or instead of (don't do it!) using an impact gun to bang the nut tight, secure the nut with a wrench and take advantage of the female hex drive on the spindle for your torque wrench. Take the time to find the torque spec in the service manual and tighten the nut correctly, instead of guessing.

cheap one, or if it's been lying around the shop for years, it may be out of calibration. Secondly, is the fastener being tightened lubricated properly? Third, is the clamping force being created suitable for the diameter and type of metal?

As I just noted, when a bolt is tightened, it begins to stretch (it's this stretch that creates the needed clamping force). What many folks don't realize is that about 90% of the torque applied during tightening is used not to create clamping force, but to overcome friction. Friction occurs between mating threads, and between the underside of the bolt head (or nut) and the parent material of the object being installed.

Excess friction can occur if galling or thread seizing takes place. This is especially common with threaded fasteners made of alloys such as aluminum, stainless steel and titanium. If galling occurs (at any level of severity), this will make your torque readings inaccurate, since the galling effect will add significant friction at the thread mating area, which will result in a severely under-tightened fastener (since your torque reading or indicator click will take place well before your desired clamping load is reached).

Several factors can affect fastener tension, including type of material, material hardness, lubrication (or lack thereof), fastener hardness, surface finish/plating, thread fit and tightening speed.

Torque Technique

So, the next time you tighten a bolt, consider the many factors at play. It's not as simple as pulling

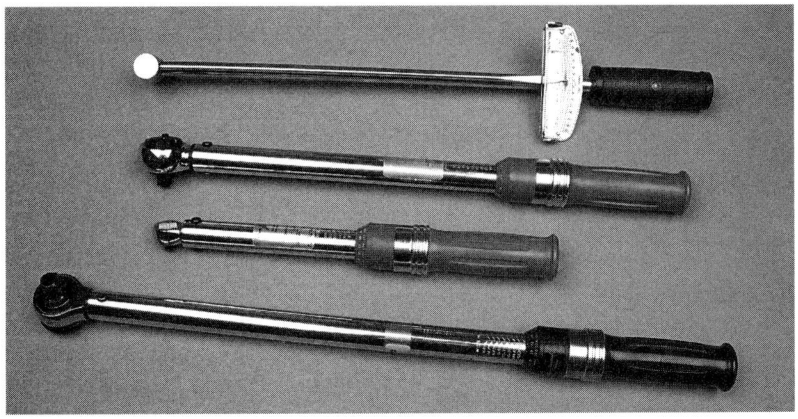

Torque wrenches allow you to tighten a fastener with the intent of achieving a target-specified torsional load.

until you hear a click. In order to give the threaded connection the best chance of performing properly, pay attention to the following:

1. Make sure the threads (both male and female) are clean.

2. Make sure the threads are in good condition, and free of deformation or burrs.

3. When necessary, apply the required lubricant to the threads before assembly (this may involve engine oil, molybdenum disulphide, an anti-seize compound, or an anaerobic thread-locking compound, depending on the situation.

4. Keep your torque wrenches clean and calibrated. Depending on the amount of use, consider sending your torque wrenches out for recalibration once per year. Also, store your torque wrenches in a safe place. Don't toss them around the shop. They're delicate instruments.

5. When tightening, whether using a common hand wrench or a torque wrench, slow down! Fast tightening creates friction and heat at the thread area, which can lead to thread galling. Fast tightening can also lead to inaccurate tightening, as the torque wrench must overcome the increased friction.

6. When you reach the torque limit (your desired torque value), approach this slowly and watch the needle or feel for the click. If you tighten too fast, you may pull the wrench past the preset limit (unknowingly adding a few more foot-pounds of torque).

7. Whenever space/access permits, always try to use

Moly lube provides a low coefficient of friction for more precise thread fastener tightening. However, pay close attention to the instructions in the vehicle service manual or in the instructions provided in the head bolt/stud kit. If only engine oil is specified in relation to the recommended torque value, use only engine oil. If the torque value is based on the use of moly, use only moly. If a choice is offered, use moly. Remember that the torque value will be based on the type of lubricant applied to the threads and underside of the bolt head or nut.

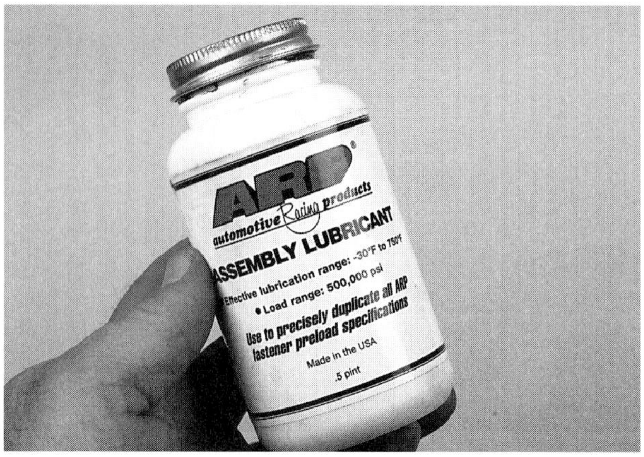

a calibrated torque wrench to tighten any threaded fastener. This will insure that you not only reach the desired clamping load per fastener, but also that each fastener on a given component is tightened at the same amount, distributing the clamping load evenly. Obviously, there are applications where the use of a torque measuring device isn't practical (when snugging down a few small bolts to secure a radio bracket, etc.), but for any engine-assembly or chassis application, it's mandatory that fasteners are tightened to a specified torque value. This isn't just theoretical mumbo-jumbo…it's vital!

The use of a torque wrench or other torque value or fastener stretch monitoring device is absolutely necessary for anything engine related, not only cylinder heads and connecting rods, but intake manifolds, carburetors, water pumps, oil pumps, rear seal housings, oil pans, timing covers, valve covers, exhaust manifolds or headers, power steering pump brackets, etc. If you own a torque wrench, take the time to look up the torque value for the carburetor fasteners, and adhere to the specs. Tightening anything on the engine by "feel" is more often than not the root cause for annoying fluid or vacuum leaks. A skilled engine builder/assembler never guesses about anything. Undertightening or overtightening can and does lead to problems, ranging from the mildly annoying to the most severe. Don't guess!

Anything chassis-related should also be installed while monitoring fastener torque values. Critical component areas include wheels, hubs, spindles, ball joints, control arms, strut rods, steering system components, axles, brake calipers, brake hose fittings, spring perches, etc. While you may or may not opt to cheat with non-critical installations such as sound systems, spoilers, consoles, etc., the performance, dependability and safety of your vehicle demand that engine, drivetrain and chassis fasteners are tightened to their required torque value. Period.

How Much Torque Is Needed?

We need to modify our thinking in this regard. It makes more sense to consider what load is required instead. After all, torquing to a specific value is simply a means to an end. The load is the important factor. Installation torque is simply a factor that needs to be considered when trying to achieve a specific load.

When torque is applied to a nut or bolt head, most of the input is spent in overcoming friction. At the end of the process, 85–95% of the energy transferred through the wrench has been lost. In other words, the clamp load itself may only represent 5–15% of your effort (this is a good example of why using a stretch gauge to tighten rod bolts makes so much sense).

Because of this frictional loss factor, slight variations in the frictional conditions can result in huge changes in the resulting preload. Variables include surface roughness, surface finish, lubricant, load range reached, dimensions, temperature and torquing sequence.

As you can see, it's just as important to ensure consistent friction conditions as it is to seek a consistent torque. If the dimensions and surface finish are fixed factors, the preload target range will depend on the lubricant and the tightening method.

The greater the friction, the higher the torsional stress in the fastener body. Since torsion is a function of the imposed friction, a given material reaches its yield strength sooner when the friction is high as opposed to low. During tightening, the apparent yield strength drops by 10–20% from the yield strength measured in tensile.

As a consequence, when regular preloads are required, good quality engine oils are sufficient for thread, washer, under-nut and under-head lubrication. In addition, the relatively higher friction will prevent any loosening under vibration. A black oxide finish may help to ensure consistent preloads (past the mandatory 3–6 pre-tightening procedure to break-in the mating surfaces).

When higher loads are needed, the inevitable torsion stress can be minimized by choosing slicker threads. Less friction will bring the yield point up a few percent, allowing the clamp closer to yield. It will also reduce the residual torsion that is present when the wrench is released.

Choose Lubes Carefully

When ultra-slick lubes are used, they require high preload to prevent backing-off under vibration. For this reason, be aware of undertorquing and make sure that your fastener is not losing clamp during

When torquing heads, position the engine on your stand to gain easy access to the head fasteners, to attain a comfortable working position. This makes monitoring torque easier.

Examine head bolts carefully if you plan to re-use them. If it's an OE (original equipment) bolt, the vehicle service manual should list a specified maximum shank length.

Torque cylinder head bolts in the specified sequence, and in steps that gradually lead to final value.

Whenever installing an aluminum cylinder head, hardened washers must be used to prevent the underside of the bolt heads (or nuts, where studs are in place) from digging into the aluminum. Always remember to lubricate the underside of the bolt head in addition to the threads.

service, which can result from compression of the mating part or stretching of the fastener.

Also, be absolutely confident in your torquing procedure. Slick lubes will emphasize the precision of any torque wrench. See Chapter 2 for more details on lubes and sealants.

Cylinder Head Torquing

In order to optimize your results, following are steps to consider:

• Pressure should be well-distributed along the joint before significant loading is applied. Towards this end, it's recommended that the first tightening step is made at 20–30% of the final desired torque value. This should prevent localized damage due to excessive pressure. In other words, perform the tightening process in multiple steps, rather than tightening a bolt to the final value in one step.

• Each additional tightening step should guarantee balanced and progressive loadings. After the first preload step, try to keep the torque targets consistent with the angle of turn needed to reach that torque. For example, the second step could be achieved in approximately two 60-degree turns, and the last step in a single 90-degree turn. This will ensure good repeatability.

• Most applications can be achieved in three steps, but more steps can be used if you're dealing with a great deal of compression.

• If you opt for a multi-step procedure, make sure that the steps are not too close to each other. Static friction is more difficult to overcome than dynamic friction, which means that if the steps are too close to each other, the wrench might click before even moving the nut or bolt head. In other words, don't try steps that are only a few ft-lbs apart.

• ALWAYS follow the tightening pattern, or sequence that's specified for your particular heads and block. If you're in a bind (2 a.m. before a race and you don't have the specs handy), it is generally

Here a torque deck plate is installed to a Ford 427 FE block prior to cylinder bore honing. Spacers are often supplied with deck plates, to allow the use of the same type and length of head bolts or studs that will be used during final assembly. The idea is to address all of the variables that might influence the stresses that will be induced during final engine assembly.

Some deck plates require the use of a used (crushed) head gasket. This better simulates the final assembly, and the gasket serves as a spacer to prevent any metal pull-out around the bolt or stud holes from interfering with the precision mating of the deck plate to the block deck.

Deck plates (also called torque plates or honing plates) stress the block to simulate the installed cylinder head. This allows a more precise creation of a round and uniform bore.

Deck plate positioned onto a Honda race block prior to cylinder honing. This block will use cylinder head studs during final assembly, so the same type studs are used when securing the deck plate. Note that some deck plates are made of cast iron, while others are made of aluminum. It's best, where possible, to use a deck plate that features the same type of material that is used in the cylinder head.

accepted that cylinder head fasteners will be tightened in a spiral pattern, starting at the center and gradually working outboard. As you're looking at the top of the cylinder head, visualize a spiral pattern that starts at one of the center fasteners. The spiral will intersect the opposite row center, then as the spiral widens in diameter, each successive fastener location is tightened. That may sound confusing at first, but if you take the time to study the bolt hole layout of the head, you'll see what I mean.

Torque Plates

Let's take a detailed look at the what, why and where of these stress-inducing cylinder shaping helpers.

Torque plates (also called deck plates) are thick slabs of metal that are fastened to the deck of a cylinder block during the cylinder bore honing process. A deck plate allows you to simulate an installed cylinder head, which places the same stresses in the block as the block will experience when fully assembled. The reason for this? Engine blocks distort when assembled. A block may look like a big, beefy chunk of iron or alloy, but as the "serious" fasteners are tightened to their respective values, the block is subjected to distortional forces that you may have a hard time imagining. This distortion is most prominently seen in the cylinder bores. As the main caps and cylinder heads are clamped onto the block, cylinder bores are subject to forces that can create tapered, barrel-shape and out-of-round dimensions in an otherwise perfectly round bore. Granted, some blocks (lighter, thinner-wall blocks) are more subject to bore distortion than others, but the forces that create bore distortion are present in all assembled blocks.

If a block's cylinder bores are properly bored and honed, they are round. However, when we begin to assemble the engine, tightening main caps and cylinder head fasteners to their required values, one or more cylinders may begin to distort. This means that when the engine operates, the pistons and rings are forced to deal with tight spots and loose spots in their bores. Once you understand this, it's not difficult to understand how this can cause increased friction and wear, as well as compression and oil blowby and loss of power. Even where these bore distortions are extremely minor and might cause no noticeable ill-effects in a street engine, it should be obvious that a race engine simply won't perform to its potential if the rings are fighting and fluttering their way home on every stroke of the crank.

In order to try to compensate for this potential bore distortion, we install a deck plate to the block deck. This theoretically creates the same block distortion that the assembled engine will see. With the block in this stressed condition, we then hone the cylinders to their final dimension, maintaining a round (consistent from top to bottom) cylindrical condition.

PS: when installing a deck plate, you may need to install a used (already crushed) head gasket (some deck plates require this, while others don't). It's also best to use the same brand and type of cylinder head fasteners that will be used on the assembled engine. And naturally, the deck plate must be

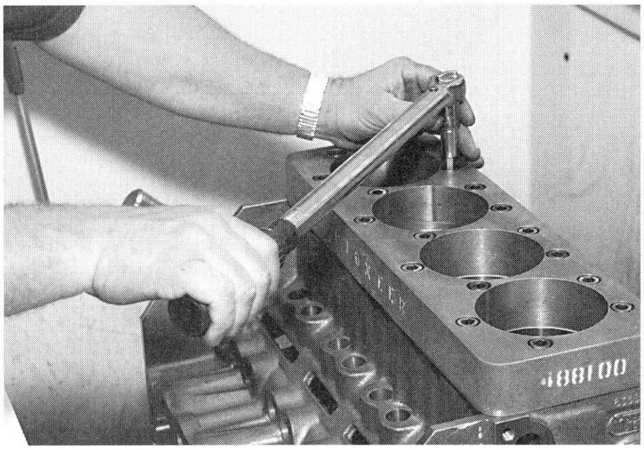

When installing a deck plate, the tightening sequence pattern must replicate that of the intended cylinder head installation.

Always follow the correct tightening sequence for the main caps.

torqued in place, following the same values and tightening pattern required for final cylinder head installation.

When a row of cylinder bores are honed to size with a deck plate installed, your chances of achieving an in-use bore that is truly round are far greater than if the cylinders are honed in a stress-free state. When plates were first used in the race engine community, this was viewed as an exercise in attaining geometrically correct bores in order to extract the last bit of power and durability from a high-stress racing engine. Over the years, the industry has realized that any block can benefit from the use of stress-honing in order to increase engine life and to create superior piston ring efficiency, and to reduce operating friction. In fact, today's OEMs use deck plates to hone new production blocks (it's nearly impossible to get the car makers to admit to anything officially, but we suspect that they're all doing it). It's certainly safe to

In some cases, such as this 1966 Ford 427 FE block, the main caps are secured in two planes, using both vertical cap bolts and horizontal "side" bolts. In the example shown here, precision main cap spacers are inserted between the cap and block before the side cap bolts are installed. Even though the crankshaft will not be installed for the cylinder honing process, the main caps must be fully tightened to specification, with all fasteners in place. Cylinder bores can easily be influenced by tightening the main cap fasteners, so in order to pre-stress the block for honing, main caps must be fully installed on any block before cylinder honing.

say that most of today's precision-cast blocks require torque plates while machining, to maintain cylinder roundness.

Head bolt torque can commonly distort cylinders from .001–.002", and in certain applications, bolt torque can result in as much as .004" distortion. On small-block and big-block Chevys, for example, the cylinder walls will pull inwards in areas immediately adjacent to the bolt locations. Big-block Fords and Pontiac 10-bolt blocks may see as much as .007" movement. When this happens, rings won't seat properly and scuffing is likely if the engine overheats. This potential for bore distortion applies to any engine, big or small. I've seen as much as 0.007" bore out-of-round on the number 4 bore in a 2.0L Dodge Neon block, as an example.

Torque Plate for Boring?—While many engine builders that use plates dedicate them for the honing operation only, it does in fact make sense to stress the block during the boring process as well. This is to minimize the potential skipping of the honing stones if gross distortion is present. If the cylinders are stressed during boring, this will help to minimize the later work that the honing machine must perform in order to correct/overcome the bore dimensional distortion. In short, if you plan to bore

Main cap fasteners being torqued to value with a digital ratcheting torque wrench.

and hone, and if you plan to use a deck plate for honing, you may benefit (in terms of cylinder bore profile) by also using the plate for the initial boring phase as well.

It simply makes sense to use a torque plate to stress the block for honing (and boring) whenever possible. Yes, this represents an investment for the machine shop (average cost is around $200–$250 or so per plate), but if you hone a certain type of block on a routine basis, the cost can certainly be justified.

When you consider the added cost, both in terms of purchasing the deck plate and the time it takes to install it), you should understand that the machine shop needs to pass this added cost along to you, the customer. If the shop doesn't charge a few extra bucks for the use of the deck plate, he's giving his investment away, and that's just not fair to the shop. So, don't complain if the shop charges you an extra $10–$30 extra for the deck plate use.

As a case in point, I recently had a Chevy 454 block bored and honed at a reputable machine shop in Toledo, OH. The block was bored +.030" (without a torque plate) and the bores were checked for roundness and taper. They seemed close to perfect at that point...nice and uniform. The block was then placed on the honing machine, and a BHJ deck plate was torqued to one bank of the block. At that point, we immediately found a series of low spots (directly across the wall meat from nearby head bolt locations) of about .002"! This is a prime example and total validation of the theory that head installation can distort cylinder walls. Would this motor run if honed without a stress plate? Of course. Would the motor run better and live longer if honed under this stress? Absolutely. In contradiction to popular belief, using a deck plate is NOT a trick reserved for race engines only. This consideration should be given to every block that runs through a shop.

Don't operate under the assumption that only X-brand or X-model of block distorts, while Y-brand or Y-model does not. There is enough deviation in casting quality and casting material rigidity among any family of blocks to warrant concern for bore shape in an assembled state. You wouldn't assume that a cylinder is sized properly without measuring, and you wouldn't assume that a valve seats properly without making the appropriate checks. Why then, would you assume that the cylinder wouldn't distort when the head is clamped down? If it sounds like I'm pushing hard for the use of deck plates, that is exactly the case. I firmly believe in their use, and of the benefits of the end result. The moral: use 'em if you got 'em.

When using a deck plate in the honing process to achieve the finished size, it's best to remove the initial couple of thousandths, then allow the cylinder to cool (as the heat buildup may pull the bore into a barrel shape). Once the cylinder has cooled, then remove the final .0005–.001". This will help you to get the most benefit of a deck plate.

Gasketed vs. Non-Gasketed Plates—In terms of stress approach, there are two types of torque plates for use on blocks: those that require the use of a crushed head gasket and those that don't. When a gasket is used on a flat-deckside plate, an air gap is created at the gasket fire ring, to simulate the fulcrum effect that will be seen when the head is attached. For production/high-volume use, special deck plates that do not require head gaskets are made by slightly spot-facing the deckside boltholes. This simulates the fulcrum effect (allowing the bolt clamping force to draw down at the bolt hole areas, which creates a similar effect of the loads that would be created if a gasket was in place). Basically, the gasketless versions are intended for high-volume production engine rebuilders who don't want to mess with keeping and using an inventory of head gaskets, while the gasketed versions are better suited to the low-volume, smaller shop that wants to simulate as closely as possible the load distribution that the block will see when assembled.

Also, counterbored plate models are available (where the top side bolt holes are recessed to create an obstruction-free top side) for those shops that want to mount a boring bar on top of the plate.

By the way, torque plates are also available to stress the cylinder head itself. This plate is bolted to the cylinder head deck, to induce torque stresses in the valve seats during grinding. Some cylinder heads show a 5–10% improvement in leakdown when the valve seats are ground under stress. Citing BHJ's cylinder head plate as an example (currently, the only source we're aware of), the plate is drilled and tapped to accept head bolts. Some clearancing of the plate may be necessary to fit large grinding stones into the combustion chamber. See the charts beginning on page 54-57 for sources of torque plates.

TORQUE-TO-YIELD

Once upon a time, it was common practice to retorque cylinder heads after initial use. This was necessitated by the expected relaxation of the compressed head gasket after initial clamping. In far too many of today's motors, this procedure of retorquing simply isn't practical, due to the extremely difficult access to cylinder head bolts, caused by an extreme overcrowding of the engine bay. While it's still easy to access the head bolts on a carbureted race engine or restored muscle car, it's a real pain in the timeclock to perform this service on a late model engine that features direct injection, a cumbersome upper intake plenum and a maze of wiring harnesses, plumbing and vacuum hoses.

The engineering answer to this was the advent of the torque-to-yield cylinder head bolt. As I mentioned earlier, when any bolt is tightened (any threaded bolt or stud, not just head bolts), it stretches by design. This is referred to as the bolt's elasticity. This stretch creates clamping force—the more the bolt stretches, the greater the clamping force becomes—to a point. When the stretch enters the bolt's yield point, the stretching (and therefore its clamping force) stops and begins to diminish.

Similar to a rubber band, the bolt will stretch to a point, retaining its elasticity, and will be able to spring back when tension is released. Once it's stretched to its yield point, it stops generating tension and beyond that point, will break.

The same phenomenon takes place whenever a bolt is tightened. Ideally, we want to stretch the bolt to a point just short of its yield. In that way, we take full advantage of its clamping force, without yielding the bolt. When loosened, it will spring back to its original length, retaining its elastic property.

A torque-to-yield bolt is designed to provide maximum clamping load on its initial tightening, since we don't expect to return to the job by retorquing the bolts in order to provide more crush for a now-compressed head gasket. With a TTY bolt, we arrive at full "long-term" clamping load during the initial tightening process. When the gasket does relax, the TTY bolts continue to provide sufficient clamping load to maintain adequate gasket seal.

TTY Torque Specs

TTY bolts are designed to be tightened within a small window of tension—just short of its yield point. The torque spec for a TTY is generally higher than that of a standard head bolt (a non-TTY bolt), because we're taking full advantage of the TTY bolt's designed elastic range. If a TTY bolt was used in an older engine, but using old, non-TTY torque specs, the TTY bolt probably won't stretch enough to provide the clamping force that it's capable of. The likely result: head gasket leaks because of low clamping force. Likewise, if new TTY specs (involving both torque and bolt angle) are used on non-TTY bolts, chances are good that you'll stretch the bolts beyond their elastic range, or break the bolts, or damage the female threaded holes in the block.

In short, TTY bolts are designed to provide more stored energy. In simple terms, they're designed to stretch more, providing a long-term clamping load without the need to retorque.

Most TTY cylinder head bolt specs will involve both an initial torque (in ft–lbs. or newton/meters), plus a specific degree of bolt head rotation (this is called torque plus angle tightening). Regardless of how silly the extra step of angle tightening may seem to some people, there is a legitimate reason for this approach.

Since the bolt engineers already know how much stretch, and therefore clamping load, will occur based on how far the bolt head rotates, they use the angle of bolt head movement to determine exactly how much load is being exerted.

A torque spec alone cannot be used to exactly determine bolt stretch because of the friction variables that come into play during tightening. A certain amount of torque loss is caused by the friction of the bolt head underside to the cylinder head contact surface, and by the friction of the thread engagement. The type and amount of oil/lubricant on the threads provides yet another variable in terms of friction.

Depending on how smooth and burr-free the bolt head contact area is, and on how smooth and uninterrupted the threads are on both the bolt and the female threaded hole, a torque reading alone really can't provide accurate and consistent clamping load information. The resistance caused by bolt head or thread friction is read by a torque wrench in the same way bolt stretch is read—it's all resistance to movement. If enough friction is created by these variables, a reading of 40 ft-lbs. on a torque wrench may in reality only provide the equivalent of maybe 20 ft-lbs. that actually works to stretch the bolt. And if the bolt was designed to stretch to its just-short-of-yield point at a true 40 ft-lbs., this inadequate bolt stretch will mean insufficient clamping load on the gasket, which means a probable gasket leak or failure down the road.

So, although some of us may not like to deal with TTY torque and angle specs, we really don't have a choice. That's the way it is.

Should TTY Bolts Be Reused?

This is an issue that will stir debate. A TTY bolt is designed to stretch to a point immediately prior to its yield point. On that basis, it is theoretically possible to reuse them. Some car makers claim that it's OK to reuse TTY bolts a specific number of times. However, that recommendation is based on the assumption that each bolt has been properly

TORQUE METHODS

Method	Tool	Error on Torque (+/-, %)	Error on Preload (+/-, %)	Notes
Feel	wrench	—	40	
Torque	impact wrench	15–30	20–40	damages surfaces
	hydraulic/air	10–20	20–35	
	click wrench	5–15	15–30	calibrate often
	dial wrench	3–10	10–20	calibrate often
	digital wrench	0.5–2	5–15	calibrate often
Torque+angle		—	5–15	requires prior testing
Torque to yield	digital wrench	—	3–10	undercut studs or "soft" bolts only
Stretch		—	1–10	access to fastener may be tricky
Ultrasonic		—	1–5	Use limited by shape and material
Strain gauge			1–2	ideal for defining torque procedures

tightened in the past. Since you have no way of knowing if a TTY bolt has been improperly tightened, perhaps past its yield, the safest course of action is to always use new TTY bolts in every single application. This recommendation is also made by the leading gasket makers, including Fel-Pro and Victor-Reinz. In fact, Fel-Pro has released a number of tech bulletins on this very subject, emphasizing the need to replace TTY bolts in every case (refer to Fel-Pro bulletins #4522-91, 4774-93, 4904S and 5052-96; as well as Perfect Circle's Machine Shop Service Bulletin CH-7). Considering the cost of the bolts, it's a cheap insurance policy to protect a costly engine rebuild.

Tightening TTY Bolts

When tightening a TTY bolt, you will invariably have to meet both a torque and angle published spec. For instance, the bolt spec may dictate that the bolt is torqued to 45 ft-lbs., then tightened further by degrees of bolt head rotation (let's say 45-degrees). Some bolt specs may ask you to reach an initial torque, followed by several steps of rotation (20-degrees, followed by 20-degrees, followed by 10-degrees, for example).

In order to apply a specified torque, obviously you'll need to use a torque wrench (needle type or click type). In order to tighten the bolt further by angle, you'll need an angle meter. These are available as separate units that are attached to the wrench. They feature indexable needles and provide a means of holding the meter base in position (so that only half of the meter moves—either the needle or the meter scale). This hold-down may be in the form of a mechanical stop built into the meter, or a remote cable secured to a convenient location on the cylinder head via a clamp or magnet. Some of these are junk, and some are very good (Lisle Corp. makes one that I like very much). On the good side, they're very inexpensive, so you can afford to try a few out in order to settle on one you like.

Precision Instrumentation—Another and very high-tech approach is to use an electronic torque wrench. A classic example of this is Sunnen's new computerized ETW-125 Sensor I. This thinking torque wrench is a high-precision fastener tightening instrument that brings laboratory accuracy into the engine builder's shop.

To make a long story short, this tool allows you to tighten a bolt to its torque-tension yield point without even knowing what the specs are. The wrench has four different tightening modes: angle control, where you can tighten to a prespecified angle (from the initial snug torque); torque control, where you can tighten to a prespecified torque; yield control, allowing you to tighten the fastener until it reaches its torque-tension yield point; and dynamic torque audit, which allows you to determine the existing torque of a tightened fastener.

The head of the tool features a strain gauge and a potentiometer that detects torque and angle respectively. The body of the tool features an easy-to-read digital display, and four control knobs. A joint switch allows selection of hard, normal and soft joint settings (joint characteristics of metal to metal with course thread, metal to metal with fine thread or gasketed). This tells the computer how often (in milliseconds) to check its torque/angle references (it will check more frequently on the hard setting). An "angle control" switch allows the operator to dial in the specific angle required by specs. A "torque control" switch allows the operator to dial in a torque value. A "mode" switch allows the operator to select the tightening mode, choosing between dynamic torque auditing (for checking existing bolt torque), torque control, yield control (measuring actual bolt stretch) or angle control.

A reset button clears the memory and readies the tool for the next tightening. A series of LED status lights provide instant information (an amber "run" light indicates that the tool is ready to use; an amber "snug" light indicates that snug torque has been reached (when set for snug torque as a first step in achieving a torque/angle procedure); a green "OK" light indicates proper completion of a tightening operation (when preset torque or angle or yield is achieved); and a red "fault" light indicates unsuccessful tightening. A built-in beeper provides another signal to the operator. A continuous tone signals an OK condition, and a beeping tone signals a fault condition.

This is more than a glorified torque wrench, since the tool is capable of actually determining where bolt yield takes place. This eliminates guesswork. The tool will not only find the precise yield point of a fastener, but it will verify if a fastener is acceptable. In other words, it'll find bad bolts. This tool eliminates the need to use a connecting rod bolt length gauge, and offers the opportunity to re-use some TTY bolts (if you can accurately verify that a TTY bolt is good, then it's safe to reuse it). Naturally though, if you don't have this tool, there's no way to determine this, so you're always better off by using new TTY bolts.

The electronic torque wrench also has the uncanny ability to determine the existing torque of a tightened bolt (by applying tightening force, it takes into account torque and bolt movement, eliminating the need to break the bolt loose and retighten). It actually audits the torque of a fastener without the need to loosen the fastener! According to Sunnen, the most effective way of auditing a fastener's torque is to tighten the fastener further by a small, precise angle, and measure the instantaneous torque. It's all a matter of complicated mathematical algorithms that my math-professor father would easily comprehend, but that I won't pretend to understand. Somehow, it knows, which is all that matters. As a laymen once remarked when a plumber attempted to explain how a water well works, "I just want to turn on the faucet and have water. I don't want to know where it comes from," we can remain blissfully ignorant of the inner workings of this tool. The bottom line is that it does the job, and it's very, very easy to use.

Precision isn't cheap, and neither is this tool. Since it goes for around $5,000, it obviously isn't a tool that every shop will pick up on a whim. However, for high-dollar race shops and high-volume production shops that depend on a high degree of accuracy and repeatability, it's a purchase that makes a great deal of sense.

Another firm, RS Technologies, offers an array of fastener verification equipment (referred to as "Torque-Angle Signature Analysis"). This sophisticated electronic equipment is compact (laptop sized), and allows you to monitor fastener torque, angle, bolt elasticity and clamping force. A special torque wrench fitted with a transducer is hard-wired to the compact computer. When a bolt is tightened or checked for tightness, a signature curve is displayed on the unit's monitor, showing the tightening curve as the bolt is snugged, drawn into an elastic state and approaches yield. Reference curves can be established (say, TTY bolt tightening on Mitsubishi 3.0 cylinder head bolts, for example) to duplicate bolt tightening or make comparisons. I saw one of these units at the Perfect Circle machine shop school, and it was pretty wild. Watching the monitor, you can see exactly what the bolt is doing, as it happens. You can also save the information, make a printout or download it onto a disc. RS Technologies makes an array of these systems. The most compact is their model 960 Hand Held Transient Recorder (we didn't have a price on this at the time of this writing).

Granted, this isn't for all shops, but we wanted to point out that instrumentation like this is available for those shops who have an interest in accurizing their fastener tightening.

THREADED FASTENER TIPS

The entire issue of TTY bolts raises a point that is by no means new or solely related to TTY technology. In any situation, wherever a threaded fastener is used (heads, water pumps, main caps, intake manifolds, etc.), attention to detail will provide its own rewards. Here are a few reminders that, although seemingly basic in nature, bear repeating.

• Inspect and clean all bolt threads before use (this applies to new and used bolts). Make sure threads

Whenever using a torque wrench, support the wrench head with your hand, only to prevent unwanted angle misalignment. This is especially important when using socket wrench extensions.

are free of burrs, dirt, rust and are not damaged.

• Lubricate threads with the appropriate oil, grease or sealer as required for the application.

• Avoid over-oiling bolt threads, since an excess of oil may seep to the bottom of a blind hole, causing a hydraulic lock as the bolt is tightened. This will provide a premature and completely incorrect torque reading and will prevent full clamping force.

• Inspect the underside of the bolt head to make sure it's free of burrs, dirt or rust.

• Inspect and clean all female bolt holes. Make sure threads are clean and undamaged. If it's a blind hole, make sure the bottom of the hole is clean and free of dirt, grease, rust, metal particles, water, oil or solvent.

• Verify the bolt's application. If the job requires a grade 8, don't use a different grade. Verify bolt length. If it's too short, inadequate thread engagement will result in potential failure. If too long, the bolt may bottom out in a blind hole, resulting in poor clamping force; or it may pass through an open hole to create interference with another moving part. In the case of head bolts, if the correct part number bolt set is used, you shouldn't have to worry about this.

• If a TTY bolt is specified, make sure the bolts to be used are indeed TTY. If in doubt, don't use them.

• Unless high-precision lab verification equipment is available for bolt inspection, never reuse TTY bolts. Play it safe and use new TTY bolts.

ALL ABOUT TORQUE WRENCHES

A torque wrench measures the amount of turning force applied to a threaded fastener (nut or bolt). Torque wrench scales usually read in foot-pounds (ft-lb) or inch-pounds (in-lb) and Newton-meters (Nm). When using the foot-pound scale, one foot-pound equals one pound of pull on one foot-long lever arm.

There are three basic types of torque wrenches commonly used for automotive use: the flex bar, the dial-indicator and the sound-indicating types. The flex-bar type (also called the scale type or beam type) features a stationary needle that runs the length of the shaft handle. The needle indicates applied torque against a printed scale that's located at the base of the handle. This type of torque wrench offers no preset limit, and there is no felt or audible "release" when a specific torque value is reached.

The dial-indicator type features a dial-indicator readout for visual display. Both the flex-bar and dial-indicator types provide visual displays of applied torque. The sound-indicating type will signal applied torque by momentarily releasing the wrench a few degrees when the preset torque value is reached. The release is usually accompanied by a "click" sound. However, there are some release-type torque wrenches that will release upon reaching the preset torque, but may not provide an audible click. The release/click type wrench is adjusted by means of a micrometer scale on the handle.

If the torque wrench releases momentarily and/or clicks, this is referred to as a "signal" type. The "indicator" type refers to the visual display units such as the flex bar or dial-indicator style.

Admittedly, this can be somewhat confusing, since there are so many different types available (some use a flex bar with no release, some use a dial indicator with no release, some use a click signal, some use a silent release, etc.). Some indicator models feature a memory pointer that remains at the maximum reading, until manually reset.

Torque wrenches are designed to permit an operator to determine applied torque on a threaded fastener. They measure torque in ounce-inches, pound-inches, and pound-feet (in addition to metric scales). Regarding foot-pound measurements, some sources will use the term pound-feet while others use the term foot-pounds. The two terms are interchangeable. Metric scale torque wrenches are available in Newton meters (Nm), meter kilograms (mKg) and centimeter kilograms (cmKg), with Nm the more common scale. Many torque wrenches provide dual scales for reading in either English or metric formats.

Any adjustable torque wrench (the commonly used micrometer-handled click type for example) should be set at its lowest torque reading when not in use. This is something that many technicians commonly forget. If left stored at a high-torque setting, the calibration may be affected over a long term. When you're done with the wrench, readjust it to the minimum setting before storing it in the

METRIC CONVERSIONS

1 Newton meter:	0.741 ft-lb.
1 Newton meter:	8.892 in-lb.
1 m-Kg:	7.25 ft-lb.
1 cm-Kg:	.870 in-lb.
1 n-Kg :	9.8 Newton meter
1 Meter:	100 centimeters
1 Meter:	39.37 inches
1 Meter:	3.2808 feet
1 Kilogram:	1000 grams
1 Kilogram:	2.2046 pounds
1 Newton:	0.2258 pounds

ENGLISH CONVERSIONS

1 in- lb.:	1.15 cm-Kg
1 ft-lb.:	1.35 Newton meter
1 in-oz.:	28.35 in-gram
1 in- lb.:	16 in-oz.
1 ft-lb.:	12 in-lb.
1 foot :	12 inches
1 pound:	16 ounces
1 pound:	453.59 grams

tool box.

Never abuse a torque wrench. It's designed as a precision instrument and should NEVER be used as a pry bar or as a disassembly/assembly tool. Handle all of your "wrenching" duties with common wrenches, and only use the torque wrench as the final-adjuster to reach a specific torque level. Don't use it as your all-around wrench.

When using an adjustable torque wrench, be careful not to overtighten by applying torque past the release point. At very low settings, the "click" may not be heard, especially in a noisy shop. It's best to become familiar with the "feel" of the release, rather than relying on the sound of a click.

When using an indicating-type torque wrench (such as a flex bar or dial indicator type), try to read the indicator while viewing it at 90-degrees to its surface. Reading the indicator at an angle will provide errors due to line of sight.

Most torque wrenches operate accurately only when held by their handle grips. Don't use cheater bars to extend your grip further away from the wrench head, and don't grab the handle closer to the wrench head. Only grip the wrench by its designated grip area.

Any busy shop should plan to have their torque wrenches recalibrated once each year at a minimum. Also, any time a torque wrench has been dropped, it should be rechecked for calibration. This is a procedure that many shops ignore, primarily because they're not reminded of the importance of this check. Torque wrenches are delicate, precision pieces, and they simply will not remain in perfect tune without care and routine maintenance.

Recalibrating Torque Wrenches

When was the last time you sent one of your torque wrenches out for a recalibration service? If the response to this question is "Huh?," you need to read the following info.

Torque wrenches, by virtue of the environment in which they operate, can be subjected to a great deal of wear. When used in an engine rebuilding shop that lives and breathes precision torque values, this wear level is at its zenith, even in the best-intentioned hands. Surprisingly, dropping a torque wrench onto the shop floor may not cause as much calibration damage as you might think. With that said, we need to strongly emphasize the need to treat any torque wrench with great care. While it's certainly not recommended to toss this precision instrument around the shop like a soccer ball, an occasional oops may not cause severe calibration changes (we use the term "may" because it's always possible for any hard impact to damage the tool). If abused badly, such as using the torque wrench as a hammer, etc., the wrench can be internally damaged if the impact force is severe enough. The bottom line: don't bandy the thing around. Treat it according to what it is: a precision instrument that deserves respect and care.

The most compelling reasons to send your torque wrenches out for recalibration include life cycle and overtorquing. If the torque wrench is used on a very occasional basis by a hobbyist, say 5 times per year, the wrench probably may require recalibration every ten years or so. If, however, the torque wrench is used on a regular basis in a professional shop, the operating cycles increase dramatically. Think about it: if you torque 32 cylinder head bolts, 10 main cap bolts, 16 connecting rod bolts and 16 intake manifold bolts per engine, that's a minimum of 74 cycles of use per engine. If you torque the main caps twice (once for align honing and once for assembly), torque the rod caps twice (once for reconditioning and once for assembly), that raises the number of torque wrench cycles to a minimum of 100 per engine. That number will rise even higher if you deal with 4-bolt main caps, and even higher if you perform a complete preassembly in addition to final assembly. If you work on two engines per week, that means that your torque

If you need to torque left-hand threaded fasteners, make sure that your torque wrench is designed to be used in a left-hand direction. If you're using a micrometer-style (click type) torque wrench that was designed for right-hand use only, trying to use it in reverse direction will result in inaccurate torque value and internal damage to the wrench. If you need to tighten left-hand threaded fasteners and you don't have a left-hand click-type torque wrench, you can always use your beam-type torque wrench, which can always be used in either direction without damage. The torque wrench shown here is clearly labeled for right hand use only.

wrench experiences somewhere in the neighborhood of at least 800 cycles per month, which equals around 9,600 cycles per year. If you rebuild a greater number of engines, and when you add in the other miscellaneous torquing applications (oil pump, water pump, exhaust manifolds, repairing shop equipment, etc.), even a small shop can easily run over 12,000 to 20,000 cycles per year. The point is that the calibration of the torque wrench has a finite lifespan. The initial calibration setting won't last forever, and as the use of the torque wrench increases, the life expectancy of the original calibration decreases proportionately. While there are no strict guidelines that can be used across the board, a routine of recalibration should be established for any working shop's torque wrenches.

If not used properly, calibration can be affected, possibly resulting in overtorque (stretched cyliunder head bolts or stripped bolt holes in the block, etc.), inadequate torque, or uneven torque. Remember, when tightening a cylinder head's fasteners, uneven or excessive bolt stretch can result in uneven or inadequate clamping. This uneven clamping force can lead to cylinder head warpage, coolant leaks, combustion leaks and even cylinder head cracking.

Torque wrench manufacturers commonly see damaged torque wrenches that are returned for repair service, with broken handles. This is a clear indication that these precision instruments are not being handled with the care that they require. If external damage (such as a broken handle) is common, it's safe to assume that it's also common for the tools to be used inappropriately as well. Granted, engine builders, as a rule, treat their tools with a much higher level of respect and care than would occur in a tire shop or general service garage, but in the heat of the day, accidents can happen which can lead to tool abuse.

When using any style of torque wrench, it is critical to stop as soon as the desired torque value is achieved. In the case of a click-type torque wrench, you must stop applying force the moment the click is heard or felt. On average, the typical user tends to stop too late, achieving about a 10% increase in torque. On a critical application such as engine assembly, this can create a real problem. For example, if the specified value is 110 ft-lbs., and the operator continually exceeds the "stop" signal, the application could experience 121 ft-lbs. or more (and often in an uneven format, when dealing with multiple fasteners on the same component). It may not seem like much of a problem at first, but uneven or excessive torquing can occur, even when a torque wrench is used, if the tool is used improperly.

When you use a click-type micrometer-style torque wrench, you're dealing with a mechanism that features an internal spring. This spring is preloaded via a screw adjustment when the wrench is first assembled and calibrated. When you feel or hear the click, you're working against the spring and a cam mechanism. When you apply a selected level of torque, the cam over-centers a bit as the wrench is pulled to the selected value. The heart and soul of this type of torque wrench is the spring mechanism, and the amount of preload on the spring can change over time. If the device is overtorqued past the selected setting, this abuses the spring and will definitely cause an out-of-calibration problem.

By the way, after using an adjustable-setting torque wrench, always back the setting off to the low side, in order to remove excessive preload from the internal spring. This will greatly extend the life of the spring. Make a habit of backing the adjuster down to a low setting before you store the tool in your box or cabinet.

Every time you click a torque wrench, that "click" represents one cycle of use. All torque wrenches probably need to be recalibrated or at least retested after 5,000 cycles (although some manufacturers recommend recalibration after as many as 10,000 cycles). Granted, 5,000 cycles may seem like an extraordinary amount, but as we mentioned earlier, that level of use can easily occur in less than six month's use in a machine shop. Most torque wrench manufacturers recommend that their products should be recalibrated at least once each year under normal use. However, torque wrench operation in a busy engine rebuilding facility hardly qualifies as "normal" use.

Even if the torque wrench is not externally damaged, wear can take its toll, because of multiple moving internal components. It's hard to understand how a few thousandths of an inch of wear can cause dramatic calibration changes.

Another aspect of torque wrench use that can seriously affect calibration is bio-directional use—using a torque wrench in a counterclockwise direction in addition to clockwise rotation. When applying left-hand force through a torque wrench, you're fighting a losing battle, because everything is stressed internally to bias the right-hand side. Once you force the torque wrench in the opposite direction, you're making it do something it wasn't intended to do. This left-hand application of force must overcome the built-in right-hand stress, moving it past neutral, toward the left.

Calibration on the left-hand side can be severely out of specification. The result is that you not only potentially damage the right-hand calibration, but you also arrive at an unknown torque value when applying left-hand force. If you require left-hand operation, this can be specified when ordering a torque wrench (or when one is being recalibrated). If calibrated for left-hand pulls, the same wrench can also then be used for right hand pulls. If your shop performs left-hand torquing on a fairly routine basis, it's best to dedicate that torque wrench as your left-hand device, and label it accordingly, although I don't see this as a common requirement for the average engine shop.

By the way, if a ratchet-type torque wrench is damaged (if the ratcheting mechanism breaks), it is a certainty that the wrench is also out of calibration and must be serviced accordingly. Never attempt to repair any torque wrench on your own. In all cases, the tool must be sent to a specialty service shop (the wrench manufacturer or an independent repair facility that is approved by the manufacturer).

Is recalibration required for all types of torque wrenches? The answer is a resounding "yes." Regardless of the design (beam type, micrometer-adjusting click-release type or dial type), they all feature moving components that can wear out. A great misconception with regard to torque wrenches relates to the ratcheting feature of the click-type version. Some users commonly believe that the tool is a ratcheting wrench that also measures torque value. As a result, some users tend to use the torque wrench to install as well as final-tighten a fastener. In reality, the torque wrench should never be used to install a fastener. The bolt or nut should be installed to initial clamping force with a conventional wrench (fixed wrench or ratchet and socket), and final-tightened to value using the

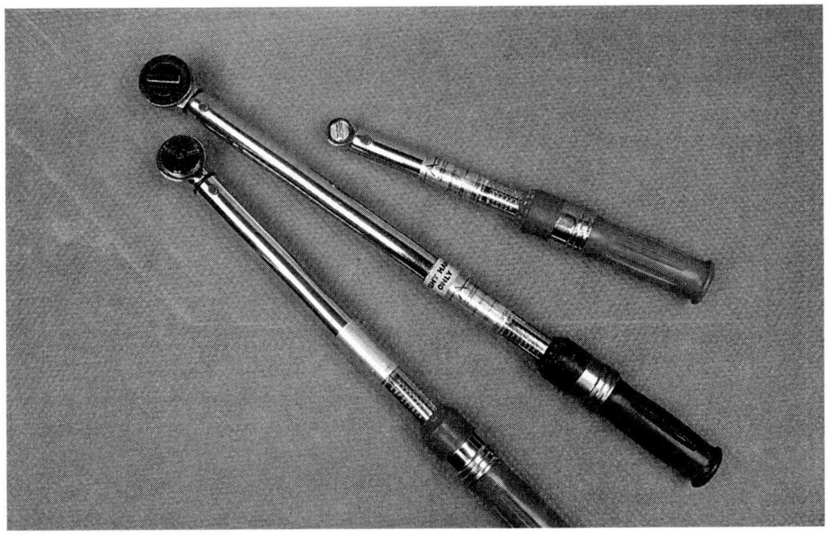

Choose the torque wrench based on its range. If a torque wrench is calibrated for a range of, say, 25 to 120 ft-lb, and the requirement at hand calls for, say, 26 ft-lb, you're better off using a torque wrench calibrated for 10 to 80 ft-lb. Try not to use any torque wrench at or very close to its minimum or maximum, for optimum accuracy. If the spec calls for, say, 24 in-lb, use an inch-pound torque wrench instead of trying to use a ft-lb torque wrench set at 2 ft-lb. Torque wrenches are available in different drive sizes (1/4", 3/8", 1/2", 3/4") and in different ranges of operation. This is why you need more than one torque wrench.

torque wrench. Don't be fooled by the convenience of the ratcheting feature. Only use a torque wrench to perform the final tightening. This will help to save the internal mechanisms and the calibration settings, which will extend the life of the torque wrench.

Use the Right Wrench for the Job

All torque wrenches are not universally adaptable to all jobs. It's important to choose a torque wrench that features your torque value requirement in the middle of its range. For instance, if you need to tighten a fastener to 100 ft-lbs, don't use a torque wrench model that has an upper range limit of 100 ft-lbs. Instead, use one that features a range of, say, 25–250 ft-lbs. If tightening a fastener to 50 ft-lbs, use a torque wrench that has an upper limit of about 100 ft-lbs, etc. Generally speaking, torque wrenches perform at their most accurate level when the application is in the mid-range area of the wrench's calibration spectrum. As a result, you may need more than one range of torque wrench for your tool inventory, depending on the type of work that enters your shop. Obviously, you need to achieve both ft-lbs and in-lbs readings, which requires separate torque wrenches. However, even when dealing in the ft-lb format, multiple torque wrenches should be used. For instance, when torquing connecting rod bolts (let's use 45 ft-lbs as

THE DON'TS OF TORQUE WRENCH USE

• Don't use a torque wrench as a general-service wrench to tighten or loosen fasteners, or as a pry-bar. Use it only for achieving final torque values.

• Don't use a ratcheting-type or dial-type torque wrench to remove fasteners (left-hand operation can damage calibration), unless the torque wrench has been specifically designed for left-hand operation.

• Don't immerse the torque wrench in solvent. Function and calibration will be affected if the internal lubricants are washed away.

• Don't apply a cheater bar to a torque wrench. This can overstress the wrench, and can lead to inaccurate torque readings.

• Use of socket extensions are permitted, as long as the extension is not positioned at an angle that is different from the torque wrench drive. For example, the use of a universal joint can cause the socket to operate at an angle relative to the drive, which will affect the accuracy of the value reading. Use of a wobble-type extension or U-joint will cause inaccurate values if the driveline angle (the line from the drive head to the fastener) is altered.

• When tightening any fasteners, try to hold the torque wrench "in-plane" as much as possible. Don't push the wrench inward toward the workpiece or pull outward away from the workpiece during final tightening. Any inward/outward pressure will dramatically affect the resulting torque value.

• Never attempt to adjust a torque wrench outside of its intended range. For example, if the range maximum is 100 ft-lbs, do not try to adjust the wrench higher and "guess" what the final value would be. For example, a "smidge" beyond the 100 ft-lb mark on a 100-max torque wrench does not mean that you're achieving 105 or 110 ft-lbs. It only means that you're off in no-man's-land, playing one heck of a guessing game.

a sample average), you should use a torque wrench that has a maximum range of, say 100 ft-lbs in order to use the "middle" of the range. When torquing main cap bolts (let's use 110 ft-lbs as our sample), you should use a torque wrench that features a 200 or 250 ft-lb maximum range. To some technicians, this may seem like a great deal of bother, but if you use torque wrenches at all, the whole point is to attain accurate a repeatable clamping values. If that's your goal, then go the extra mile and maintain several torque wrenches, each with a different low-to-high value range. In that way, you're covered for whatever job enters the shop. Examples include 1/4-drive in a range of 5–50 in-lb, a 1/4"-drive in 30–200 in-lb; a 3/8"-drive in 10–100 ft-lb, 3/8"-drive in 50–250 in-lb, a 3/8"-drive in 100–1000 in-lb; a 1/2"-drive in 10–150 ft-lb and a 1/2"-drive in 25–250 ft-lb. If you also perform heavy-duty engine work, or take in the occasional hub or axle job, you may need a 3/4"-drive wrench in 100–600 ft-lbs. The moral of the story: if you only own one torque wrench that you try to use for a wide variety of work, you will not be able to achieve accurate clamping loads for all jobs.

Use of Adapters

As long as the adapter (socket extension, etc.) is inline with the torque wrench drive, no compensation is required. However, if an adapter is used that effectively lengthens the wrench (such as a crow's-foot wrench), a calculation must be made in order to achieve the desired torque value.

For those occasions when a straight socket can't be used, a special attachment might be needed (such as a crow's foot). The use of an offset adapter changes the calibration of the torque wrench, which makes it necessary to calculate the correct torque settings. Following is a method of calculating this change:

E = Effective length of extension, measured along the centerline of the torque wrench
L = Lever length of the wrench (from center of the wrench drive to the center of the adapter's grip area)
TW =Torque setting on the torque wrench
TE = Torque applied by the extension to the fastener

Formulas:
TW = L ÷ (L+E) x Desired TE
(where the adapter makes the wrench longer)
or:
TW = L ÷ (L-E) x Desired TE
(where the adapter makes the wrench shorter)

If you want to know where to set the torque wrench when using an adapter that alters the effective length of the wrench, you must calculate to compensate for the adapter. If the distance from the wrench drive to the center of the bolt makes the wrench longer, the final wrench setting must be adjusted to a lower value in order to compensate. If the distance from the wrench drive to the bolt center makes the wrench shorter, the wrench must be set to a higher value in order to compensate.

Let's say that you want to torque a bolt to 140 ft-lbs, but you're using a crowsfoot adapter. Let's say the length of the torque wrench is 12" (from center of the handle to center of the drive). Let's also say that the crowsfoot is aiming away from the wrench drive, making the distance from the center of the wrench drive to the center of the bolt 1". This makes the wrench 1" longer. In this case, you would divide the length of the torque wrench (L, from the center of the handle to the center of the drive) by L+E, then multiply that ratio by the desired value.

In this example, the formula would be as follows:

12 ÷ (12+1) x 140 =
12 ÷ 13 x 140 =
.923 x 140 = 129

So in this case, where the crowsfoot adapter has made the torque wrench 1" longer, the wrench would be set at a value of 129 ft-lbs, in order to actually achieve a value of 140 ft-lbs.

If the crowsfoot is aimed towards the handle (turned 180-degrees from the prior example, and we still want to achieve 140 ft-lbs of torque, we know that the adapter has now made the wrench shorter (because the center of the bolt is now closer to the center of the wrench handle)...

12 ÷ (12–1) x 140 =
12 ÷ 11 x 140 =
1.0909 x 140 = 152.7

So in this case, where we want to achieve 140 ft-lbs, but the wrench has been made 1" shorter, we adjust the wrench to a setting of 152.7 ft-lbs in order to actually achieve the desired 140 ft-lbs of torque.

If the adapter makes the wrench longer, we must back off on the adjustment setting. If the adapter makes the wrench shorter, we must raise the adjustment on the setting.

UNDERCUT FASTENERS

An overloaded fastener will always yield and potentially break at its smallest cross-section. Traditionally, course threads are the weakest link. The goal of a stud undercut is to provide a degree of relief to the threads by balancing the stress distribution along the length of the fastener. An undercut stud will elongate slightly more than a full-bodied stud, serving as a "spring." As an example, a 3.00" undercut stud at 10,000 lbs force stretches barely 0.002" more than a full-body stud, but that minute amount of extra stretch can be enough to save a stud from failing.

An undercut stud elongates in an "intelligent" manner. As the cylinder head warms and reaches full temperature, it expands. The undercut stud will accommodate this cylinder head expansion, protecting the threads from premature yield.

The stretch is a function of the length of the fastener and the depth of the undercut. The stud set can be calibrated and optimized to the specific gasket load requirements. In other words, no single undercut spec for all applications. The undercut studs must be selected according to the specific engine application.

A second benefit of an undercut comes into play when the stud's reduced-diameter section acts as a calibrated "breaker" which yields at a load that is defined by the diameter of that undercut.

In a full-body stud, stress will increase as the cylinder head continues to expand. The gasket will then reach its compression strength and may lose its sealing capability. The undercut section is programmed to yield at a certain load in an effort to save the head gasket from excessive crushing.

Of course, when the undercut is yielded, part of the initial clamp load will be lost as soon as the head cools. A light retorquing may be required, but the increased plasticity of the undercut will alloy the stud to handle reasonable stretch without creating problems.

Tremendous elongation is required to break an undercut stud. Plus, if it does break, the break will occur in the undercut section, and not in the threads. This means that you won't be faced with the painful task of trying to exorcize a busted stud that snapped flush or below the deck surface.

For high-RPM engine applications, an undercut stud will also handle fatigue better and will "filter" or dampen dynamic bending caused by harmonic vibrations.

Torquing Undercut Studs

Undercut studs theoretically achieve more consistent clamp loads. However, consider the gain as negligible in comparison to the precision of most commonly-used torque wrenches.

Keep in mind that the undercut stud is not the solution to every head gasket problem. Some high-horsepower engines require the stiffness of a full-body stud to prevent head lifting. In that case, step-studs can also be an attractive solution to consider.

TIGHTENING CONNECTING ROD BOLTS

Published torque specifications aside, race engine

builders have long realized that the correct approach to tightening connecting rod bolts is to stress the bolts into their "working" range of stress, but not beyond. Since OEM connecting rod bolts may vary in terms of their ideal torque by as much as 10 ft-lbs from batch to batch due to variations in heat treating and materials, if the concern is to arrive at both peak bolt strength as well as maintaining concentricity of the rod big-end, the rod bolts should be measured for stretch instead of simply tightening until the torque wrench hits its mark.

In simple terms, in order to measure bolt stretch, first measure the total rod bolt length (from the head surface to the tip of the shank) in the bolt's relaxed state. Then measure the bolt again after the nut has been tightened to value.

The difference in length indicates the amount of stretch the bolt experiences in its installed state. For the majority of production rod bolts, stretch will likely be in the 0.005" to 0.006" range. If the stretch is less, the bolt is probably experiencing too much friction that is preventing the proper stretch (requiring lubricant on the threads). If stretch is excessive, the bolt may have been pulled beyond its yield point and is no longer serviceable.

After measuring the rod bolt's free length, zero the gauge and carefully install the gauge onto the bolt.

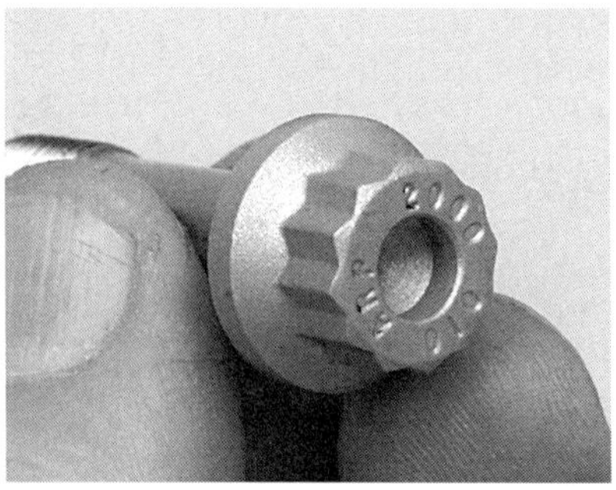

High quality performance aftermarket connecting rod bolts (like this one from ARP) feature a dimple at each end to accommodate the use of a rod bolt stretch gauge. Here you see the dimple on the bolt head.

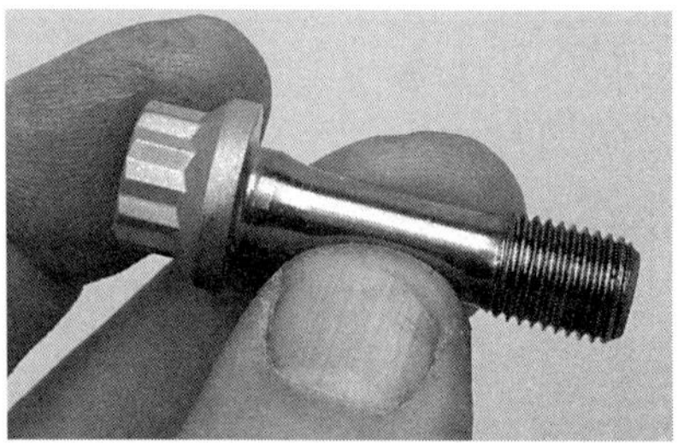

Connecting rod bolts are precision crafted to provide the proper diameter, length and tensile strength required for this specific application. NEVER substitute a generic bolt for connecting rod use!

Here we see the dimple at the bolt tip. The smoothly-chamfered surface allows precision fit of a stretch gauge.

Note how the stretch gauge dial indicator plunger nestles into the rod bolt head dimple.

While an outside micrometer may be used to measure the rod bolt length, the most accurate method is to use a specialty fixture that is outfitted with a dial indicator. Excellent examples of this gauge include units from GearHead Tools, ARP and Goodson Shop Supplies. GearHead's bolt stretch gauge features a heat-treated aluminum frame (with a very handy thumbhole) with a specially modified dial indicator with sufficient spring tension to hold the gauge firmly to the ends of the rod bolt. The indicator can be rotated for right- or left-hand operation, and the lower anvil is adjustable to accommodate various bolt lengths. Goodson Shop Supplies also offers a rod bolt stretch gauge, P/N RBG-4, featuring spherical points for consistent and repeatable readings, and can also be rotated for right- or left-hand operation. Also, ARP offers its own bolt stretch gauge, P/N 100-9941, designed with 0.0005" increments, with a heavy spring and ball tips.

There is a debate among some engine builders regarding the validity of measuring rod bolt stretch, due to potential compression of the rod material as the rod cap is clamped to the rod. While this can occur, the use of a stretch gauge remains the BEST practical method of accurately determining bolt load.

Connecting rod bolts can be viewed as high-tensile springs. The bolt must be stretched short of its yield point in order for accurate, and most importantly, repeatable, clamping of the rod cap to the rod. Improper or unequal bolt clamping force can easily result in a non-round rod bore.

Stock, or production, rod bolts typically offer a tensile strength of approximately 150,000–160,000 psi. However, due to variances in bolt production, tolerances can be quite extreme, with peak bolt stretch occurring anywhere from, say 0.003" to 0.006". If the installer uses only torque in the attempt to achieve bolt stretch, he runs the risk of unequal rod bolt clamping loads, due to the potential inconsistencies between bolts.

High-performance rod bolts are manufactured to much tighter tensile strength tolerances. ARP, for instance, calculates each and every rod bolt for stretch, and the bolt packages include reference data to that effect. The instructions actually recommend that a specific amount of bolt stretch should be achieved on each bolt (ARP cites 190,000 psi as their nominal, or base tensile rating, with actual ratings much higher in some applications).

How can unequal/inadequate rod bolt tightness affect the connecting rod big end bore shape? Let's cite an example: If one technician reconditions the connecting rods using torque value alone to tighten the rod bolts, and another technician who is responsible for final assembly uses the bolt stretch method, the final result can be out-of-round bores. This is because of frictional variances that will be encountered. As a result, the assembler using the stretch method may achieve a higher clamping load on one or more bolts as compared to the loads imposed when the rod reconditioner torqued the nuts without regard to actual bolt stretch. When a bolt is tightened with dry threads, as much as 80% of the torque can be exerted because of friction as opposed to bolt stretch.

In a high-volume production rebuilding facility, technicians may not have the time to measure for bolt stretch. However, a slower-paced operation that is attempting to obtain maximum accuracy (for a race engine, as an example), is far better off using the stretch method instead of relying only on the torque method.

A set of connecting rod bolts' instructions may list both a torque value and a stretch range, effectively giving you a choice of methods. Yes, tightening only to a specified torque value is quicker, and measuring bolt stretch requires more time, but the best results will be achieved by measuring bolt stretch. So, unless you're in a rush, take the time to measure stretch, tightening each rod bolt to the recommended stretch range. It's all about the quest for precision.

FRICTION FACTOR

If a bolt is tightened using straight torque, you may not necessarily achieve the desired preload due to the variable of friction. Since we can't predict the frictional loss, measuring rod bolt stretch provides the most accurate method of ensuring that the

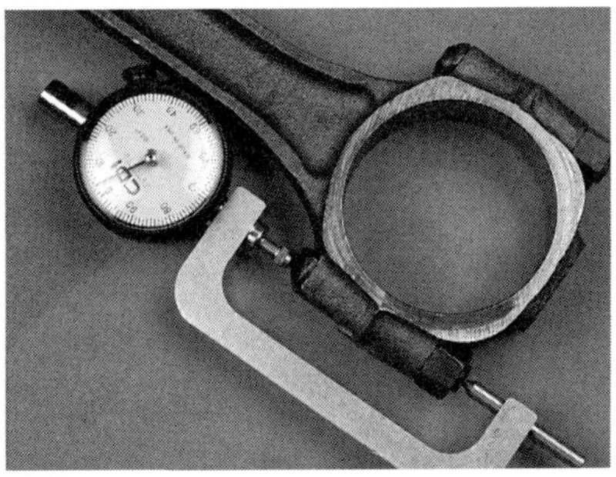

You can see how the stretch gauge engages onto the rod bolt. The indicator plunger and the anvil tip must both be centered onto the bolt.

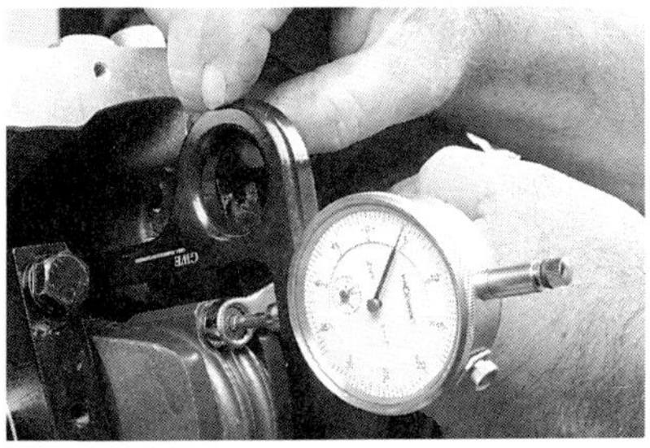

Once you know the bolt free length and with the gauge zero'd to that free length, you can begin to tighten the nut, while monitoring the gauge. The rod bolt maker will supply a specification sheet with the rod bolts, with a recommendation of the amount of bolt stretch for a given application. This stretch amount will likely be in the 0.003 to 0.005" range, but always follow the recommendation provided by the bolt maker, as this will vary depending on bolt material and bolt diameter.

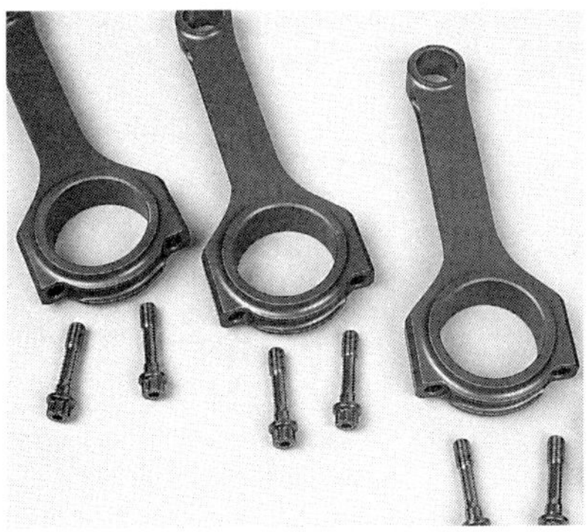

At all times during handling, regardless of the assembly stage, make a concerted effort to keep all rod bolts clean and free of grit or other contaminants.

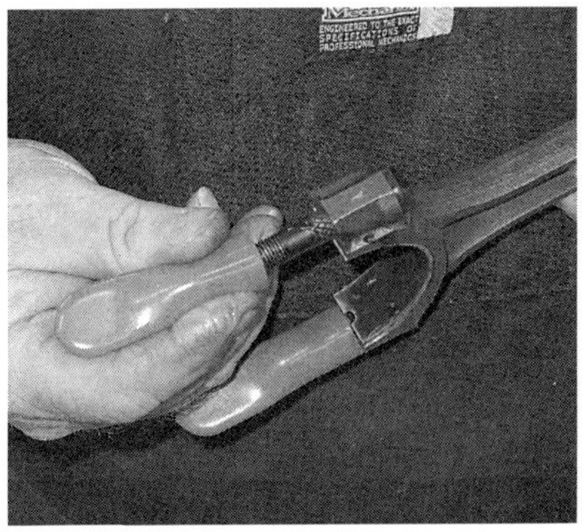

If rod bolts are press-fit into the rod big end, install protective sleeves onto the exposed bolts to prevent nicking the crankshaft journals during rod installation.

This closeup shows the distinct dimple on an ARP rod bolt head.

Depending on the type of engine block you're dealing with, access with a connecting rod bolt stretch gauge can be a challenge. This Honda block provides a tight fit. Using a stretch gauge is worth the time and effort, so be patient.

If connecting rod bolts are to be loosened or tightened when the connecting rod is not attached to the crankshaft, the rod big end must be properly supported to avoid twisting the big end bore and beam. This connecting rod vise is bench-mounted and clamps the big end of the rod to allow bolt service without harming the rod. Do not stick the rod into your conventional bench vise!

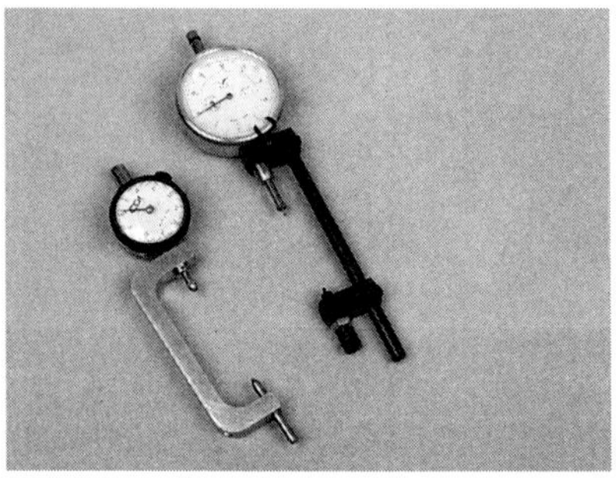

Two examples of rod bolt stretch gauges. A variety of manufacturers offer these tools. The bottom anvil (stationary pointed pin) is adjustable to accommodate bolt length. The tool on the left is adjustable by loosening a small set screw amd sliding the anvil. The tool on the right features an adjustable sliding base plus a fine-tune adjustment of the anvil pin itself.

When tightening connecting rod bolts (either as the final method or as part of a torque/angle method), stabilize the ratchet head and pull with a slow, steady motion. Never jerk or pull quickly, as this will result in inaccurate torque value, possibly pulling past the desired point.

This rod bolt stretch gauge from Gearhead features a handy fingerhole, which makes handling easier.

Carefully locate the stretch gauge to insure that the top and bottom gauge anvils are centered in the bolt head and top dimples.

This angle gauge connects between the socket and a ratchet. Set the pointer to zero and watch the needle while slowly tightening the bolt. The flex clamp shown here is part of the angle gauge shown in this example. This prevents the gauge base from accidentally rotating.

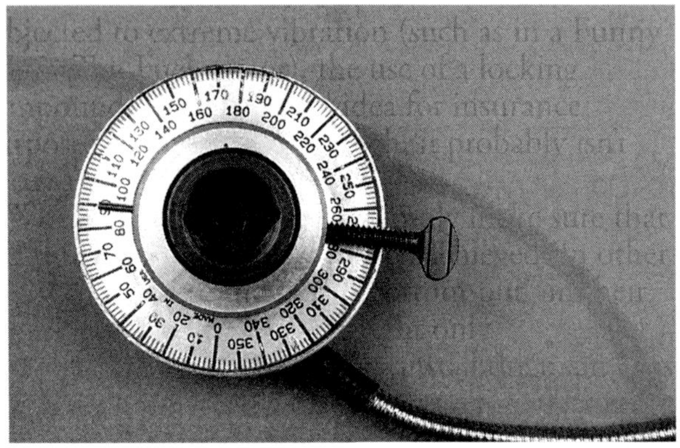

Here's a Lisle angle gauge. The compass is clearly marked in degrees. Simply apply gentle pressure on the wrench only to take up any slop in the wrench, position the pointer to the zero mark and tighten the tool's thumbscrew. As you tighten the bolt, watch the pointer, and stop when the pointer reaches the target degree number.

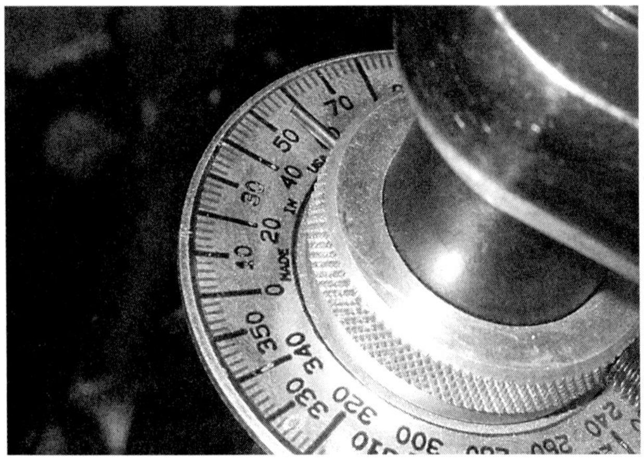

An angle gauge features a pointer to make it easy to monitor bolt rotation.

clamping loads will be both sufficient for the task and that each pair of rod bolts will achieve EQUAL loads.

Bolt stretch is generated by a number of factors, including tensile strength and mass (the length of the bolt being stretched). The effective diameter of the bolt contributes to this. For example, let's consider two 3/8" x 1" bolts. One features a 1"-long shank, with threads on the full 1" of the shank length. The other bolt features 3/4" of shank length that is full-diameter and smooth, and only 1/4" of thread length at the tip. The bolt with partial thread will stretch less, because the shank area between the head and nut engagement area has a thicker cross-section. The partial-thread bolt will have a .375" diameter shank, while the all-thread bolt will have only a .324" shank (due to the smaller root diameter inside the thread path).

ARP, to cite one example, calculates the stretch

number for every bolt. On the spec sheet that is included with every bolt set, this stretch goal is listed, in addition to a torque value, but the torque value should be used as a guide only. ARP does not want the installer to use a torque value as the final indication of bolt stretch, Rather, the bolts should be individually measured for stretch, to assure that each bolt is installed at its optimum strength.

While we cannot control the reaction of the connecting rod base material, at least consider the potential compression of the connecting rod material itself during bolt clamping. As the bolt is tightened, the head of the bolt will tend to embed itself into the rod, slightly compressing the stock material of the connecting rod under the bolt head. Production rods are typically softer, allowing the head of the rod bolt to sink into rod, until the material under the bolt heads "work hardens" under compression. ARP recommends that the bolt stretch is based on the bolt itself, and not on the compression of the rod, since we can't accurately predict what the rod does in this state. Since too many variables exist in terms of rod bolts and connecting rods, we can't draw any generalized conclusions regarding ideal connecting rod bolt stretch. However, to use the Chevy small block 350 as but one isolated example, ARP typically looks for an installed bolt stretch of .0063". Since each engine/rod/bolt application differs, we cannot assume that ideal bolt stretch would be the same for any given application.

SPECIAL TORQUE/ANGLE TORQUE WRENCHES

Thanks to advancements in technology, torque wrenches are now available that allow you to achieve both torque value and applied rotation angle without the need for a separate angle gauge.

One example is Snap-On's Techangle series of wrenches. Featuring sensor electronics and digital control and readout, you can preset (program) the torque value you want; or both torque value and final applied angle, depending on your needs. The electronic control allows you to also choose between in-lb, ft-lb or Nm, plus angle.

An internal gyroscope provides the desired angle sensing. Here's how it works: you program the desired torque value, and tighten the fastener. When you reach the programmed torque value, the wrench beeps and vibrates.

Then, if you need additional angle rotation, you program the desired angle.

When you continue to apply pressure and reach the programmed angle, the wrench again beeps and vibrates.

Snap-On's Techangle torque wrench handles both torque value and angle value at the press of a button.

Press the "U" button to enter the ft-lb mode. When you're in the mode you want, simply press the + or—buttons to dial-in exactly the torque value you want. Press the "U" button to also toggle between Nm and in-lb. modes.

Press the "U" button until you enter the angle mode. Use the + or— buttons to program the desired angle. Notice the small speaker holes to the left of the display screen. The wrench provides both audible (beep) and vibration alerts when the desired torque or angle preset has been reached.

The preset angle range is 5 to 360 degrees, with a resolution of 1 degree and accuracy of +/- 1 degree. Unlike the use of a separate angle gauge, where you can't ratchet (with an angle gauge you must start and continue the angle rotation in a steady one-direction sweep), this tool allows you to ratchet without "losing" the angle memory. Pretty cool.

Two models are currently available, including ATECH2FR100, with a torque range of 5 to 100 ft-lb; and ATECH3FR250, with a torque range of 12.5 to 250 ft-lb.

DECK PLATE SOURCE/APPLICATION LISTINGS

Following is a listing of torque plate applications by manufacturer.

TORQUE PLATE APPLICATIONS (B-H-J)

P/N	Application	Size
CH-1	Chevy small block	265, 283, 302, 327, 350 CID
CH-2	Chevy small block	400 CID
CH-3	Chevy big block	396, 400, 427, 454 CID
CH-4	Chevy big block	with extra valley studs
CH-5	Chevy big block	symmetrical port head
CH-V6	Chevy V6	200, 229 CID
CH-6	Chevy straight 6	250, 292 CID
FO-1	Ford small block	261, 289, 302, 351, 400 CID
FO-3	Ford big block	352, 390, 406, 427, 428 CID
FO-4	Ford big block	429, 460 CID
FO-1.6	Ford 4-cyl.	1600cc
FO-1.6E	Ford Escort	1.6, 1.9L
FO-2.0	Ford 4-cyl.	2000cc
FO-2.3	Ford 4-cyl.	2300 cc
FO-2.8V6	Ford V6	2.6, 2.8L
FO-3.8V6	Ford V6	3.8L
FO-4.0V6	Ford V6	4.0L
FO-6	Ford straight 6	240, 300 CID
MP-1	Mopar small block	273, 318, 340, 360 CID
MP-2	Mopar big block	361, 383, 400, 413, 426, 440 CID
MP-3	Mopar big block	426 Hemi (top studs not supplied)
MP-4	Mopar big block	392 Hemi
MP-6	Mopar 6 cyl.	225 CID
MP-2.2	Mopar 4-cyl.	2200 cc
MP-2.5	Mopar 4 cyl.	2.5L
MP-V6	Mopar V6	3.9L (238 CID)
MP-B1	Mopar T-5 V8	V8 (all)
GM-3.8V6	GM V6	3.0 (181 CID), 3.8L (231 CID)
GM-4.1V6	GM V6	4.1L (252 CID)
GM-2.0	GM 4-cyl.	1.8, 2.0L
GM-2.5	GM 4-cyl.	2.5L (151 CID) (1985 & later VIN U short deck block)
PO-SD	GM 4-cyl.	2.5L (151 CID) (fits long deck block, 1997 thru current; 1985 & later VIN R)
GM-112	GM 4 cyl.	1.8, 2.0L OHV
GM-2.8V6	GM V6	2.8L
GM-3.8/4.1	V6GM V6	3.8, 4.1L
GM-6.2	Detroit	6.2L
PO-1	Pontiac	389, 400, 421, 428, 455 CID
PO-2	Pontiac	350 CID
PO-3	Pontiac, other GM	301 CID
OL-1	Oldsmobile	350, 400, 425, 455 CID
OL-3	Oldsmobile	Special WJ engine (alum. only)
OL-Q4	Oldsmobile 4-cyl.	Quad 4

Torque Plate Applications (B-H-J, continued)

P/N	Application	Size
BU-1	Buick	455 CID
BU-2	Buick	350 CID
AM-1	American Motors	360, 390, 401 CID

Note: Honing plates for all engines produced by foreign and domestic manufacturers are available, including Toyota, Datsun, Nissan, VW, Honda, BMW, Mercedes, Ferrari, Fiat, etc. Custom honing plates are available on special order. Plates are available, depending on application, in cast iron or 356-T6 heat-treated aluminum. Plates are Blanchard ground on both sides within precision commercial tolerances. Bored to .090–.095" over maximum stock bore size. This maintains full gasket compression. Boltholes are precision drilled to factory tolerances. Clearance holes are drilled for locating dowels. "R" models are specially machined and/or supplied with spacers and washers to allow use of OEM head bolts or aftermarket studs, at user's discretion. Heat-treated inserts under-head bolts are supplied to protect aluminum plates. O-ring receiver grooves can be cut into honing plate surface, to customer specifications. Use suffix "O" to order this option.

TORQUE PLATE APPLICATIONS (GOODSON)

P/N	Application	Size
HCH-3-R-S	Chevy big block Bowtie (up to 4.530" bores)	454, 502 CID
HFO-46-R	Ford	4.6L
HGM-1	Chevy small block	265–350 CID
HGM-1A	Chevy small block	400 CID
HGM-2	Chevy big block	396–454 CID
HCR-4	Chrysler small block	273–360 CID
HCR-6	Chrysler big block	361–440 CID
HCR-22	Chrysler 4-cyl.	2.2 & 2.5L
HFO-7	Ford small block	260–302 CID
HFO-7	Ford small block	351–400
HFO-9	Ford big block	352–428 CID
HFO-9A	Ford big block	429–460 CID
HFO-12	Ford V6	3.8L
HFO-6	Ford Ranger 6-cyl.	240–300 CID
HFO-23	Ford Ranger 4-cyl.	2.3L
HOL-1	Oldsmobile V8	350, 400, 425, 455 CID

Note: Plates feature recessed deckside bolt counterbores to allow for distortion. All plates include special non-coated bolts for accurate torquing.

TORQUE PLATE APPLICATIONS (SUNNEN)

P/N	Application	Size
DP-10	Chevy small block	283–350 CID
DP-11	Chevy small block	400 CID
DP-12	Chevy big block	396–454 CID
DP-13	GM 4-cyl. VIN R	2.5L (151 CID)
DP-14	GM 4-cyl. VIN U	2.5L (151 CID)
DP-15	GM 4-cyl. pushrod motor	1.8–2.0L
DP-16	GM 4-cyl. OHC	1.8–2.0L
DP-17	Oldsmobile	350, 400, 425, 455 CID
DP-18	GM, Buick, Olds, Pontiac	301 CID
DP-19	Chevy V6	specify motor
DP-20	GM V6	3.8–4.1L
DP-21	GM V6	2.8L (60-deg bank)
DP-22	Pontiac	389–455 CID
DP-23	Pontiac	350 CID
DP-24	Buick	455 CID
DP-25	Ford small block	260–400 CID
DP-26	Ford FE	352–428 CID
DP-27	Ford	429–460 CID
DP-28	Ford V6	3.8L
DP-29	Ford V6	2.6–2.8L
DP-30	Ford 4-cyl.	2000 cc
DP-31	Ford 4-cyl.	2300 cc
DP-32	Mopar small block	273–360 CID
DP-33	Mopar big block	361–440 CID
DP-34	Mopar V6	3.9L
DP-35	Mopar 4-cyl.	2.2L
DP-36	Chevy straight 6	250 & 300 CID (steel blank)
DP-37	Ford straight 6	240 & 300 CID (steel blank)
DP-38	Ford Escort	1.6 & 1.9L (aluminum blank)

Note: Torque plates include hardened bolts and washers. According to Sunnen, these plates do not require gaskets when mounting.

TORQUE PLATE APPLICATIONS (PETERSON)

P/N	Application	Size	Bore
XT 62035	Chevy small block	265, 305 CID	3.840"
XT 62033	Chevy small block	283 CID	3.975"
XT 62011	Chevy small block	350 CID	4.100"
XT 62014	Chevy small block	400 CID	4.200"
XT 62029	Chevy big block	396 CID	4.190"
XT 62017	Chevy big block	427, 454 CID (18 bolt)	4.350"
XT 62002	Ford Pinto	2300cc	3.879"
XT 62005	Ford	260, 289, 302, 351, 400 CID	4.100"
XT 62039	Ford	352, 390 CID	4.150"
XT 62037	Ford big block	406, 428 CID	4.225"
XT 62008	Ford big block	427 CID	4.330"
XT 62020	GM (19" deck, VIN R)	151 CID	4.088"
XT 62020A	GM (18 3/8" deck, VIN U)	151 CID	4.088"
XT 62026	GM V6 (11mm bolts ordered sep.)	173 CID	3.600"
XT 62023	GM V6	231 CID	3.895"
XT 62031	GM V6	252 CID	4.065"

Note: All torque plate kits include bolts and washers. Plates are counterbored to recess bolts for easy mounting on all types of boring and honing machines. Bolt lengths should be checked for each application. All plates feature integrally cast lifting pads for easy handling.

RECOMMENDED TORQUE TO ACHIEVE OPTIMUM PRELOAD (CLAMPING FORCE)

Note: the torque values listed here refer to clamping loads achieved with specific type of fasteners. Use this as a guide only, as these will not necessarily reflect the torque value specification for a given engine application. Always follow the specifications listed by the engine manufacturer or the fastener supplier for a given application.

Note: all torque values listed below are represented in ft-lbs. Always lubricate threads prior to torque to assure accurate readings! The preload/torque chart information shown here was provided courtesy of ARP.

FASTENER TENSILE STRENGTH 170,000 (PSI)

Thread Size And Type	Torque w/30 wt. Oil	Torque w/Moly	Preload
1/4" stud	12	10	3,804
1/4-20	13	10	3,804
1/4-28	14	11	4,344
5/16" stud	25	20	6,264
5/16-18	26	21	6,264
5/16-24	28	22	6,948
3/8" stud	45	35	9,276
3/8-16	46	36	9,276
3/8-24	50	39	10,512
7/16" stud	71	56	12,720
7/16-14	73	58	12,720
7/16-20	80	62	14,220
1/2" stud	108	84	16,992
1/2-13	111	88	16,992
1/2-20	122	95	19,164
9/16" stud	156	122	21,792
9/16-12	159	126	21,792
9/16-18	174	136	24,312
8mm stud	25	20	6,264
10mm stud	54	42	10,680
11mm stud	80	63	14,220
12mm stud	97	77	15,540

Note: Torque values are in ft-lbs.

FASTENER TENSILE STRENGTH 190,000 (PSI)

Thread Size And Type	Torque w/30 wt. Oil	Torque w/Moly	Preload
1/4" stud	14	11	4,280
1/4-20	14	11	4,280
1/4-28	16	13	4,887
5/16" stud	28	22	7,047
5/16-18	29	23	7,047
5/16-24	32	25	7,817
3/8" stud	50	39	10,436
3/8-16	51	41	10,436
3/8-24	57	44	11,826
7/16" stud	80	63	14,310
7/16-14	90	70	15,998
1/2" stud	122	95	19,116
1/2-13	125	99	19,116
1/2-20	137	107	21,560
9/16" stud	175	137	24,516

Thread Size And Type	Torque w/30 wt. Oil	Torque w/Moly	Preload
9/16-12	179	142	24,516
9/16-18	196	153	27,351
8mm stud	28	22	7,047
10mm stud	61	48	12,015
11mm stud	90	71	15,998
12mm stud	109	86	17,483

FASTENER TENSILE STRENGTH 220,000 (PSI)

Thread Size And Type	Torque w/30 wt. Oil	Torque w/Moly	Preload
1/4" stud	15	12	4,756
1/4-20	16	13	4,756
1/4-28	18	14	5,430
5/16" stud	32	25	7,830
5/16-18	32	26	7,830
5/16-24	35	28	8,685
3/8" stud	56	44	11,595
3/8-16	57	45	11,595
3/8-24	63	49	13,140
7/16" stud	89	70	15,900
7/16-14	91	72	15,900
7/16-20	100	78	17,775
1/2" stud	135	105	21,240
1/2-13	138	110	21,240
1/2-20	152	119	23,955
9/16" stud	195	152	27,240
9/16-12	199	158	27,240
9/16-18	217	170	30,390
8mm stud	32	25	7,830
10mm stud	68	53	13,350
11mm stud	100	79	17,775
12mm stud	122	96	19,425

Chapter 4

THREAD REPAIR

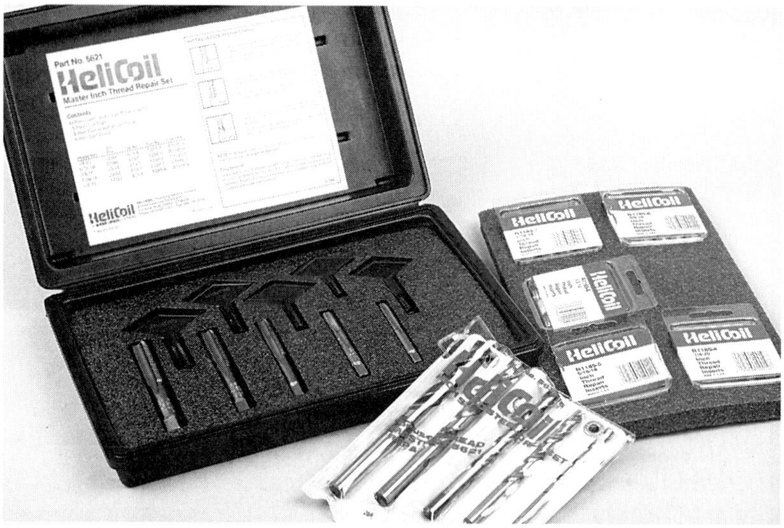

Helically wound thread insert kits (like this one from HeliCoil) are available for all inch and metric sizes. A full kit will include a selection of inserts (inch or metric), drills, dedicated taps, insert installation tools and a selection of inserts. The kit shown here includes sizes 1/4" x 20, 5/16" x 18, 3/8" x 16, 7/16" x 14 and 1/2" x 13 (coarse thread inch). Other kits offer fine thread inch, while other kits offer metric sizes.

Threaded holes can usually be repaired, depending on the integrity of the parent material. If female threads (threads inside a hole) are only slightly deformed, a thread chaser, or forming tap may be used. This restores the threads without removing a substantial amount of material. If the threads are too damaged (stripped out), the hole can usually be repaired with the use of a thread insert. Thread inserts are available in the desired internal thread size (the size of the original threads). The exterior of the insert also features threads, either a common thread diameter and pitch; or with an external thread that's unique to a particular brand of insert. The external threads of the insert allow you to tap an oversize hole and install the insert.

INSTALLING THREAD INSERTS

Installation of a thread insert requires drilling the hole out to a specific oversize, cutting new threads that will accept the insert's external threads, and installation of the insert.

Various styles of thread inserts are available, the most common being the helically wound stainless steel insert that is often generically referred to as the Heli-Coil (Heli-Coil is actually a brand name, but other makers have copied their design, so this style is available under various brand names). The helically wound insert is made of a tightly wound angular thread (before it's installed, it resembles a little coil spring). The insert kit will provide instructions regarding what size hole to drill to accept a specific size insert. A complete kit will also provide the needed drills, taps and the required installation driver tools.

Note: When using a helically wound thread insert, you MUST use the specific drill specified in the instructions, and you MUST use only the tap provided in the insert kit. These taps are unique to the requirements of the insert. You must also install the insert only with the unique installation tool included in the kit. You can purchase inserts alone, but you'll need the maker's unique taps and installation tools anyway, so if you're just starting out, buy an entire kit. You can always buy additional inserts when you run out.

Basically, a helically wound insert installation is as follows: drill the damaged hole out to the proper oversize, cut threads into that hole with the unique tap supplied in the kit, then install the insert. The insert features a small tang at the bottom of the insert, which engages to the driver tool tip. As the insert is threaded into the hole, keep an eye on your installation depth. Proper installation requires that the top of the insert be installed about 1–2 threads below the top of the hole (to avoid leaving a coil above the surface). Once the insert has been inserted at the desired depth location, the driver tang must be removed. The tang features a small notch to facilitate snapping the tang free. The tang can be removed by inserting a straight (not tapered) drift and a hammer. One quick tap against the tang will break the tang loose. Or, needle-nose pliers can be used, employing a twisting/wiggle motion to fracture the tang at its relief notch. If the hole is open and the workpiece is off the car, the tang will fall away. If however the workpiece is on the car, you'll need to avoid allowing the tang to drop into the engine. In those cases, needle-nose pliers will be best (or even better, skinny needle-nose vise grips), to maintain a solid grip on the tang).

Helically wound thread inserts are made of stainless steel. It resembles a smal coil spring.

The bottom of a helically wound insert features a driver tang. This engages with the driver tool for installation, and is removed after installation.

This close-up shows the driver tang. Notice the small notch near the tang. This is the break-off point that allows the tang to be easily snapped off for removal.

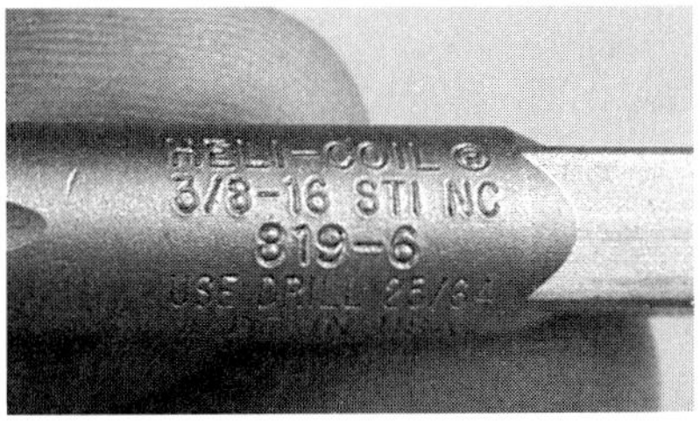

While the drill sizes provided in a Heli-Coil kit are common, the taps are not. Heli-Coil taps are unique to the kit and must be used to cut compatible threads for the inserts. Each tap is marked for the size of the insert's internal thread. The marked size does not denote the size of the tap threads.

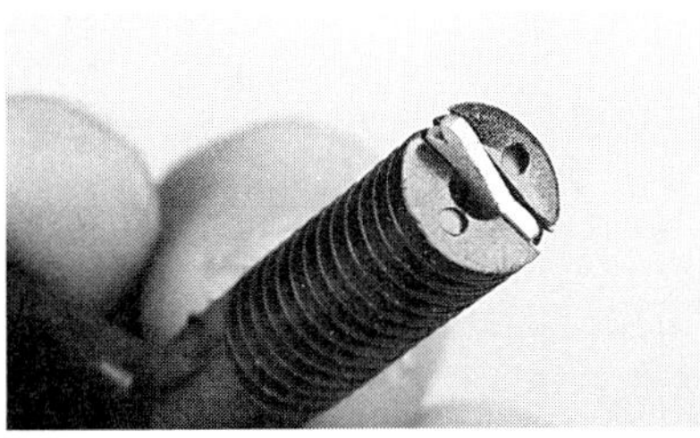

The bottom of the tool features a driver tang. This will engage onto the insert tang.

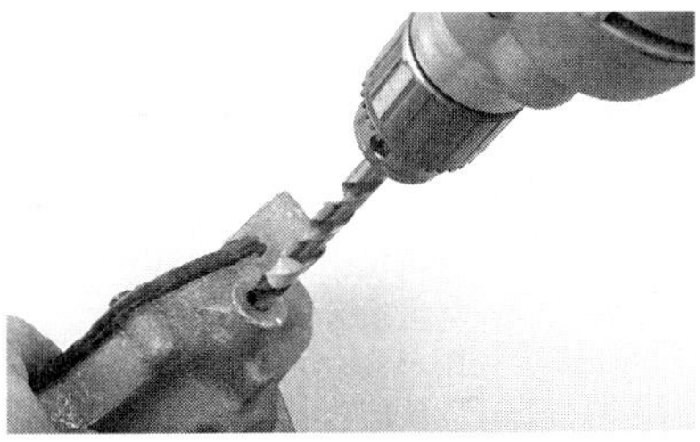

Drill out the damaged threaded hole using the recommended drill size. Use extreme care to make sure that you drill straight into the hole. If you can secure the piece on a drill press, by all means do that. If the piece is hand-held, on a vise or on the car, do your best to prevent drilling off-angle.

After blowing the hole free of chips, carefully tap the hole using only the recommended tap.

A Heli-Coil insert installation tool. This is a simple finger-operated screw that accepts the insert.

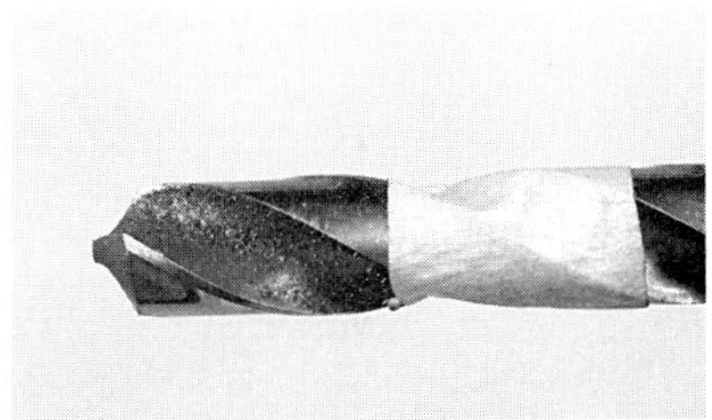

Before drilling a blind hole, first measure the depth of the hole, then mark the drill as a reference guide. This will eliminate guesswork, preventing you from drilling too shallow or too deep.

Be careful not to drill too deep in a blind hole. Over-depth drilling could result in opening the hole to a fluid jacket.

It's also a good idea to reference-mark the tap, to prevent tapping too short.

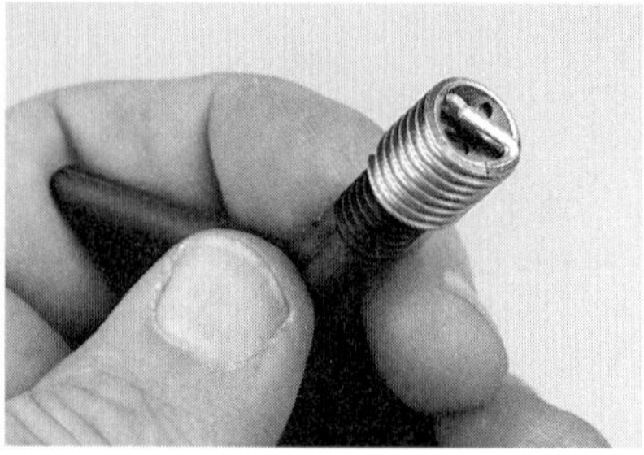

The insert easily finger-threads onto the installation tool.

This close-up shows the insert tang stopped against the driver.

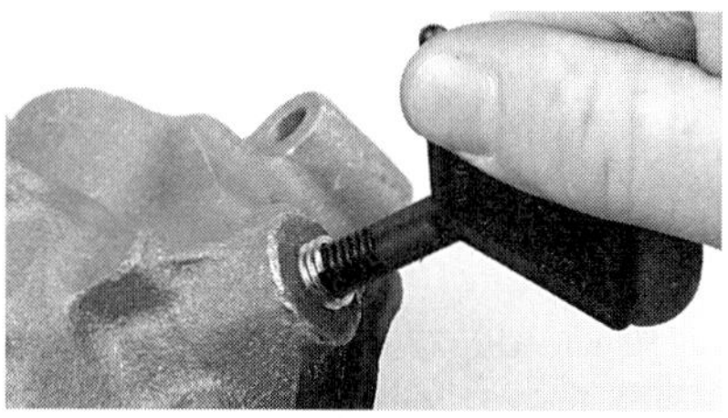

Once the insert has entered the hole, it begins to expand slightly. You won't be able to remove the insert (if you change your mind in the middle of the job), since the design provides a great deal of retention. The only way to remove the insert is to grab it with pliers and strip it out. This will destroy the insert, and possibly damage the fresh threads that you created. So make sure you're ready before starting to install the insert.

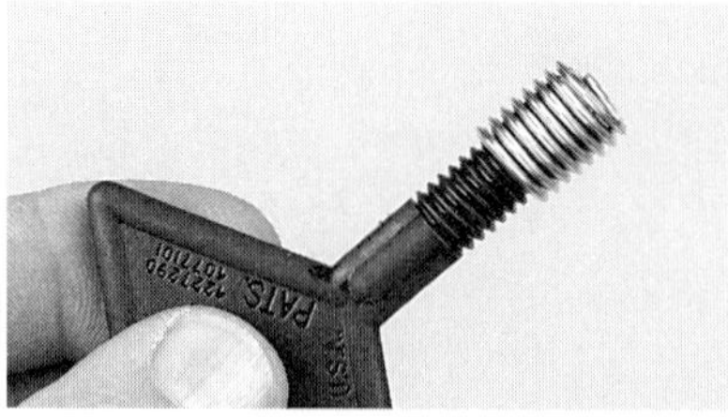

Thread the insert fully onto the tool until the driver tang engages. There's no need to force the insert. You'll know when it wants to stop.

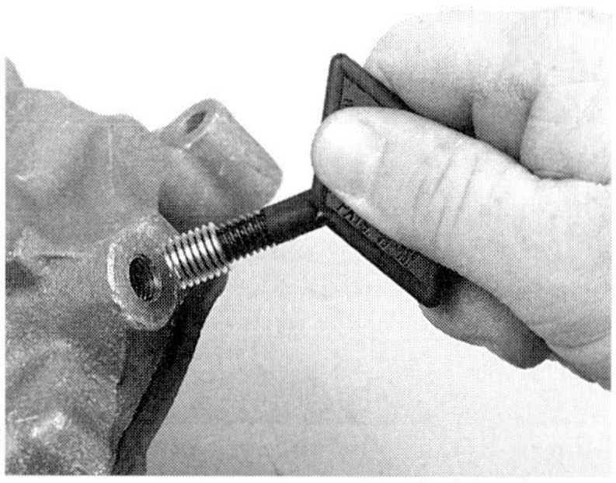

Blow the hole clean, making sure no chips or debris are present. The insert will thread into the hole easily.

Carefully thread the insert into the hole until the top of the insert is slightly below the parent material surface, then remove the tool by simply unthreading it from the insert (turning the tool counterclockwise).

The top of the insert must be fully captured in the hole. Do not leave any insert material above the hole.

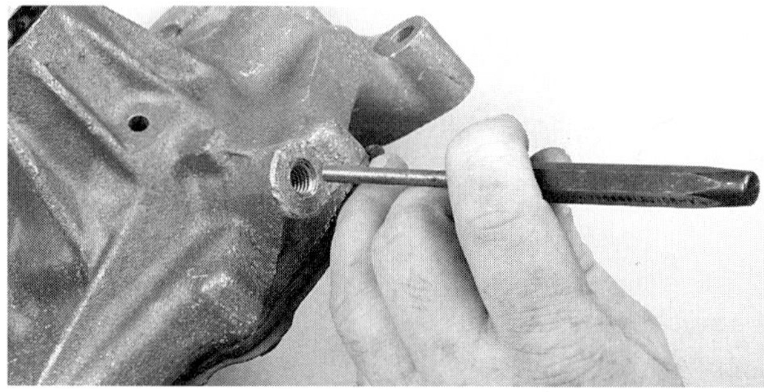

Once the insert is fully seated, you must remove the insert's tang. If you're dealing with a blind hole, use a flat-tipped drift and a hammer. If the part being repaired is on the engine (and you don't want the tang to drop into the motor), use needle-nose pliers. If pliers are used, grip the tang securely and wiggle the tang back and forth to break it loose, then extract the tang.

Using a drift and hammer in a blind hole application such as shown here, place the drift against the tang (as close to the tang's break-off notch as possible) and give it one good whack. This will fracture the tang off of the insert.

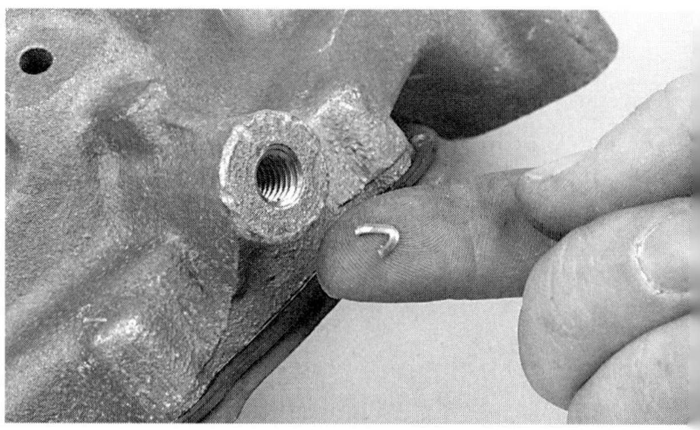

Here's the tang we broke off of this 3/8" x 16 insert in a blind hole. The tang must be removed for two reasons: to get it out of the way of the bolt, and to prevent the insert from threading its way further into the hole as the bolt is installed. Remember: even though the insert can't be unthreaded, this type of insert can always be threaded deeper if the depth allows and if the tang is left in place.

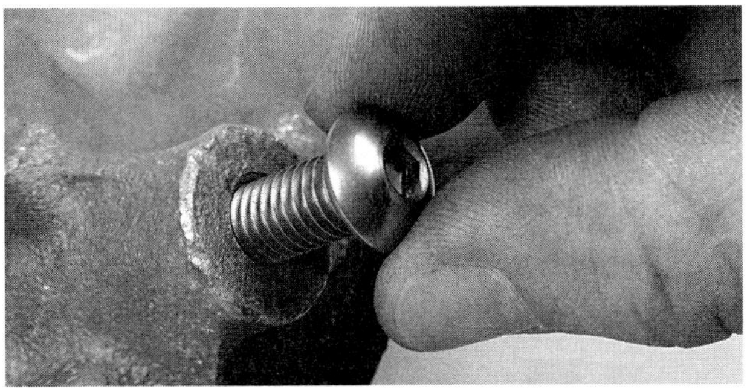

Once the tang has been removed, install a screw to verify the job.

A stainless steel insert will provide excellent fastener retention, and in the case of an aluminum component, the insert will provide threads that are stronger than the original threads in the parent material. Also, because of the unique threads that are cut to accept the insert, the insert is actually spread and pre-loaded somewhat, creating a solid mating between the insert and the parent hole. In other words, it can't loosen and fall out.

Other styles of threaded inserts include steel or stainless steel inserts that are machined from solid material (not coil-wound), featuring small stake pins that are driven into the parent material once the insert is in position. These small stakes lock the insert in place, preventing loosening or further rotation.

Threaded inserts are available in all common sizes in both fractional and metric formats, so no matter what size thread you're dealing with, inserts will be available.

Stake-type threaded inserts are available in kits, like this KeenSert kit from Tridair Industries. Kits are available in both coarse and fine inches as well as metric versions.

This thread insert features a solid threaded body. Internal threads provide the desired thread diameter and pitch that you need to restore your hole, while the exterior threads allow insert installation. The stakes (captured in outer wall grooves) are driven down once the insert is in place. This type of insert does not require a unique-thread tap. The kit instructions tells you what drill size and tap size to use for any given insert.

The stake-in type insert uses a driver tool to seat the retaining stakes.

Example of two inserts and their corresponding stake driver tool.

1. First, drill out the damaged threaded hole, using the specific diameter drill listed in the kit instructions. Note that the stake-in type insert usually recommends a slightly large drill than would normally be used for the exterior thread size of the insert. For instance, when using a KeenSerts 1/4" x 28 insert (featuring 1/4" x 28 inside thread), that insert's outer threads are 3/8" x 16. Normally, you would use a 5/16" drill to prepare the hole for a 3/8" x 16 tap. However, the KeenSert requires a Q drill (0.332"), which is a tad larger than 5/16". This is to allow easy finger installation of the insert. In other words, don't guess. Always follow the exact recommendation for drill or tap size with any given insert kit.

2. After we drilled the damaged 1/4" x 28 hole with a 0.332" drill, we tapped the hole with a 3/8" x 16 tap.

3. Once the hole is tapped and blown clean, finger-install the thread insert. You can apply thread-locking compound to the exterior insert threads if you like, but it isn't necessary.

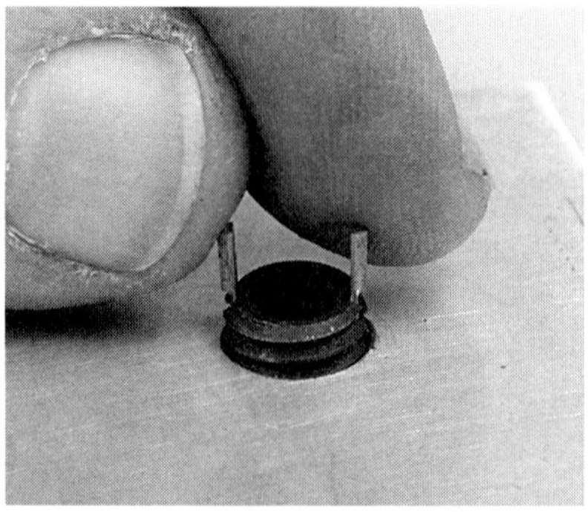

4. The insert should thread easily by hand. If you feel too much resistance, remove the insert and reclean the hole. If you try to use a prybar across the stakes to drive the insert, you'll only damage the stakes.

5. Install the insert so that the top of the insert is slightly below the deck surface. Don't allow the top of the insert to protrude above the surface.

6. Once the insert is in place, position the stake driver tool's guide nose into the insert.

7. With the tool in place, the bottom of the driver tool's outer jacket contacts the top of the stakes.

8. Using a hammer, tap the driver tool to drive the stakes into the parent material.

9. Here's the finished 1/4" x 28 insert, fully staked into place. The stakes will prevent the insert from rotating.

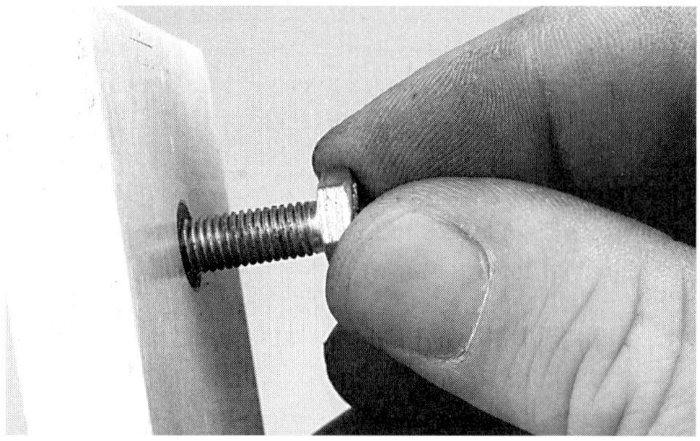

10. Whenever you finish installing any insert, run a bolt into the threads to verify the job.

Metal Shavings

If you need to repair spark plug holes, or any other hole on a component that is already attached to the engine, care must be taken to avoid dropping metal chips (caused by drilling or tapping) into the engine. The best course is always to remove the component, fix the hole, then reinstall the part. If the part must stay on the engine (let's use a spark plug hole as the example), you'll need to take some precautions to prevent engine damage. First of all, most damaged/stripped spark plug holes occur on aluminum cylinder heads, because the material was weak in the first place, or more likely because someone installed spark plugs too tightly, with no thread lubricant, or both. Here are a few tips to keep contamination to a minimum: apply some grease to the drill bit before drilling (this will help to keep much of the aluminum debris captured in the drill bit flutes). Do the same with the tap—apply grease to the flutes to try to capture chips. On one hand, it's sometimes a good idea to bring that cylinder's piston to TDC, to try to contain any chips in as small an area as possible (after the repair, you can try to suck debris out of the chamber with a vacuum, or you can use a long cotton swab (apply grease to the swab) in an attempt to retrieve any

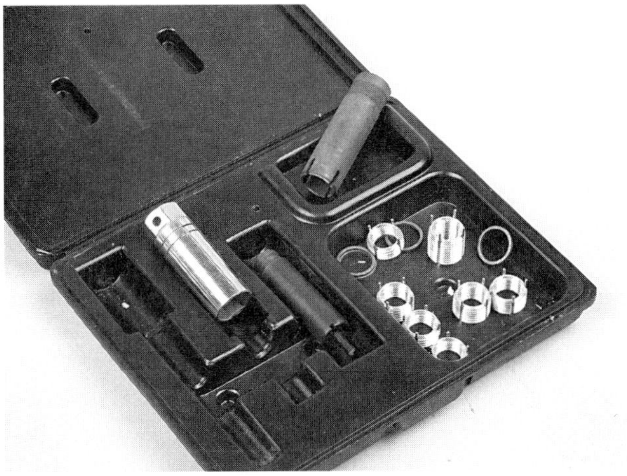

1. Both helically wound and staked inserts are available for spark plug hole repair. Shown here is a stake-type insert kit that features aluminum inserts (for repairing aluminum cylinder head spark plug holes). Different people have different views regarding which material to use on aluminum heads. Using stainless steel inserts in an aluminum head provides stronger-than-original threads, but some folks believe that the dissimilar metals and their potentially different rates of thermal conductivity adversely affect heat transfer. To address this concern, aluminum inserts are available. Personally, I've used stainless steel inserts on aluminum heads for years with no problems.

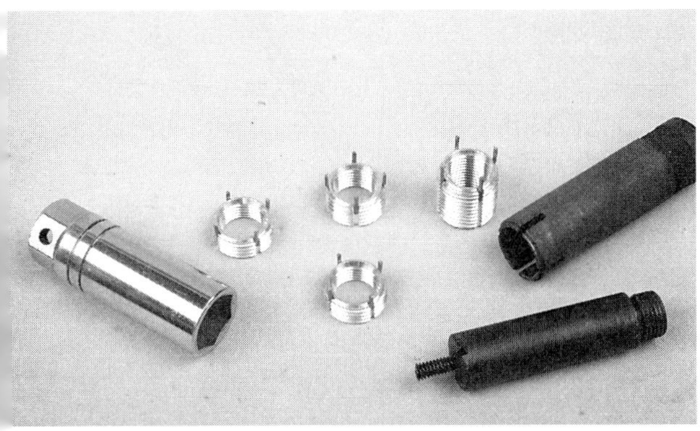

2. The stake-type spark plug thread repair kit shown above includes a selection of aluminum inserts (of various depths, depending on application), installation and driver tools and a spark plug socket wrench.

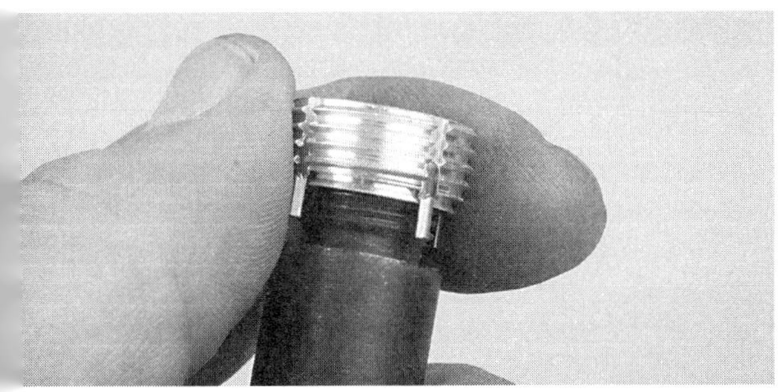

3. Using this kit, drill out the damaged hole and tap it according to the instructions, then thread an insert onto the installation tool with the stakes facing the tool.

4. The insert is fully threaded onto the installation tool.

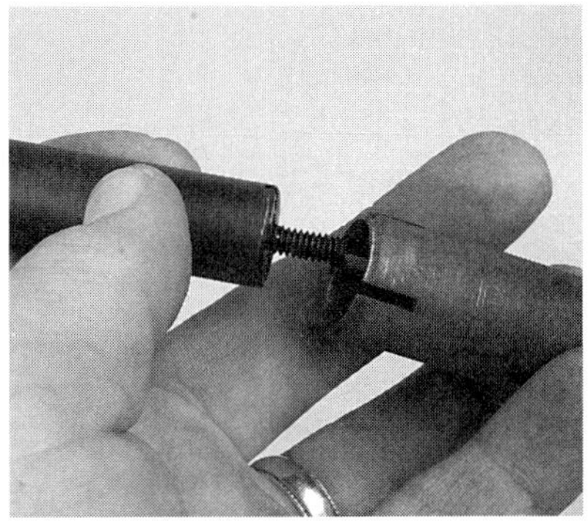

5. The driver sleeve is slipped over the tool body.

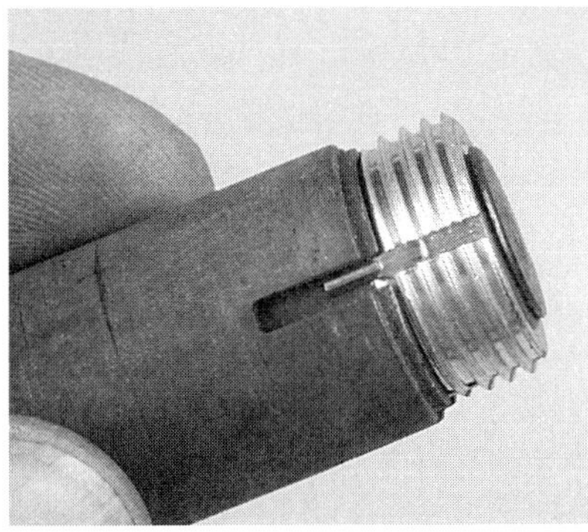

6. The sleeve is fully seated, with the insert stakes engaged into the sleeve's grooves. This allows you to hand-thread the insert into the spark plug hole.

7. Once the insert has been fully threaded into the spark plug hole, lift the sleeve away to clear the stakes, rotate the sleeve and drop the sleeve edge onto the stakes. Using a hammer, tap the tool to drive the stakes into place, flush with the perimeter of the spark plug hole.

chips that may be lying on the piston dome. Regardless, always check piston location before drilling, to avoid drilling a hole in the piston dome!!!! You can understand why it's always best to remove a cylinder head for this type of repair. Performing this work on the engine can lead to severe engine damage if you're not extremely careful.

REPAIRING BOLTS OR STUDS

The best way to "repair" a damaged bolt or stud is to simply replace it with a new unit.

If you must repair male threads (on a stud or bolt), a forming die is the answer. Avoid using conventional cutting dies, since this will remove much needed material, possibly decreasing the strength of the stud or bolt. The only reason to attempt repair of a bolt is if you have no replacements available (maybe you're at the track or off-road in the boonies), or if the bolt is unique and an exact replacement cannot be located easily. If a stud is damaged, again the best course is to replace it, unless reforming the threads is mandated by time and expense. The point is to avoid removing thread material. When you remove material from the threaded surfaces, you simply weaken the fastener.

AN INVENTIVE THREAD REPAIR THAT FIXES CRACKS

While it is not our intent to promote one manufacturer's product over another, when I find a product that I feel will inform, educate and assist our readers, I'm all for it. After reviewing Loc-N-Stitch's new Full-Torque thread repair system, featuring Spiralhook thread inserts, I thought you'd enjoy learning about this unique repair system.

Previously, if someone had told me that a block's damaged main cap or cylinder head bolt hole could be repaired using a threaded insert, I would have dismissed their banter as wishful thinking. However, the recently introduced full-torque system incorporates the proven spiral lock external thread design and offers inserts specifically designed for a variety of engine applications including critical high-torque holes such as head and main thread locations. In short, it sounds as though this thread repair system could salvage what would otherwise be a trashed block.

Cracked cylinder head bolt holes and main cap bolt holes have been considered unrepairable due to the nature and location of the damage. Obviously, if we subscribe to the view that anything is repairable, we still have to face reality, in that some repairs simply are not worth the time, trouble and expense. With that said, the inventive folks at Loc-N-Stitch have developed a repair insert that is designed to fix not just thread damage, but cracked bolt holes as well. The really interesting aspect of the system is that the "insert" is capable of pulling the separated thread walls together, creating enough strength and material integrity to accommodate even the high-torque-value applications involving head and main cap clamping.

When bolt hole walls crack, it's due to the radial force that generates outward from the threads as a result of applying torque. The force results from the interaction of the 60-degree included angles of the bolt threads to the bolt hole threads.

According to Gary Reed of Loc-N-Stitch, "It is actually such a big problem that all engine manufacturers are forced to make special compensations to try to prevent cracking. This is an inherent flaw in the design of bolts with standard threads that have plagued the fastener industry since day one." The most common solution to crack prevention is to recess the threads below the

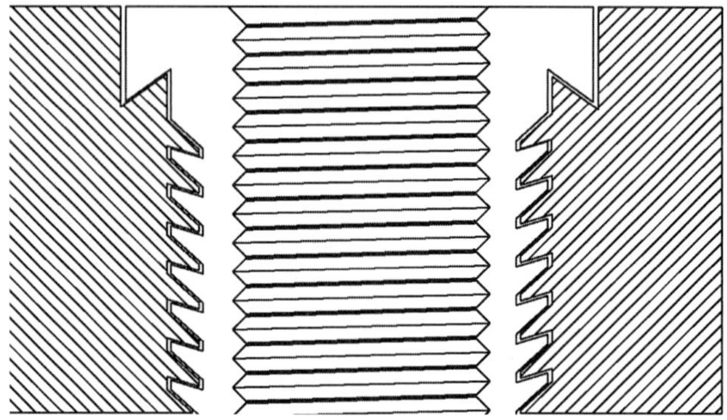

The Spiralhook thread repair system features an upward-angled outer thread that grips the surrounding wall and actually pulls the parent metal inward.

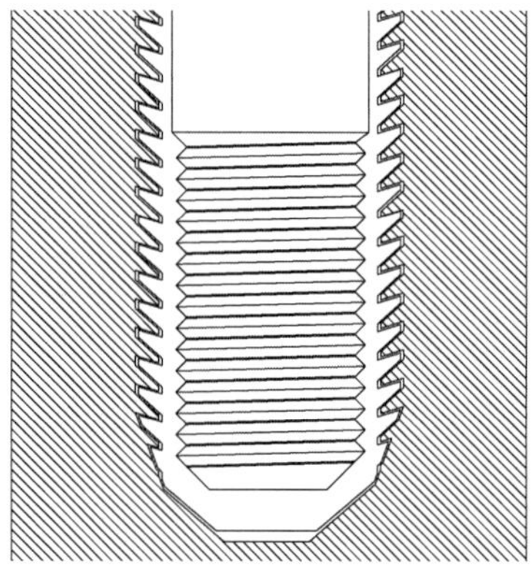

Cross-sectional view of a Spiralhook repair insert in a blind hole.

surface. Other remedies include heavy bosses surrounding the bolt hole and spacing holes far enough away from edges and openings to resist cracking. In late-model engines, adding a torque plate before honing the cylinders helps compensate for bolt hole stretch to ensure a round bore in spite of the distortion caused by the radial spreading force of the bolts.

Now that aluminum is being used with greater frequency for heads and blocks, the frequency of stripped, cracked and damaged tapped holes is higher than ever. In aluminum cylinder heads, for instance, a stripped spark plug hole is a very common occurrence.

While the use of stainless steel inserts offers a leve of installation ease and convenience, "the problem in using stainless steel coils and thin-wall steel inserts in aluminum heads for spark plug holes," noted Reed, "is that they do a very poor job of transferring heat away from the spark plugs, and th plugs can overheat. Stainless steel coils act as a heat dam and restrict heat transfer because stainless transfers heat very slowly. Steel inserts fail to expanc at the same rate as the aluminum, thereby losing contact and thermal transfer. The Full-Torque spar plug thread repair inserts for aluminum heads are made of aluminum and hard-anodized to maintain

Note the crack running from the threaded hole to a water jacket hole on a block deck. Red dye was applied to make this crack more visible.

Once the damaged hole has been drilled out and tapped using the Loc-N-Stitch dedicated drill and tap, the Spiralhook insert is installed. A special driver tool allows insert installation by engaging onto the insert head (notice the small dowel pin hole in the insert head).

Here's a Spiralhook insert fully installed in a cylinder head bolt hole on the deck of a Chevy bigblock engine block.

uniform heat transfer even in air-cooled heads. In addition, these new inserts will not back out, due to a locking pin that is installed through the flange of the insert."

The problems relating to the installation of the old-style solid inserts is that they remove valuable material surrounding the bolt hole, weakening the casting. This can result in failure of the insert or cracking of the casting. "Full-Torque inserts," according to Reed, "solve most of the problems caused by the bolt itself. The insert's Spiralhook external threads pull the surrounding metal into the threads. This radial drawing force improves with bolt torque. It's like putting Spiralhook threads on the bolt itself."

The accompanying photos illustrate the crack repair on a 1969 4-bolt 427 big-block Chevy. The crack originates from a cylinder head bolt hole and extends down into the cylinder about 5/8". The crack is first stitched from the bottom of the crack up onto the deck surface, so that the last pin will be intersected by the Full-Torque thread repair insert. The bolt hole is then drilled out, spot faced and tapped to accept the F1 series thread insert. A small amount of thread sealant is applied to the threaded hole, and the insert is then torqued in place to 10% over the required head bolt torque specification. After the installation bolt is removed, the locking pin is driven into place to prevent the insert from backing out (simply a failsafe feature). If the insert should ever be removed, simply drill out the locking pin and install a new insert. The insert is peened around the edges to produce a flawless finish once the excess insert material is machined flush to the deck surface. The tops of the stitching pins in the cylinder are ground close, followed by cylinder boring and honing. Cylinder sleeve installation is not recommended when using this repair.

When repairing cracked main bearing cap bolt holes, it's not necessary to fix the crack in most cases. You only need to drill, tap and install an F3 series insert and the hole will be better than new. The more force that is applied to the bolt, the tighter the crack is held together.

Once the Spiralhook insert has been fully installed, the protruding driver head is milled off flush with the deck. The block deck is then resurfaced. (Spiralhook photos courtesy Loc-N-Stitch)

Damaged pipe threads can be repaired with these inserts as well, even if a crack extends from the hole. According to Reed, simply stitch the crack and install a Full-Torque insert. Special plugs for Caterpillar injector chamber threads, Detroit Diesel injector holes and many other special application inserts are now available to fix specific problems on specific engines.

The Full-Torque inserts are available in four different styles. The F1 is a through-wall through-hole insert for repairs such as spark plug holes, pipe thread, diesel injector and automotive cylinder head bolt holes. F2 is a through-wall blind hole insert designed for repairs that require a dry, pressure-tight hole such as motor mount, oil pan rail and some cylinder head bolt holes. F3 is a solid-wall blind hole insert intended for repairs such as diesel head bolts and main cap bolt holes. The F4 is a general-purpose solid insert for use on any badly damaged threaded hole.

Note: This Spiralhook thread/crack repair system is intended for professional engine rebuilders only.

THREADED NUT INSERTS

If you need to screw one part to another, but the parent part's material is too thin, or if a material is selected that won't accept tapping (like sheet metal, wood or fiberglass), the solution is often to install a metal insert commonly and generically referred to as a "nutsert." Simply drill an appropriately sized hole in the material to accept the insert, insert the nutsert, and using a special tool, expand the body of the nutsert to lock it in place. As you tighten the insert, the main body (which is now located within or behind the material) will expand, much in the same way a rivet body expands to hold it in place. Use of this type of female thread insert provides an easy and handy way to create threads were non-existent before, but they're not for critical or high-torque applications. Nutserts (they come in various shapes, sizes and brands) are available, at least to my knowledge, in sizes ranging from 6x32 all the way up through 1/2". Metric sizes are also available (I've seen them in 3mm to 10mm).

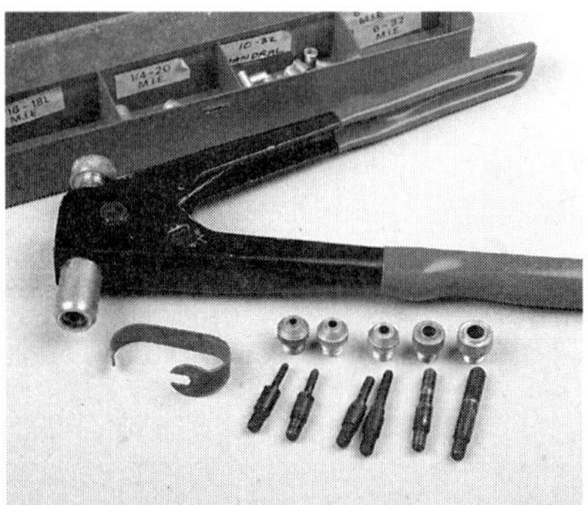

1. This nutsert kit includes the installer tool, a selection of mandrels and nosepieces (to accommodate different thread diameters) and a selection of nutserts. A manual tool like this will run anywhere from $40 to $60. Hydraulic tools can cost around $250 or so, cordless electric tools can run around $1000 or so, and pneumatic tools can run upwards of $1200. If you're not performing production runs and you don't mind flexing your hand muscles a bit, stick with the hand-operated tool. The dough you save can be used for better things, like engine or chassis parts.

2. Three typical insert nuts. From left: the A-T series nutsert and two rivet-type nut inserts. They all do the same thing: once in the hole and drawn by the installation tool, the threaded portion is drawn towards the top lip, expanding the body in the process, which locks them in place.

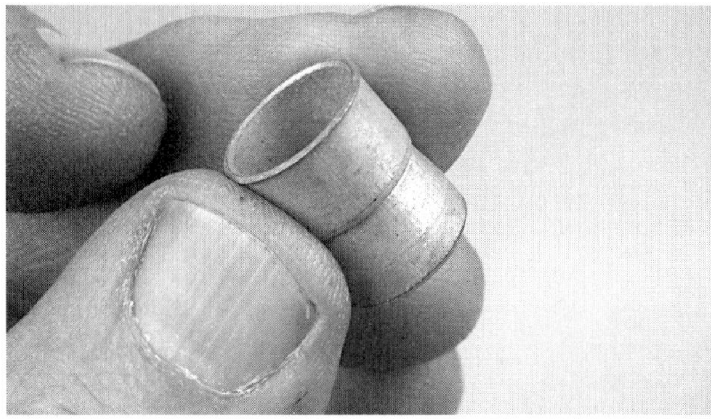

3. The A-T series nutsert provides strong threads in almost any material 0.030" or thicker. As the nutsert is drawn by the installation tool, the threaded portion is drawn into the sleeve, creating pressure around the hole (nutsert diameter increases to obtain a tight interference fit). These are available in cadmium-plated steel, stainless steel and aluminum. This style is my personal favorite.

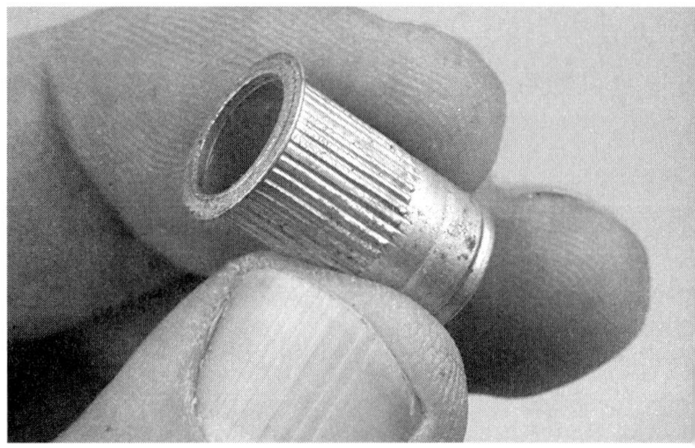

4. Serrated rivet type insert. This can be "staked" in plac by a tight fit into a hole (allowing the serrations to bit into the material). Use the nutsert tool to expand th insert.

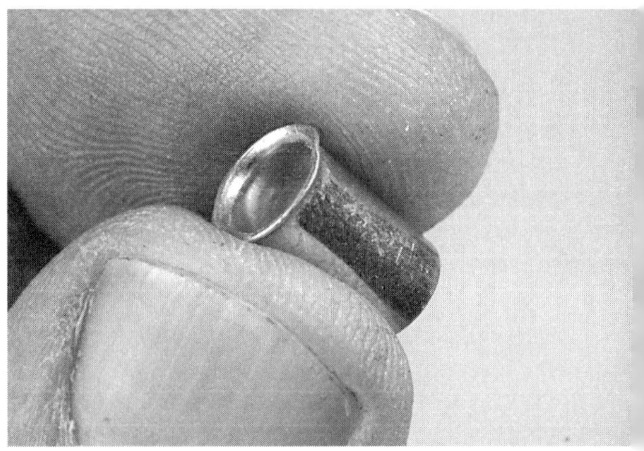

5. This is the cheapest type of rivet nut. Again, using nutsert tool, insert into a hole and squeeze the tool gri to draw the threaded area up, expanding the backside f retention. These are good for light-duty applications.

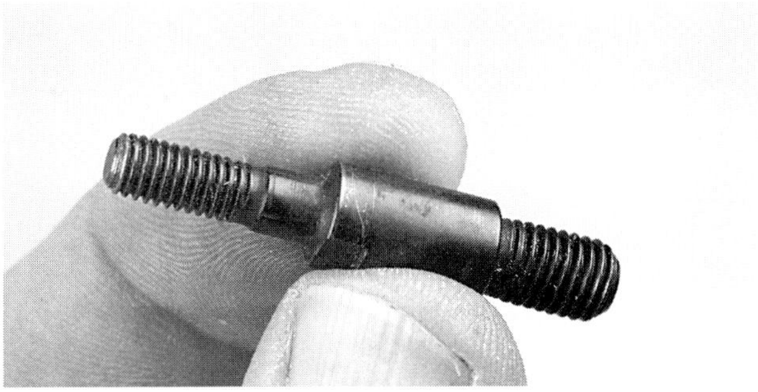

6. Select the appropriate size mandrel. One end features left-hand thread (the end that threads into the tool) and the opposite end features the desired thread size (the nutsert threads onto this).

7. Thread the mandrel into the tool (remember...it's left-hand thread, so install the mandrel counterclockwise).

8. Snug-tighten the mandrel to the tool using the wrench supplied with the tool.

9. Select the nosepiece that fits the exposed mandrel thread. The bore in the nosepiece is smooth...it slips over the mandrel. The male threads on the nosepiece thread into the tool (also left-hand thread).

10. Install the nosepiece to the tool and hand-tighten.

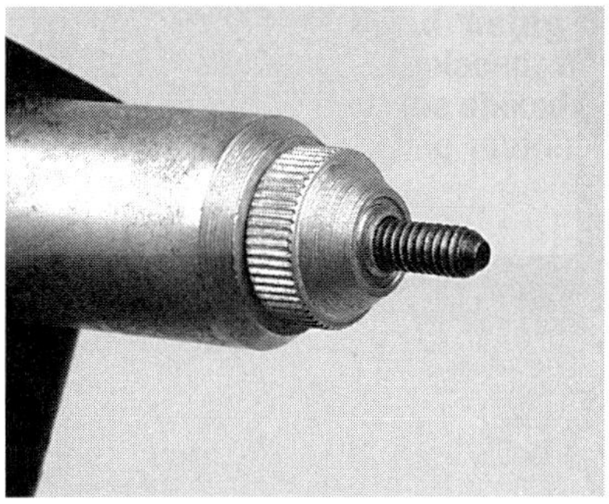

11. The tool is now ready to accept the nutsert. The nutsert threads onto the exposed threads of the mandrel.

12. Insert the nutsert into the hole (don't apply grip pressure to the tool yet). Here, we're installing an 8x32 nutsert. This will accept an 8x32 machine screw.

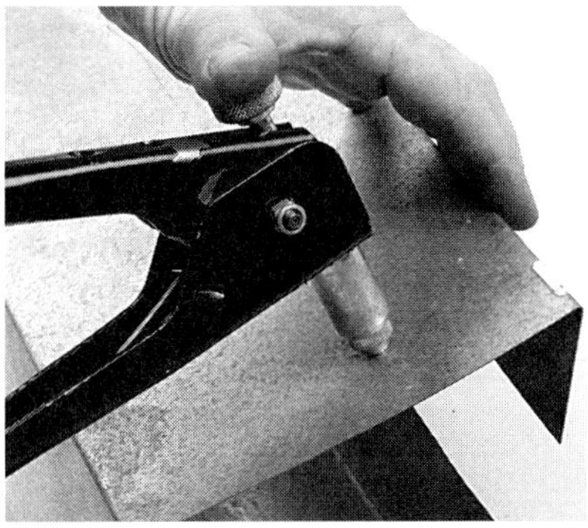

13. Once the nutsert is fully inserted into the hole, Begin to squeeze the tool grips.

14. Squeeze until the tool stops.

15. To extract the tool from the workpiece, turn the thumbscrew counterclockwise to unthread the mandrel from the nutsert.

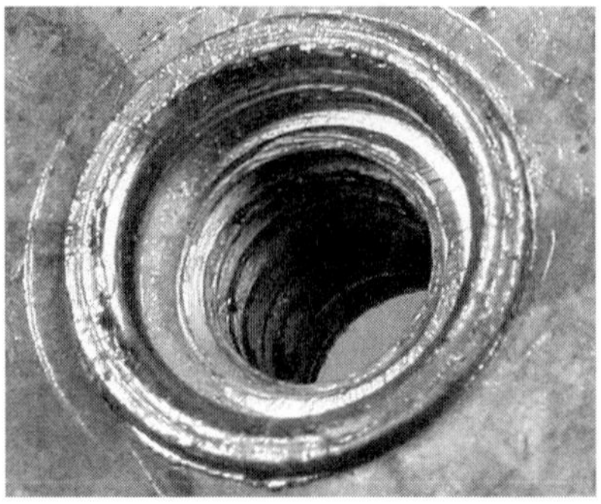

16. Closeup of an installed nutsert.

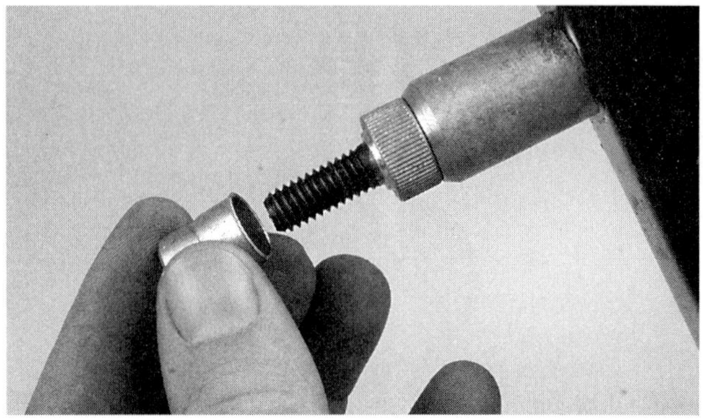

17. Here, we selected a 5/16" x 18 nutsert (the A-T type). We simply installed the appropriate size mandrel and nosepiece to the tool.

18. The 5/16" nutsert on the tool, ready for use.

19. This 5/16" nutsert requires a 1/2" hole.

20. With the nutsert positioned and flush, use the tool to draw the threaded section up into the sleeve. The expansion snugs the nutsert in the hole.

21. Installed 5/16" x 18 nutsert.

22. This side view shows how the rear threaded section was drawn up into the nutsert sleeve, creating sleeve expansion.

Chapter 5

CREATING THREADS WITH TAPS & DIES

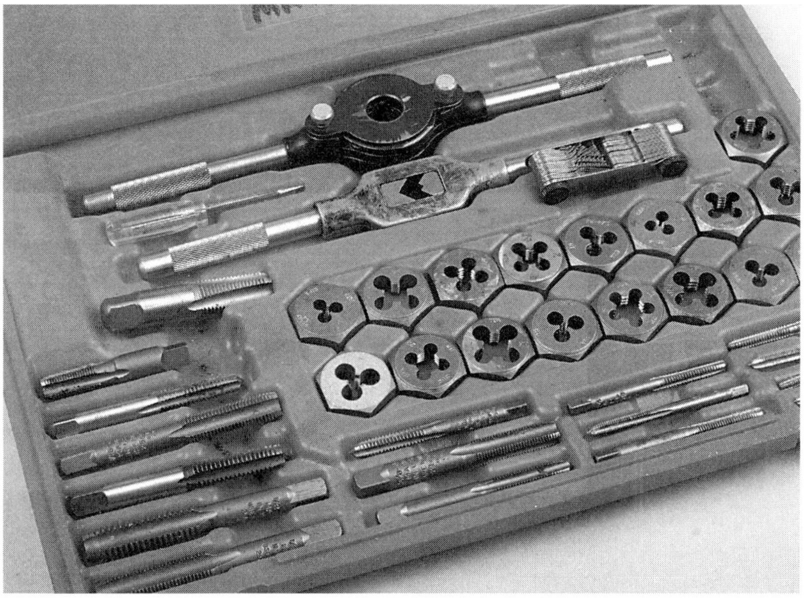

Consumer-grade tap and die sets are commonly available in both inch sets (shown here) and metric sets. This inch set includes sizes 6x32 all the way up to 1/2". The 1/4" through 1/2" sizes feature both "fine" and "coarse" thread taps and dies.

Taps and dies allow you to create threads. Taps are male cutting tools that create threads in a female hole. Dies are female cutting tools that create threads on a male rod or shank.

Note: The vast majority of taps are designed for creating female threaded locations. By their nature, these tools cut the parent metal in order to create the threads. When you need to "restore" or clean-up "boogered" threads in a critical engine component, such as the cylinder head threaded holes in an engine block, avoid using a normal cutting tap. nstead, use a "forming" or "chaser" tap that is specifically lesigned to chase and reform the existing threads instead of cutting thread material away. By using a regular cutting tap n a block's cylinder head bolt holes, you could easily remove oo much thread material, weakening the threads and possibly resulting in cylinder head gasket failure down the road (along with a resulting warped head, oil/coolant cross-contamination, etc.).

If you're faced with aluminum or steel threaded holes that re only slightly damaged (burrs, etc.), you may be able to lean and reform those threads by inserting a high-grade hard) bolt (grade 8 or 9). The bolt's metal will be harder han the parent material, and will often restore damaged hreads. This avoids the use of a cutting tap where too much netal might be removed from the threads.

Also, for spark plug hole thread repair, a special thread chaser is available that will clean and reform the existing spark plug hole threads without cutting (so no metal chips will be dropped inside the cylinder). An example is the 14mm spark plug thread chaser offered by Goodson Shop Supplies as P/N SPB-14. For more on thread repair, see Chapter 4.

THREAD CUTTING AND REPAIRING DON'TS

• If a bolt or stud's threads are stripped, galled or otherwise mangled, replace the bolt or stud. If the bolt or stud doesn't feature enough material to provide full-depth threads, you'll waste your time.

• Never be in a hurry when running a tap or a die. Cutting threads is a precise process. Rushing the job will only cause problems.

• Never guess or eyeball the size of the drill required to create a threaded hole. Take the time to look at a reference chart to determine the correct size drill for a given hole.

• If you're dealing with a shallow blind hole, you must use a bottoming tap in order to create threads as close to the bottom as possible. However, start the threads with a chamfered tap, following up with a bottoming tap.

• Handle and store your taps and dies carefully. Always clean them when you're done, and store them away from moisture. Before using a tap or die, check and clean the threads.

• Remember that taps and dies are designed for cutting and

Chaser dies are available individually or in sets. This set includes popular inch and metric sizes.

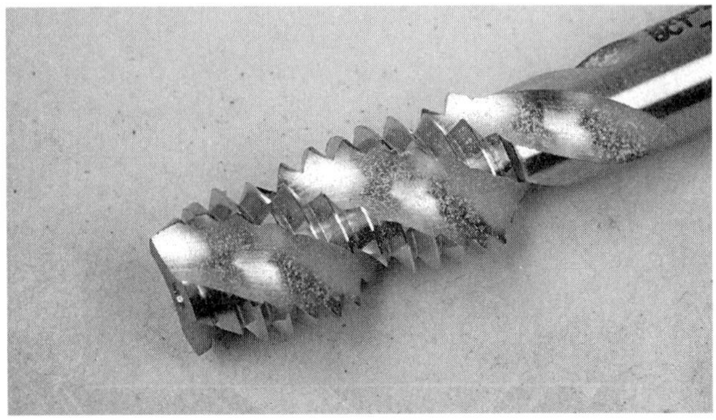

This 7/16" x 14 chaser tap (from Goodson) is specifically designed to *chase*, or follow female threads, cleaning and restoring thread shape, as opposed to cutting threads. The tap shown here is of extremely high quality.

creating threads. Whenever possible, use a following, or forming tap or die to clean or repair existing threads, in an effort to remove as little metal as possible from existing threads

Types of Taps

Spiral point taps are also referred to as *gun taps* because of the angular point that shears and shoots chips out ahead of the tap. This minimizes the chance of the flutes being clogged with chips, which in turn reduces the chance of tap breakage. This type of tap can be run faster, with less power required. To be used only for through-holes or holes that are at least two diameters deeper than the required thread length. For instance, if you're using a 1/2" diameter tap (cutting, say 1/2x20 threads), if the hole is blind, you'll need at least 1" of extra depth where threads are not required.

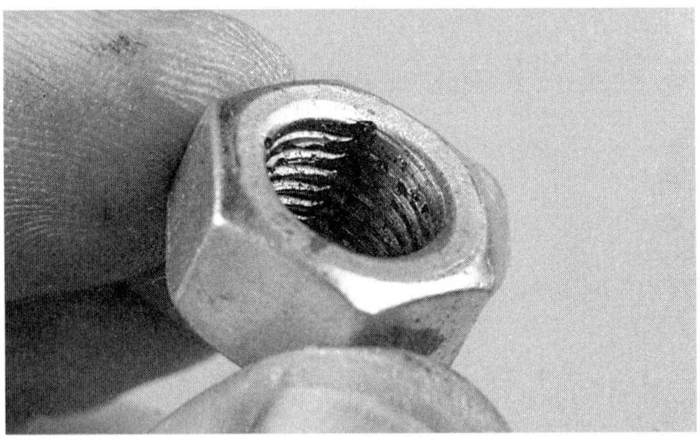

Inexpensive chaser dies are hardened steel nuts with flutes. Run the nut onto a damaged bolt or stud, then back it off. If you're about to cut (to shorten) a bolt or stud, you can thread a chaser die onto the threads, past the cutting area. Make the cut, deburr the edge of the cut, then back this nut off, restoring threads as you go. This type of chaser die is a good idea, but be careful to avoid the cheap imported junk. Some of these are so poorly made that the threads aren't even cut properly, and you'll damage the bolt or stud instead of restoring it. Buy only respected brand-name products. I tossed this one into the trash after I took this shot. As with any tool, if the price is too good to be true, avoid it. Spend some dough and buy quality tools. Cheap tools cause more grief than they're worth.

Hand taps feature straight flutes. Hand taps can be used either by hand or by machine (for example, on a lathe).

Spiral flute taps feature spiral flutes (also called helical flutes), and are best used on soft materials such as aluminum, brass, copper or magnesium. The flute spiral offers a lifting action that removes stringy chips that form when tapping soft metals. Good for deep holes, blind holes and holes that feature a keyway or a gap.

Pipe taps, as the name implies, are designed to create the specific diameter and pitch required for pipe threads. Note that not all pipe taps are tapered.

Tap Chamfer Styles—Three basic styles of tap tips are available based on the tip's chamfer. These include tapered chamfer, plug chamfer and bottoming chamfer. Tapered chamfer taps feature a noticeable taper, or chamfer at the tip, featuring about 7 to 10 threads in this tapered area. This eases tap entry and alignment. Plug chamfer taps feature only about 3 to 5 chamfered threads at the tip area. Both tapered and plug chamfer taps are OK to use in through holes.

Bottoming chamfer taps allow you to "bottom" the threads in a blind hole (creating threads as close to the bottom of the hole as possible). Bottoming chamfer taps feature 1 to 2 chamfered threads (the slight chamfer offering initial entry into the hole).

Pay attention to the type of tap chamfer. When tapping a through-hole, any of these types will do. However, when tapping in a blind hole, if it's critical to use up all of the hole, you'll need a bottoming tap. And if you need to tap all the way to the bottom, it's best to use all three types in

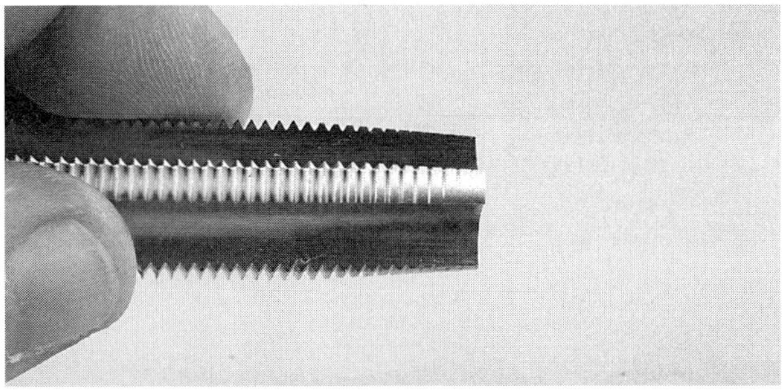

A tapered chamfer tap provides a lengthy taper at the tip, which eases starting into a drilled hole. Always begin to cut threads with a tapered chamfer tap.

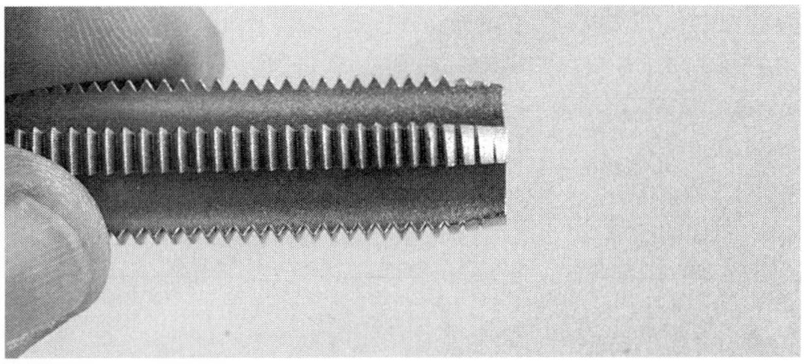

A plug chamfer (commonly called a plug tap). This features a smaller taper area. Good for general use, but a tapered tap will provide a better start. If you're dealing with a through-hole, you can cut all of the threads with a tapered tap. If you're dealing with a blind hole, but threads are not required all the way down to the bottom, you can use a plug tap.

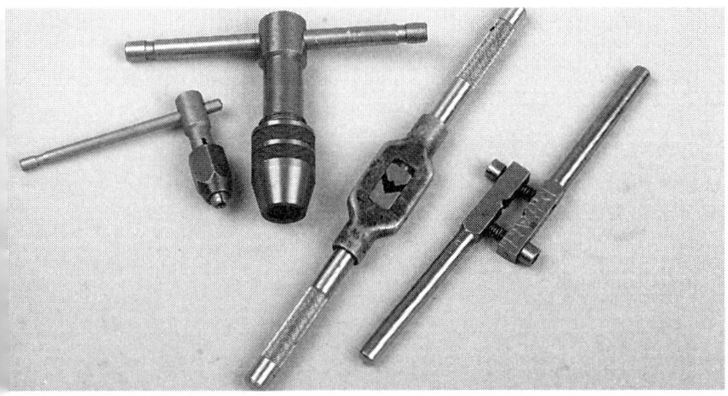

Tap wrenches, or tap handles, are available in several styles, including T-handle tools with chuck jaws, straight-bar style with sliding jaw and 2-piece clamping bar type.

progression—start with the taper chamfer tap, followed by the plug chamfer tap, and finishing with the bottoming tap.

Tap Coatings & Treatments

If you plan to use your taps often, and especially if you plan to machine tap, you might want to consider the tap surface coatings/treatments in order to extend tap life. Coated taps can enhance tap life by as much as a whopping 800% as compared to uncoated (bright) taps. TiN (Titanium Nitride) coating, which appears as a gold color, is suitable for a variety of both ferrous and non-ferrous materials (but ironically is not appropriate for cutting threads in titanium alloys). TiCN (Titanium Carbonitride) coating is harder and more wear-resistant than TiN. These taps feature a blue/gray/violet color.

Chrome-plated taps feature harder threads with improved lubricity for cooler cutting and less loading. These taps are silver colored and are good for ferrous and non-ferrous materials. Oxide-treated taps prevent buildup or welding (where the tap may weld to the parent material if high force/speed is used). The oxide treatment (blue/black color) reduces friction and offers a slightly porous surface that holds tapping lubricant. These taps are good only for ferrous

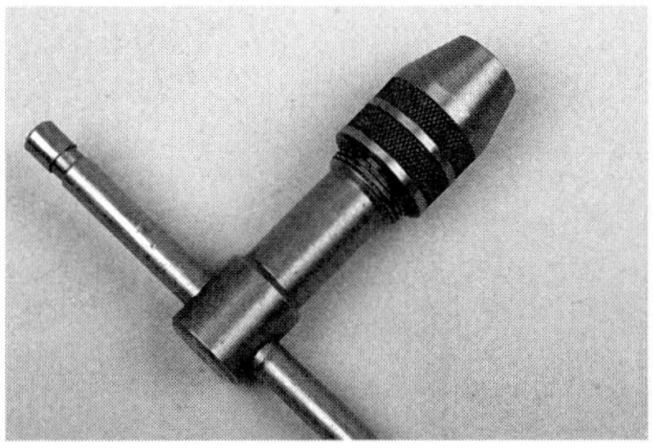

This is a good quality tap wrench. The body is machined steel, as is the chuck head. Periodically, remove the head and clean and lube the threads.

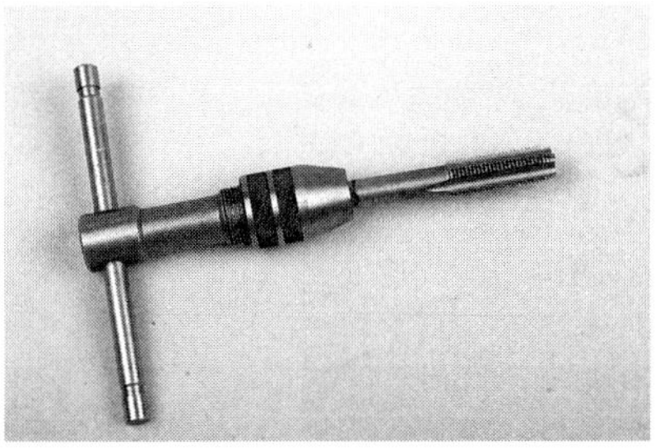

The T-handle tap wrench shown here (2-jaw) is a heavy duty tool that you can depend on. I've used this one for about five years in my shop with no problems. Go to an industrial tool supply house and invest in a decent tap wrench.

The cheap T-handle chuck-type tools feature four jaw fingers that compress together as the head is tightened. These are not as rigid or secure. The one shown here was included in a cheap kit, and it's a piece of junk.

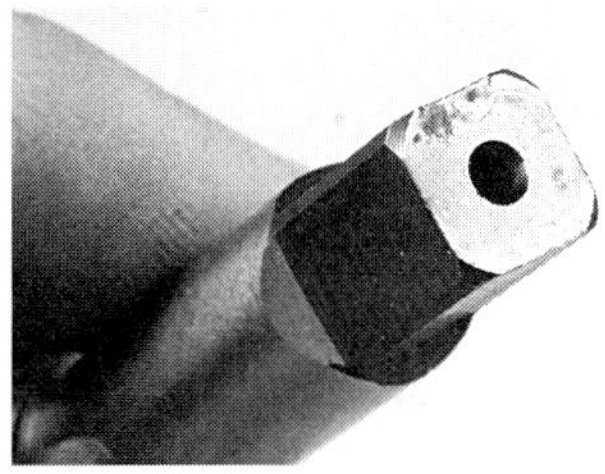

Industrial grade taps feature a centering dimple, unlike consumer versions. This allows you to use the tap on a lathe, if you so desire. A centering pilot tool keeps contact with the dimple, assuring that the tap remains straight during setup and operation. If you plan to use taps on a lathe, make sure you buy the industrial-style taps.

The sliding-jaw type tap wrench (the type that usually comes with a cheap kit) works, but it won't secure the tap for long. Always recheck the tightness. This style is OK for general use, where you have room to swing the long handles. The T-handle style features a sliding bar that allows you to push the handle out of the way as you rotate the tap.

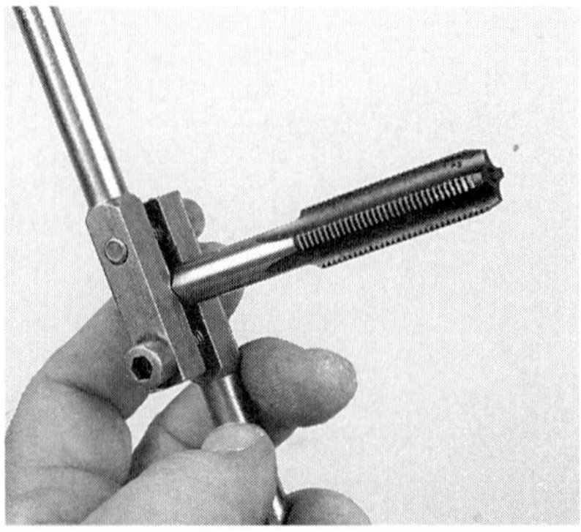

The 2-piece clamp type tap tool is simply made of two identical rods that attach with a pair of screws. The V-cuts face each other. Loosen the two socket head cap screws, insert the tap and tighten the screws to secure the tap. Of course, you need a hex wrench to service the screws.

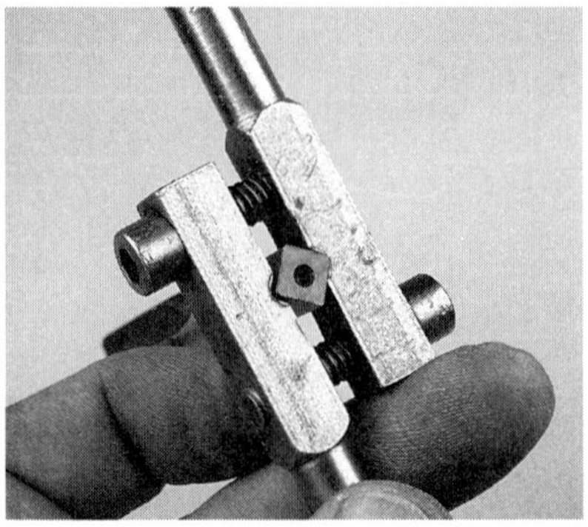

If you'll be using the same tap for a while during a given job, the clamp style does a good job of securing the tap, since you can tighten the screws for a stable connection.

materials. Oxide-over nitride taps are good for ferrous materials, high-temperature alloys, titanium, titanium alloys, iron, high-tensile steel and stainless steel. These taps also feature a blue/black color. Nitride taps are suitable for use on ferrous and non-ferrous materials that are abrasive (materials that tend to dull taps). Use on aluminum, cast iron, brass, bronze, magnesium, zinc, copper and die castings.

Tap Accuracy Tolerance

Not all taps of the same size are equal. Depending on how precise your threads must be, taps are available in a range of tolerance limits. Seven tap diameter tolerance limits are offered, as follows. (Note that the letter H designates inch sizes and D designates metric sizes.) The higher the number, the less "precise" the tap:

H1: +0.0005"
H2: +0.0005 to 0.0010"
H3: +0.0010 to 0.0015"
H4: +0.0015 to 0.0020"
H5: +0.0020 to 0.0025"
H6: +0.0025 to 0.0030"
H7: +0.0030 to 0.0035"

Tapping Fluids

For best tapping results, in terms of thread quality and reducing wear and tear, tapping fluids are highly recommended. For best results, use a lubricant that is specifically designed for tapping/cutting threads.

Special Taps

Taps are available in longer lengths for cutting threads in deep locations. I've seen extra-long taps of up to 6" in overall length. Naturally, these long-shank taps will cost more.

If the part (being tapped) is to be chrome-plated or powder-coated, and you're concerned about plating or coating buildup on the threads, special 0.005" oversize taps are available that cut the thread diameter 0.005" larger.

Want to create left-hand threads? No problem, as long as you can locate the size you need in LH thread. Your local sources will probably not carry them at all, or may be able to special-order them. If you need LH taps, try a good industrial tool supply house, or try one of the big mail-order suppliers. Outfits like McMaster-Carr and others usually carry LH taps in stock in both inch and metric sizes.

As you would expect, the "normal" tap features a right-hand thread. However, left-hand taps are available but probably not on-the-shelf items for your local suppliers, so expect to special-order these. Left-hand threads have their place. For instance, if you're fabricating a tie-rod adjuster sleeve and plan to use a left-hand thread outer tie rod or spherical rod end, you may want to make an adjuster sleeve with right-hand thread on one end for the rack tie rod and left-hand thread on the other end for the outer tie rod end (this allows length adjustment without the need to disconnect one end). The same goes for making a RH/LH adjuster sleeve for stabilizing and adjusting belt tension on an alternator or A/C compressor in a street rod or other custom application. Left-hand threaded taps may seem like a waste of money now, but when you need one, you need one.

As I mentioned in Chapter 1, left-hand taps and dies may not be easy to find at your local auto parts or hardware stores, but they're readily available from any full-line machine tool or fastener supplier. If you can't locate what you need locally, jump onto the Internet. See the Source Index for suggested sources.

Drilling A Hole to Be Threaded

Don't blindly drill a hole that "looks" like the right size in preparation for tapping. Consider the bolt size you plan to use, then refer to the drill chart (usually included with any decent tap and die set). Or refer to the charts listed here. The drilled hole must be large enough to allow tap use, but small enough to allow creation of properly sized thread diameter. Basically, we want to drill a hole that equals the root diameter of the desired threads. This will provide enough meat in the hole to allow the tap to cut the major diameter.

It's a really good idea to keep a precision caliper (dial caliper or digital caliper) with your tap and die set. In some cases, your tap and die set sizing chart may call for an alpha- or a numeric-coded tap. You can refer to the charts here to determine what decimal-inch equivalent is represented by an alpha or numeric drill, but you'll need a caliper to measure the drill O.D. in order to identify the bit.

After drilling the hole, it's a good idea to slightly chamfer the hole entry. This will remove any raised edge burr around the hole perimeter, which will aid in tap insertion as well as bolt insertion down the road. The hole edge can be chamfered with a tapered chamfering bit, or with a hand-operated deburring tool.

Using A Tap

When you need to create threads in a hole, a tap will be required. Threads can be tapped by hand (securing the tap's square drive end in a tap handle, or tap wrench, and turning the tap by hand; or the workpiece may be secured in a lathe chuck (where the workpiece rotates) and the tap secured in a drill head that is stationary on the lathe. The lathe must be adjusted down to its lowest speed, and the drill head must be allowed to easily slide on the lathe horizontal rails. With the lathe head turning, slide the drill head towards the workpiece until the tap enters the hole to be threaded. The tap threads will begin to cut into the parent material, and will self-feed (give the drill head a slight push to help it along).

Note: This will work only if the workpiece can be properly secured and centered on the lathe head; and if the female hole is open (allowing the tap to pass through). If the hole is blind, or needs to be only partially threaded, the lathe must feature a quick shut-off. Otherwise, the tap will feed through the hole uncontrolled. When the tap bottoms out, disaster will strike, with the tap breaking, the workpiece breaking, or the workpiece being knocked out of the lathe head. Either way, it won't be pretty.

However, you can use a tap on a lathe by working the tap manually, using the lathe only as a fixture. A centering tool can be mounted to a lathe-mounted drill head, which will locate the hole's dead center. With the tap secured in a hand-tap tool, locate the centering tool's point into the small dimple featured on the rear end of the tap (**Note:** common hand taps found at retail outlets will normally not feature this centering dimple. If you need dimpled taps, these must be purchased from an industrial supply house). This will be slow-going, but it'll keep the tap straight and centered. Tap a bit, and push the drill head closer to the workpiece to keep contact between the centering tool and the tap, etc. The problem that many people face

Here we're about to drill and tap a hole to serve as a typical example. For accuracy of hole placement, after marking the hole center, use a pilot bit. This will allow you to place the hole center more precisely (without the need to use a small-diameter flexible bit that can flex or break).

This hole has been center-drilled with the pilot bit and chamfered.

For this sample hole, we intend to create a 3/8" x 16 thread. In order to use a 3/8" x 16 tap, a 5/16" diameter hole is required. Here we drill the hole using a 5/16" drill.

The sturdy pilot bit features a chamfer cutter. Take advantage of this to create a slight chamfer at the hole entrance. This will aid in centering the final-sized drill.

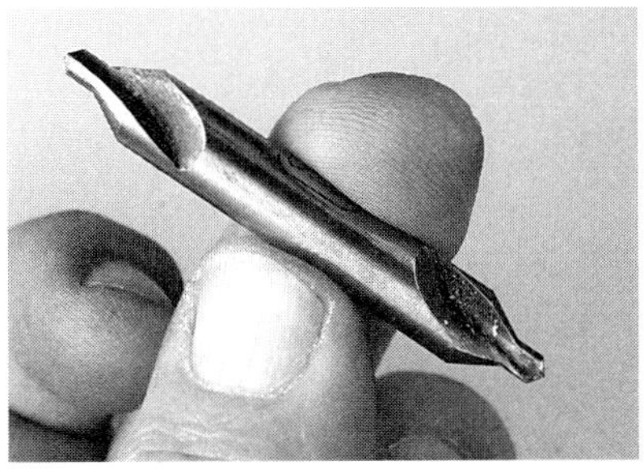

Pilot bits are short and sturdy, so they won't flex (which could result in an off-center hole or fracture of a skinny wire drill). Pilot bits also feature two identical tips (when one wears out, you can flip it over).

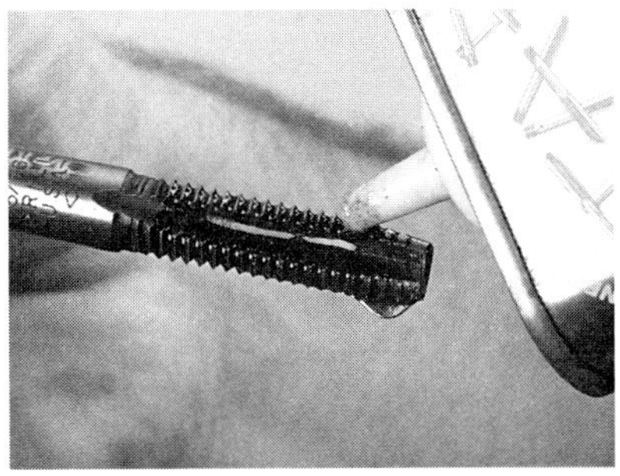

With the tap secured in the tap wrench, we apply tap lubricant. In this particular case, we're about to tap a piece of aluminum. Tap lube designed for aluminum will prevent galling and tearing of the parent material, and will result in easier, faster and cleaner creation of threads. Since this sample hole will be a through-hole, there's no reason for us to use a bottoming tap, so we're starting our threads with a tapered chamfer tap, and following with this plug chamfer style tap.

In this case, we decided to use our drill press to maintain a straight shot for our tap. We installed a centering tool (we grabbed one of our chamfering bits) in the drill chuck, engaging the bit's point into the dimple on the tap wrench head.

Once the tap was driven through about half of the workpiece, we continued to rotate the tap wrench T-handle with both hands.

Once the hole has been tapped, blow it clean of debris, then insert an appropriate-sized bolt to verify the new threads. If the threads have been cut properly and cleaned, you should be able to easily run a bolt into the hole with your fingers.

Once the threads have been cut, we then gently kissed the top edge of the hole with a chamfer bit, to eliminate the raised burr around the hole perimeter.

The hole has been tapped and deburred. The slight chamfer not only removes the raised burr, but the chamfer also allows easier entry for a bolt.

when trying to tap a hole is getting the tap to begin a straight entry. If the tap enters and begins to cut threads off-plane from center, you'll end up with a hole that has deeper-cut threads on one side of the hole and shallower threads on the opposite side of the hole.

Another option is to use a drill press. If you're uncomfortable in allowing the drill press to power the tap, you can tap manually with the workpiece secured in the drill press vise. Once the hole has been drilled, without disturbing the drill press table or vise, chuck a pointed pilot or pointed chamfering bit into the drill chuck, and use this as a centering device (with the centering bit engaged into the center dimple of the tap or tap wrench), holding the tap in a centered and straight position. Turn the tap wrench with one hand while applying gentle pressure on the drill press up/down arm. Once the tap has entered about halfway into the hole, you can continue to drive the tap with both hands on the tap wrench. The idea is to keep the tap at a straight angle relative to the hole. If you begin to tap at an off-angle, you'll create a sloppy, angled threaded hole that will be useless.

In most do-it-yourself cases, the tap will be completely hand-operated. Do your best to keep the tap as straight as possible in both planes.

Take advantage of tapping fluid (not just any lube, but dedicated tapping fluid) to ease cutting and to avoid galling. This is especially important when cutting threads in aluminum. Assuming the threads will feature a right-hand twist, rotate the tap handle clockwise relative to the hole entry until you begin to feel noticeable resistance. At that time, back the tap out a little bit and then resume cutting threads. Occasionally, it may be wise to remove the tap completely and blow out any chips from the hole. Clean the tap threads with a stiff brush (an old toothbrush works fine), and resume tapping.

If you're cutting threads in a blind hole, before you begin to cut threads, first carefully measure the hole depth, and place a reference mark on the tap to indicate when the tap is approaching the hole bottom. The mark can be a magic-marker mark, or you can wrap a piece of masking tape to the tap at the depth mark. If you're using a bottoming-style tap, you'll usually be able to cut threads within about 1/8" from the hole bottom (depending on hole and tap diameter). Don't continue to turn the tap into the hole once the tap bottoms out, since this will only result in thread damage, or possible seizing or failure of the tap.

Once the hole has been threaded, run a rifle brush through the threads (helping it along with a few turns), followed by blowing the hole clean with compressed air. Always wear eye protection whenever applying compressed air into any hole.

Note: If by chance you blast or glass-bead the workpiece prior to threading, make sure that the hole is free of any blast media. By the same token, if you blast the piece after the hole has been tapped, make sure that you clean the threads of blast media

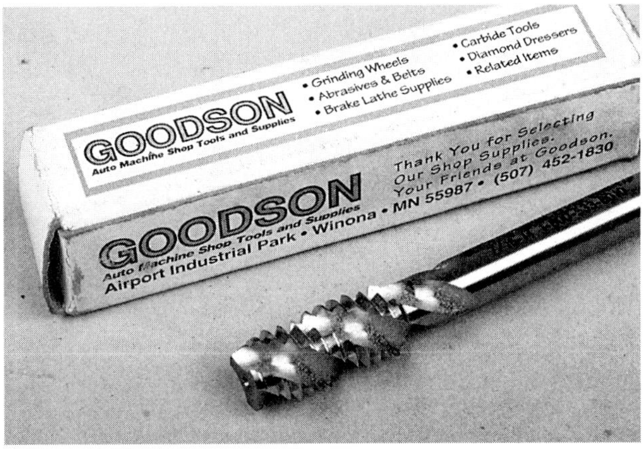

An example of a chaser tap. The sample shown here is size 7/16" x 14, from Goodson Shop Supplies, their part number BCT-7/16 "Block Clean Out Tap." This will clean and re-form the existing threads without removing excess metal from the threads. This professional-grade chaser tap is not cheap, but it's worth it. If you plan to restore an old engine block, buy one these in the block's cylinder head bolt size and clean out every head bolt hole. You won't regret it. Dirty, cruddy and boogered head bolt holes will prevent proper cylinder head clamping.

Each die will be marked for thread size and pitch.

One side of the die will feature a chamfer entry. It will also be marked "start this die" as well. Always use the chamfered end as the entry side.

Run the chaser tap using your tap wrench. Be sure to run the tap all the way through to clean the entire length of existing thread. The block clean out tap shown here is a bottoming-type tap, enabling us to clean threads at the very bottom of a blind hole. Once the tap is removed, blow the hole clean with compressed air, clean the hole with a rifle brush, then blow the hole clean again.

Typical cheap die handles will feature a hex recess that accepts the hex-sided die. The die is secured with a set screw. The die can be installed in either direction (relative to the chamfer).

If your die wrench features cam-type centering fingers, take advantage of them by first installing the die's chamfer side facing the cam fingers. You can then contact the rod to be threaded with the cam-side of the wrench facing the rod. Adjust the cam fingers to contact the rod and tighten the cam set screws. This will help guide the die onto the rod while starting the thread cut with the chamfered side of the die.

before painting, plating or inserting a bolt. Abrasive blasting media can really mess up your threads if they're forced into the threads with a tap or bolt. Keep them clean!

Chasing Burred Female Threads

If you want to clean up existing female threads (maybe the threaded hole has burrs or a bit of corrosion, etc.), it's best to use a chaser, or follower tap instead of a common cutting tap. A chaser tap will try to reform the threads, as opposed to a cutting tap, which will cut its way through, possibly removing too much thread material. This is especially important when dealing with critical-torque-load threaded holes such as cylinder head bolt holes in an engine block's deck. These taps that are designed specifically for cleaning cylinder block threads are also called block clean out taps.

USING A DIE

A threading die will allow you to create threads on a male workpiece. Using a die is a bit trickier than using a tap, since it's very easy to start threading crooked when using a die by hand. Don't assume that the die will automatically self-center, because it won't. It's easy to accidentally start the die off-angle. The die MUST be kept at a 90-degree angle to the workpiece, especially when starting the cut.

Pay attention to the die itself, since the cutting threads on one side will be chamfered (this side will be stamped "start with this side" indicating the side of the die that needs to make contact with the

Here we cut threads with the die secured in the die wrench. Notice that the centering cam side of the wrench is facing the workpiece. If you need to cut threads very close to the base of a rod where the cam finger set screws will get in the way, you can loosen the die set screw, remove the die wrench, flip it around and reinstall it onto the die to continue the thread cutting.

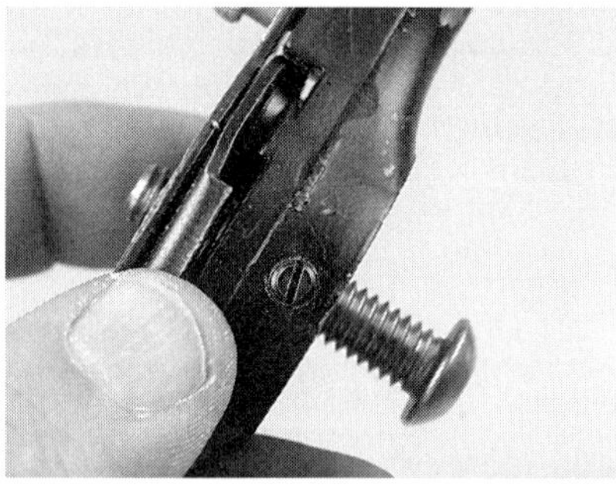

To illustrate, we threaded a screw into a die. With the two set screws loose, the centering cam fingers are easily adjusted by hand to contact the screw shank or rod being threaded.

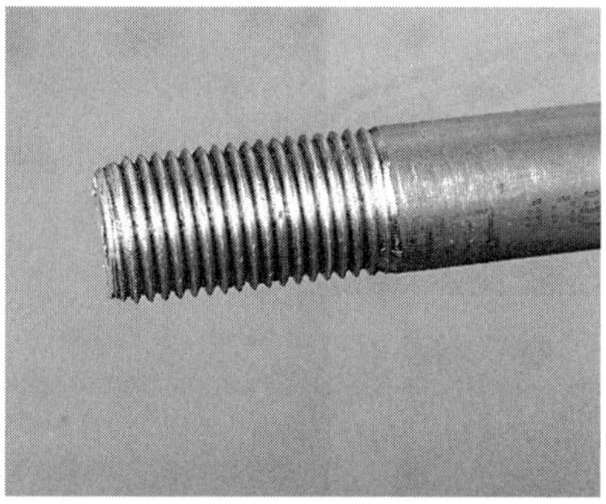

Newly cut threads on an aluminum rod. Clean the threads thoroughly using compressed air and a brush. Avoid using a stiff steel wire brush, as this may damage the threads.

This closeup (using a screw for illustrative purposes) shows the centering cam fingers. When you're about to cut threads onto a blank rod, adjust the centering fingers onto the rod surface and tighten the two set screws. This will help guide the die and will aid in keeping the die straight.

Before backing the die off of the newly-cut threads, use a blow gun or brush to remove as much debris as possible.

Run an appropriate-sized nut onto the new threads to verify thread condition. The nut should easily run on/off with your fingers.

workpiece upon initial entry. This chamfer will help to center the die, but you'll still need to pay close attention to keep the die level (perpendicular to the rod being threaded). **Note:** Just as when you're tapping, apply a good tap oil to the threads as you cut, and keep your thread cutting to a slow speed. Especially when cutting threads on aluminum or stainless steel, you'll need to cut slowly to avoid galling and smearing the parent material.

Most dies that you'll buy at a hardware or auto parts store will feature a hexed outer face. This allows the die to be driven by a dedicated hex-type die wrench, or with a hex wrench or socket from your tool box. Round-outside dies are also available for use in industrial equipment such as a lathe. Before you begin to use a die on your project, it's a good idea to practice on some scrap rod of the same material in order to get accustomed to the die's use. If you've never used taps and dies before, believe me...using a die is trickier. It's easy for the novice to make mistakes.

SHORTENING A BOLT OR STUD

It's not unusual to discover (usually at night or on weekends, when the fastener store is closed) that you need a bolt that's a bit shorter than those you have on hand. We've all been there...due to clearance issues, you need a 5/16" x 18 bolt with a shank length of 1.25", but the only bolts you happen to have handy are 1.5" long. You don't want to (or can't) wait until Monday morning, so you decide to whack 1/4" off of the threaded length.

If you need to trim the threaded shank length, the shank can be cut using a hacksaw, a reciprocating metal saw or an abrasive cut-off wheel. Obviously, be sure to secure the bolt in a vise before cutting. If you can't secure the bolt in your vise without damaging the head (maybe you're dealing with a polished stainless button head), you can install two nuts onto the shank, jamming them together. You can then secure the bolt in the vise by allowing the vise to grip the flat sides of the nuts.

However, once you make the cut, you'll then need to address the threads at the newly-cut tip, which are sure to be deformed. Instead of trying to clean up the threads with a thread file, simply install a threading device onto the threads before you make the cut. Depending on the hardness of the bolt, you may be able to use a grade 8 or grade 9 nut, or you might need to use a die. Simply thread the nut or die onto the threads, past the cutting area (placing the nut or die between the cutting area and the bolt head). Then make your cut to shorten the shank. Follow this by using a file or a hand-held pneumatic mini-belt sander to carefully radius

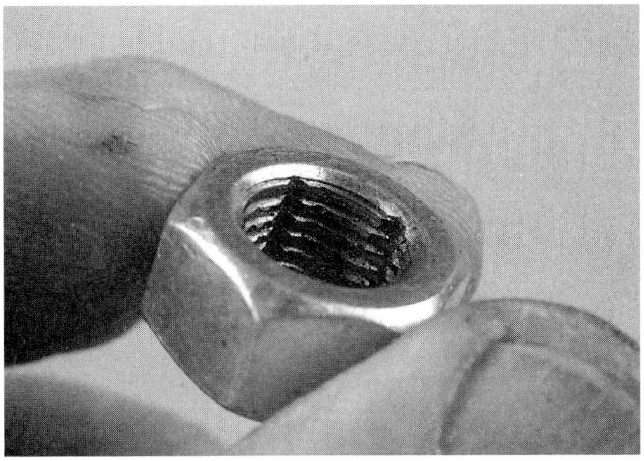

Here's a option. If you need to repair damaged threads, chaser dies are available. These features a series of flutes cut into a hardened steel nut. The theory is that you'll reform the threads rather than shaving material from them. These work OK, but they'll still often remove some thread material, depending on how badly deformed the threads were at the outset. Apply a bit of tap cutting fluid onto the threads, and run one of these nuts on/off. It's a good way to save a bolt or stud. Just make sure that you buy good quality chasers, as some of the cheap imports don't work very well.

(clean up and slightly round) the edges of the new tip. Then remove the nut or die, which will repair/form the damaged threads as it runs over the cut area. Test-install a fresh nut, to verify that the bolt tip won't cause any starting problems.

DETERMINING TAP DRILL BIT SIZE FOR FRACTIONAL BOLTS

Bolt diameter - (1/threads per inch) = Tap drill size

First, divide the threads per inch (the thread pitch) into 1. Then subtract that number from the bolt diameter. For example: If you plan to drill and tap a 3/8 x 16 hole, divide 16 into 1 (for a result of .0625). Subtract the .0625 from 3/8 (or its decimal equivalent of .3750).

So, .3750 - .0625 = .3125". A drill bit measuring .3125" (or its fractional equivalent 5/16") would be correct.

DETERMINING TAP DRILL BIT SIZE FOR METRIC BOLTS

Bolt diameter - thread pitch = Tap drill size

Example: If the desired bolt hole is 10mm x 1.50, subtract the thread pitch (in this case 1.50) from the diameter (in this case 10). The resulting number (in this case 8.5) is the correct drill size to use if you plan to tap a 10mm x 1.50 hole.

DRILL SIZES REQUIRED FOR TAPPED HOLES

Before drilling a hole that you plan to tap (cut threads into), you'll refer to a chart provided in your tap and die set. Following are typical drill sizes required for tapping specific size holes. Always refer to the tap set's instructions when provided.

Inch Format

Tap Size	Drill Size Required
4-40	43
6-32	36
8-32	29
10-24	25
10-32	21
12-24	16
1/4-20	7
1/4-28	3
5/16-18	17/64
5/16-24	9/32
3/8-16	5/16
3/8-24	11/32
7/16-14	3/8
7/16-20	25/64
1/2-13	27/64
1/2-20	29/64

Metric Format

Tap Size	Metric Drill Size Required	Equivalent Decimal Drill Size Required (Inches)
4mm x 0.70	3.3mm	0.1299
4.5mm x 0.75	3.7mm	0.1457
5 mm x 0.50	4.5mm	0.1772
5mm x 0.80	4.2mm	0.1654
6mm x 0.75	5.25mm	0.2067
6mm x 1.00	5.0mm	0.1968
7mm x 1.00	6.0mm	0.2362
8mm x 1.00	7.0mm	0.2756
8mm x 1.25	6.7mm	0.2638
10mm x 1.00	9.0mm	0.3543
10mm x 1.25	8.7mm	0.3425
10mm x 1.50	8.5mm	0.3346
12mm x 1.00	11.0mm	0.4331
12mm x 1.25	10.8mm	0.4252
12mm x 1.50	10.5mm	0.4134
12mmx 1.75	10.2mm	0.4016
14mm x 1.50	12.5mm	0.4921
14mm x 2.00	12.0mm	0.4724
16mm x 1.50	14.5mm	0.5709
16mm x 2.00	14.0mm	0.5512

TAP DRILL SIZES

Most tap and dies kits provide a handy reference table for determining the appropriate drill size for an intended tap size. Here's a reference chart that was included in a commonly available tap and die set.

Inch Format				Tap Drill Sizes			
Machine Screw Size		Threads	Minor	Alum., Brass, Plastics		Stainless Steel, Iron	
No. or Diameter	Major Diameter	Per Inch	Diameter	Drill Size	Decimal Equiv.	Drill Size	Decimal Equiv.
0	0.600"	80	0.0447"	3/64"	0.0469"	55	0.0520"
1	0.0730"	72	0.0560"	53	.0.0595"	52	0.0635"
1	0.0730"	64	0.0538"	53	0.0595"	1/16"	0.0625"
2	0.0860"	64	0.0668"	50	0.0700"	48	0.0760"
2	0.0860"	56	0.0641"	50	0.0700"	49	0.0730"
3	0.0990"	56	0.0771"	45	0.0820"	43	0.0890"
3	0.0990"	48	0.0734"	47	0.0785"	44	0.0860"
4	0.1120"	40	0.0813"	43	0.0890"	41	0.0960"
4	0.1120"	48	0.0864"	42	0.0935"	40	0.0980"
5	0.1250"	40	0.0943"	38	0.1015"	7/64"	0.1094"
5	0.1250"	44	0.0971"	37	0.1040"	35	0.1100"
6	0.1380"	32	0.0997"	36	0.1065"	32	0.1160"
6	0.1380"	40	0.1073"	33	0.1130"	31	0.1200"
8	0.1640"	32	0.1257"	29	0.1360"	27	0.1440"
8	0.1640"	36	0.1299"	29	0.1360"	26	0.1470"
10	0.1900"	24	0.1389"	25	0.1495"	20	0.1610"
10	0.1900"	32	0.1517"	21	0.1590"	18	0.1695"
12	0.2160"	24	0.1649"	16	0.1770"	12	0.1890"
12	0.2160"	28	0.1722"	14	0.1829"	10	0.1935"
12	0.2160"	32	0.1777"	13	0.1850"	9	0.1960"
1/4"	0.2500"	20	0.1887"	7	0.2010"	7/32"	0.2188"
1/4"	0.2500"	28	0.2062"	3	0.2130"	1	0.2280"
1/4"	0.2500"	32	0.2117"	7/32"	0.2188"	1	0.2280"
5/16"	0.3125"	18	0.2443"	F	0.2570"	J	0.2770"
5/16"	0.3125"	24	0.2614"	I	0.2720"	9/32"	0.2812"
5/16"	0.3125"	32	0.2742"	9/32"	0.2812"	L	0.2900"
3/8"	0.3750"	16	0.2983"	5/16"	0.3125"	Q	0.3320"
3/8"	0.3750"	24	0.3239"	Q	0.3320"	S	0.3480"
3/8"	0.3750"	32	0.3367"	11/32"	0.3438"	T	0.3580"
7/16"	0.4375"	14	0.3499"	U	0.3680"	25/64"	0.3906"
7/16"	0.4375"	20	0.3762"	25/64"	0.3906"	13/32"	0.4062"
7/16"	0.4375"	28	0.3937"	Y	0.4040"	Z	0.4130"
1/2"	0.5000"	13	0.4056"	27/64"	0.4219"	29/64"	0.4531"
1/2"	0.5000"	20	0.4387"	29/64"	0.4531"	15/32"	0.4688"
9/16"	0.5625"	12	0.4603"	31/64"	0.4844"	33/64"	0.5156"
9/16"	0.5625"	18	0.4943"	33/64"	0.5156"	17/32"	.0.5312"

ALPHA/NUMERIC DRILL SIZES

Some drill sizes are identified by alpha (letter) designations, while others are identified by numeric (number) gauge sizes. Quite often, when referring to the required drill size in a tap set's chart, alpha or numeric drill sizes will be noted. This can be confusing for someone who isn't familiar with these sizing designations. The following list provides the decimal equivalent (in inches) of both alpha and numeric drill sizes.

Alpha Drills

Code	Decimal Equivalent
A	0.234"
B	0.238"
C	0.242"
D	0.246"
E	0.250"
F	0.257"
G	0.261"
H	0.266"
I	0.272"
J	0.277"
K	0.281"
L	0.290"
M	0.295"
N	0.302"
O	0.316"
P	0.323"
Q	0.332"
R	0.339"
S	0.348"
T	0.358"
U	0.368"
V	0.377"
W	0.386"
X	0.397"
Y	0.404"
Z	0.413"

Numeric Gauge Drills

Number	Decimal Equivalent
1	0.2280"
2	0.2210"
3	0.2130"
4	0.2090"
5	0.2055"
6	0.2040"
7	0.2010"
8	0.1990"
9	0.1960"
10	0.1935"
11	0.1910"
12	0.1890"
13	0.1850"
14	0.1820"
15	0.1800"
16	0.1770"
17	0.1730"
18	0.1695"
19	0.1660"
20	0.1610"
21	0.1590"
22	0.1570"
23	0.1540"
24	0.1520"
25	0.1495"
26	0.1470"
27	0.1440"
28	0.1405"
29	0.1360"
30	0.1285"
31	0.1200"
32	0.1160"
33	0.1130"
34	0.1110"
35	0.1100"
36	0.1065"
37	0.1040"
38	0.1015"
39	0.0995"
40	0.0980"

Number	Decimal Equivalent
41	0.0960"
42	0.0935"
43	0.0890"
44	0.0860"
45	0.0820"
46	0.0810"
47	0.0785"
48	0.0760"
49	0.0730"
50	0.0700"
51	0.0670"
52	0.0635"
53	0.0595"
54	0.0550"
55	0.0520"
56	0.0465"
57	0.0430"
58	0.0420"
59	0.0410"
60	0.0400"
61	0.0390"
62	0.0380"
63	0.0370"
64	0.0360"
65	0.0350"
66	0.0330"
67	0.0320"
68	0.0310"
69	0.0292"
70	0.0280"
71	0.0260"
72	0.0250"
73	0.0240"
74	0.0225"
75	0.0210"
76	0.0200"
77	0.0180"
78	0.0160"
79	0.0145"
80	0.0135"

DRILL SIZES WITH DECIMAL EQUIVALENTS

FRACTIONAL

Drill Size	Decimal
1/64"	0.0156"
1/32"	0.0312"
3/64"	0.0469"
1/16"	0.0625"
5/64"	0.0781"
3/32"	0.0938"
7/64"	0.1094"
1/8"	0.1250"
9/64"	0.1406"
5/32"	0.1562"
11/64"	.0.1719"
3/16"	0.1875"
13/64"	0.2031"
7/32"	0.2188"
15/64"	0.2344"
1/4"	0.2500"
17/64"	0.2656"
9/32"	0.2812"
19/64"	0.2969"
5/16"	0.3125"
21/64"	.0.3281"
11/32"	.0.3438"
23/64"	0.3594"
3/8"	0.3750"
25/64"	0.3906"
13/32"	0.4062"
27/64"	0.4219"
7/16"	0.4375"
29/64"	0.4531"
15/32"	0.4688"
31/64"	0.4844"
1/2"	0.5000"
33/64"	0.5156"
17/32"	0.5312"
35/64"	0.5469"
9/16"	0.5625"
37/64"	0.5781"
19/32"	0.5938"
39/64"	0.6094"
5/8"	0.6250"
41/64"	0.6406"
21/32"	0.6562"
43/64"	0.6719"
11/16"	.0.6875"
45/64"	0.7031"
23/32"	0.7188"
47/64"	.0.7344"
3/4"	0.7500"
49/64"	0.7656"
25/32"	0.7812"
51/64"	0.7969"
13/16"	0.8125"
53/64"	0.8281"
27/32"	0.8438"
55/64"	0.8594"
7/8"	0.8750"
57/64"	0.8906"
29/32"	0.9062"
59/64"	0.9219"
15/16"	0.9375"
61/64"	0.9531"
31/32"	0.9688"
63/64"	0.9844"
1	1.0000"

METRIC

Drill Size	Decimal
1.00	0.0394
1.05	0.0413
1.10	0.0433
1.15	0.0453
1.20	0.0472
1.25	0.0492
1.30	0.0512
1.35	0.0531
1.40	0.0551
1.45	0.0571
1.50	0.0591
1.55	0.0610
1.60	0.0630
1.65	0.0650
1.70	0.0669
1.75	0.0689
1.80	0.0709
1.85	0.0728
1.90	0.0748
1.95	0.0768
2.00	0.0787
2.05	0.0807
2.10	0.0827
2.15	0.0846
2.20	0.0866
2.25	0.0886
2.30	0.0906
2.35	0.0925
2.40	0.0945
2.45	0.0965
2.50	0.0984
2.60	0.1024
2.70	0.1063
2.80	0.1102
2.90	0.1142
3.00	0.1181
3.10	0.1220
3.20	0.1260
3.30	0.1299
3.40	0.1339
3.50	0.1378
3.60	0.1417
3.70	0.1457
3.80	0.1496
3.90	0.1535
4.00	0.1575
4.10	0.1614
4.20	0.1654
4.30	0.1693
4.40	0.1732
4.50	0.1772
4.60	0.1811
4.70	0.1850
4.80	0.1890
4.90	0.1929
5.00	0.1968
5.10	0.2008
5.20	0.2047
5.30	0.2087
5.40	0.2126
5.50	0.2165
5.60	0.2205
5.70	0.2244
5.80	0.2283
5.90	0.2323
6.00	0.2362
6.10	0.2402
6.20	0.2441
6.30	0.2480
6.40	0.2520
6.50	0.2559
6.60	0.2598
6.70	0.2638
6.80	0.2677
6.90	0.2717
7.00	0.2756
7.10	0.2795
7.20	0.2835
7.30	0.2874
7.40	0.2913
7.50	0.2953
7.60	0.2992
7.70	0.3031
7.80	0.3071
7.90	0.3110
8.00	0.3150
8.10	0.3189
8.20	0.3228
8.30	0.3268
8.40	0.3307
8.50	0.3346
8.60	0.3386
8.70	0.3425
8.80	0.3465
8.90	0.3504
9.00	0.3543
9.10	0.3583
9.20	0.3622
9.30	0.3661
9.40	0.3701
9.50	0.3740
9.60	0.3780
9.70	0.3819
9.80	0.3858
9.90	0.3898
10.00	0.3937
10.20	0.4016
10.50	0.4134
10.80	0.4252
11.00	0.4331
11.20	0.4409
11.50	0.4528
11.80	0.4646
12.00	0.4724
12.20	0.4803
12.50	0.4921
13.00	0.5118
13.50	0.5315
14.00	0.5512
14.50	0.5709
15.00	0.5906
15.50	0.6102
16.00	0.6299
16.50	0.6496
17.00	0.6693
17.50	0.6890
18.00	0.7087
19.00	0.7480
20.00	0.7874
21.00	0.8268
22.00	0.8661
24.00	0.9449
26.00	1.0236

Chapter 6

ENGINE STUDS

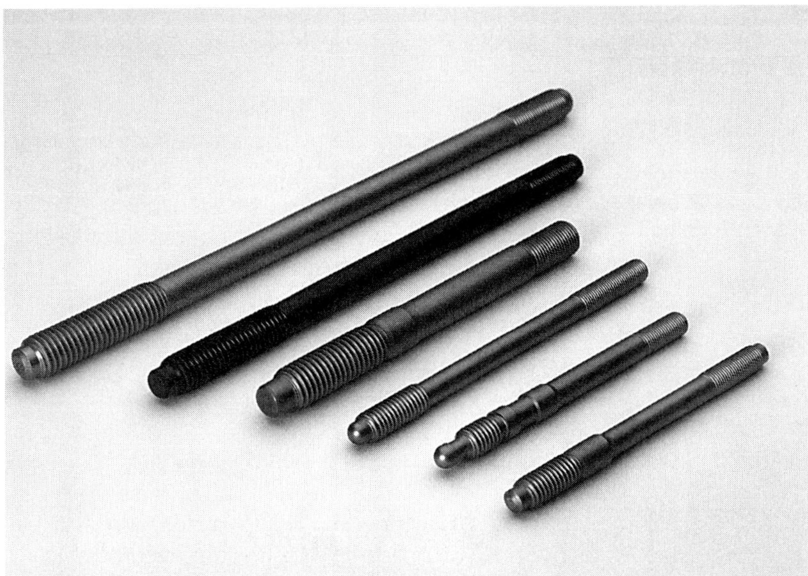

Studs, for engine applications, are available for most popular engines, especially in terms of cylinder head applications. The use of studs, as opposed to bolts, for cylinder head installation provides a more accurate and uniform clamping load pattern throughout the cylinder head.

In far too many cases, the engine builder's attention focuses on selection of the proper components for a specific race engine build, determining proper clearances, intake and exhaust volume and flow, surface finishes and precision assembly. Threaded fasteners, the vital link that secures everything together, are often taken for granted. Here, we'll discuss a variety of issues and concerns directly relating to the most important components of any engine—studs and bolts and their handling.

MAIN STUDS

For a performance application, studs should be used whenever possible instead of main cap bolts, in those instances where a choice is available. Studs provide the ability to obtain much more accurate torque values because the studs don't twist during tightening as do bolts. Because the studs remain stationary during nut tightening, the studs stretch in one axis alone, providing much more even and accurate clamping forces. Also, because the use of studs results in less force applied to the block's threads, this extends the life of the threaded holes in the block. This is especially important when dealing with alloy blocks. The use of studs also eases main cap installation, and contributes to main cap alignment. There is less chance of main cap walking because the studs remain stationary during cap clamping.

Main Cap Stud Installation Tips

Before installation begins, clean the block, the caps and the studs thoroughly. Any debris on the threads can easily affect thread engagement quality and can cause incorrect torque wrench readings. In order to ensure that the block's female threaded holes are clean and uniform, they should be chased. Never use a cutting tap to perform this task, as this will cut, shave and remove thread material, which can reduce thread strength. Instead, use only dedicated thread chasing taps, which will restore threads by "forming" instead of cutting. Also make sure that your torque wrench is properly calibrated. All torque wrenches should be checked for calibration at least once each year (more frequently for heavy use). Even new torque wrenches should be checked, as it's not uncommon for even some new wrenches to be out of calibration by as much as 10 ft-lbs.

Use consistent tightening techniques. In other words, don't turn quickly on some nuts and slowly on others. The best method is to slowly "creep" towards the preset value. Quick-banging to reach the "click" on a ratcheting click-type torque wrench can result in uneven and inaccurate values.

In the majority of cases, screw the studs into the block FINGER-TIGHT ONLY. Do not double-nut the stud and tighten severely. Remember...the torque value given for the installation refers to the tightening of the nut only, not the stud itself! If you desire a "permanent" installation, apply a small amount of anaerobic compound to the "course"

A set of main studs installed in a Honda race block. This will provide superior and more dependable clamping for the main girdle cap assembly. Plus, by using studs threaded into the aluminum block, only the nuts will experience on/off during any subsequent assembly and future service. This saves the aluminum threads from wear.

Always check the edges of main fastener holes for raised edges that would prevent proper main cap flush mating. While the use of main cap bolts might draw the upper threads up, creating a raised edge, studs are not installed at the high torque values that bolts would experience, so creating raised edges around the holes is eliminated by using studs.

The front and rear main caps, along with the center cap girdle, are secured with moly-coated washers and nuts. Since any race engine build requires numerous installation and removal of main caps during block prep and clearance checks, the use of studs as opposed to bolts avoids thread wear and potential thread damage in the aluminum block's threaded holes. This is a great benefit whenever an aluminum block is involved.

Obviously, main cap studs also serve as helpful locating guides when installing main caps.

threads that engage in the block. Loctite 242 or a similar grade is acceptable. If you use Loctite 271 or similar high-strength compound, the studs will be extremely difficult to remove at a later date without the application of high heat. If a locking compound is used, be sure to immediately install the main cap before the compound sets, to avoid any possible misalignment of the studs in relation to the cap. This means installing the cap and tightening the nuts to specs, before the thread compound hardens. This allows the necessary preload to be placed on the stud-to-block thread engagement before the compound hardens. Use of a locking chemical really isn't necessary, unless you want to ensure stud position during repeated and hurried main cap changes (such as quick changes between heats in a drag race application, etc.). Also, if the block will be subjected to extreme vibration (such as in a Funny Car or Top Fuel motor), the use of a locking compound may be a good idea for insurance purposes. In most cases though, it probably isn't necessary.

When installing the studs, simply make sure that full thread engagement has been achieved. In other words, make sure the studs "bottom-out" in their holes. Again, tighten finger-tight only.

Once the studs are installed, install the main caps and check for stud-to-cap alignment (check for binding—the caps should slip over the studs smoothly).

Lubricate the exposed (fine) threads of the studs, and be sure to also lube the washers and nuts. **Note:** Torque specifications WILL vary depending on the lubricant used. Moly lube offers a more consistent torque reading than various oils. If Moly is used, be sure to follow the torque specs for Moly. If oil is used, follow the specs for oil. They WILL be different! With caps in place, the nuts should be tightened to the specified torque value three times. As noted earlier, if an anaerobic compound is applied to the stud-to-block threads, the nuts must be torqued before the compound sets.

To achieve best overall (and optimal long-term) results, the block should be align-honed after stud and main cap installation.

CYLINDER HEAD STUDS

Again, as in the case of main studs, the use of studs is always preferred over bolts. However, this may vary depending on application. On a street engine application, or in a situation where the heads would be difficult or impossible to remove with the engine in place, studs may not be the best choice from a cylinder head service standpoint. For example, if a master cylinder or other component prevents a cylinder head from being removed or installed with the engine mounted in the vehicle, bolts may be a better choice simply in terms of practicality. However, if the situation allows, the use of studs will usually provide a superior assembly.

The use of head studs will aid in cylinder head installation, simply from a standpoint of gasket and head alignment. This is especially helpful in a race engine application where frequent head removal will occur (between rounds in drag race applications, for potential engine service in an endurance race engine, etc.).

In terms of function, the use of studs provides much more accurate and consistent torque loading. When a bolt is installed, the act of tightening results in both twisting (torsional load) and stretching (vertical, or axial load). This results in the

Cylinder head studs were selected for the build of this Ford 427 FE engine. For thin-wall blocks where cylinder bore distortion is a concern, studs make sense because they provide more uniform and precise clamping force.

Some cylinder head studs, like this ARP stud, feature a female hex drive. This drive socket is provided to aid in stud installation or removal. Don't get carried away and tighten the willies out of the stud. Finger-tight is usually recommended. The stud simply needs to obtain full thread engagement. The clamping load will be achieved as the nut is tightened.

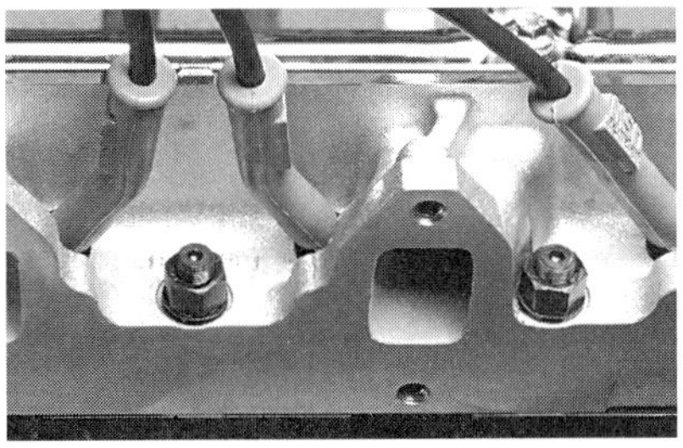

An aluminum cylinder head installed using studs. The use of a flat washer is highly recommended in all stud applications (to provide a bearing surface for the nut), but in the case of clamping an aluminum part in place, hardened washers must be used to both provide a bearing surface for the nut, and also to prevent damage to the softer alloy component. If hardened washers are required, the stud kit should include these.

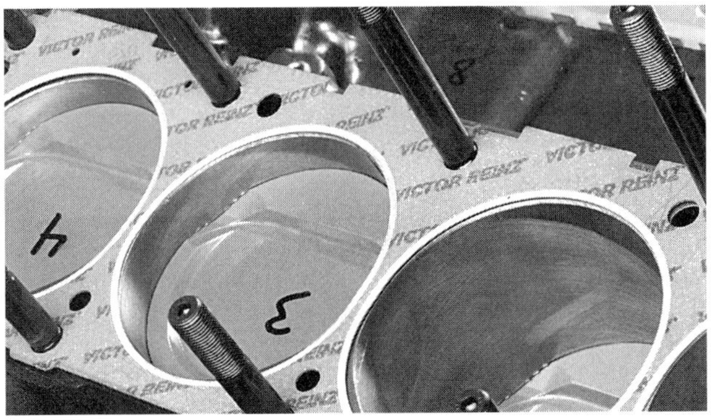

A cylinder head stud kit will include the correct stud lengths for the specific application, so you won't end up with too-tall studs exposed above the nuts. By the way, NEVER create your own studs by grabbing lengths of all-thread rod, especially for critical locations such as main caps or cylinder heads. Use only studs that are designed specifically for each application. Just because a stud is the correct thread diameter and length doesn't mean it will provide the required tensile strength.

bolt being exposed to two forces at the same time, as well as experiencing frictional loads at the thread engagement. When a stud is installed (finger-tight in the block), the stud stretches on its vertical axis only. The exposed end (top) of the stud features "fine" threads, which allow more precise and therefore accurate, torque readings when the nut is torqued to spec.

Another advantage of using studs is potentially increased working clearance at the nut. For instance, a smaller-head 12-point nut may be used, which could provide additional needed clearance at the top of the cylinder head (as opposed to conventional hex heads found on bolts, which may (in some cases) provide a tight fit for the outer diameter of a socket (if close to a boss, etc.).

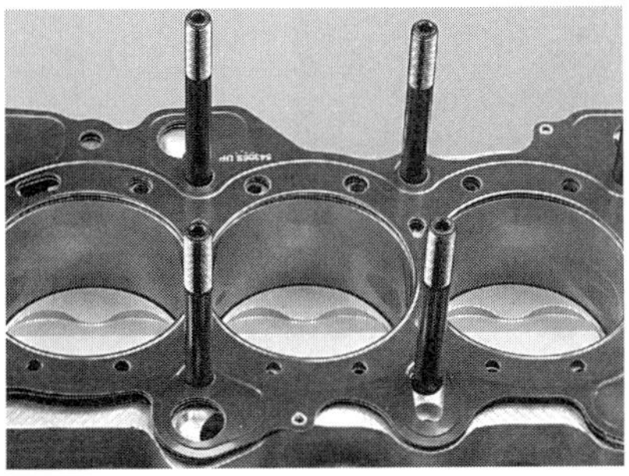

An MLS (multi-layer steel) head gasket on a Honda race block. Since the use of studs guides the gasket into place, you eliminate the chance of damaging the gasket, which could occur if the gasket is misaligned while you attempt to force head bolts into place.

With the block oriented on the stand with the deck facing up, the cylinder head can be lowered over the studs with no worries about hole alignments.

Cylinder head studs also provide handy "grabbers" for rotating a short block on the engine stand (just kidding). Pay attention to all cylinder head bolt holes, checking to see if they are open to waterjackets. If a threaded hole is open to water, a quality thread sealant must be applied to the block-end of the stud threads before stud installation. If in doubt, apply thread sealant to all studs (again, only to the threads that enter the block).

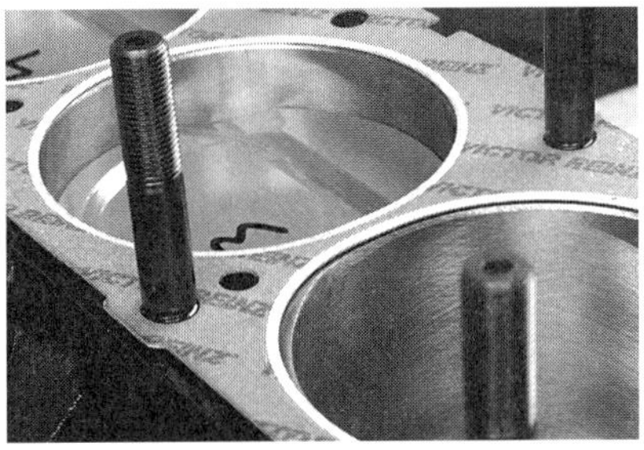

Positioning the head gasket over a set of studs also allows you to easily verify that the gasket's cylinder holes and fire rings are properly located over their respective bores.

Hardened washers and nuts installed on cylinder head studs. Lubricant must be applied to the exposed stud threads, to both sides of the washer, and to the underside of the nut. While either engine oil or moly lube may be used, read the instructions included with your head stud kit, since torque value will be different depending on which type of lubricant is used. A lower torque value will be specified with the use of moly, due to this lubricant's lower friction coefficient.

Here, cylinder head studs are used for deck plate installation (prior to cylinder bore honing). Whenever a deck plate will be used in order to pre-stress the block for the honing process, always use the same style and brand of cylinder head fasteners that you plan to use for the final assembly, whether you've decided to use bolts or studs.

The extended chamfered tip of this head stud (lower tip shown here) is designed to bottom-out inside the hole. The installer is then required to tighten the stud, using a torque wrench, to create a very slight preload, theoretically to stabilize the stud against the threads for a superior thread engagement. Remember: this is not true of all head studs. Always pay close attention to the instructions.

Moly-graphite lubricant provides less thread friction than engine oil. Carefully read the instructions included with your studs. Depending on the source of the studs, the use of engine oil may be specified in accordance with the recommended final nut torque value, or two distinct torque values may be listed, one for oil and one for moly. The use of moly is preferred, since this is what the stud maker may have used during their research and testing to achieve ideal clamping loads. While moly lube is available from a variety of brand sources, it's always best to use the specific moly recommended or offered by the maker of the studs.

Head Stud Installation Tips

Make sure the threads in the block and on the studs are clean to provide accurate torque readings.

Since many applications feature cylinder threaded holes that are open to water jackets, it's advisable to coat the (course) end of the studs with a quality thread sealer. If you are certain that the holes are not open to water, you may skip this step, but if in doubt, apply the sealer. It can't hurt. As with main cap studs, if you prefer that the studs are installed more permanently, you may apply an anaerobic compound to the threads. In either case, whenever a sealant or locking compound is applied, the cylinder head must be installed immediately following stud installation so that the nuts can be torqued to value before the compound sets (the studs must be preloaded to allow set-up to occur in the proper tensioned state, to assure stud alignment). Install the studs into the block finger-tight only. DO NOT double-nut the studs and tighten severely. As long as full thread engagement occurs (the "course" end of the stud is

Make sure the threads are clean, and apply a small amount of moly to the threads.

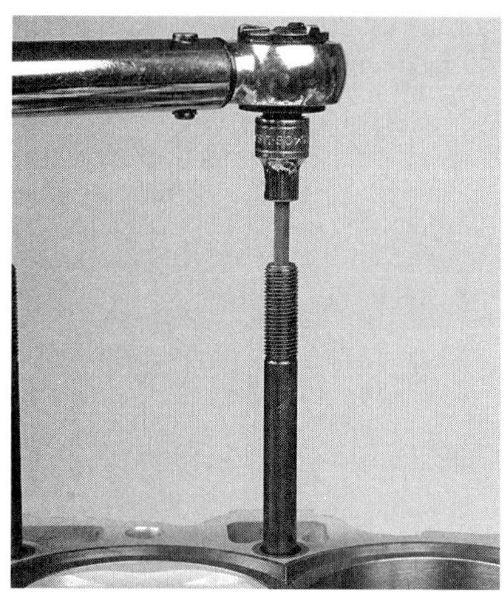

Using a torque wrench equipped with a hex wrench bit, this cylinder head stud is tightened into the block, to a value of 10 ft-lbs as per the stud maker's instructions.

screwed into the hole fully), clamping load will be achieved by tightening the nut. If the studs are too tight in the block, they may splay and cause a misalignment with regard to the head gasket and the head.

Prior to installing the nuts, lubricate the exposed "fine" threads on the studs, the washers and the nuts and torque the nuts in the proper sequence and at the values specified. Remember: torque value may vary depending on the lubricant used (oil or Moly lube), so always pay close attention to the torquing instructions supplied by the stud manufacturer. The nuts should be torqued three times.

There are some exceptions. Some cylinder head studs are designed to be installed with a slight preload to create superior stability. As a case in point, I recently built a couple of Honda-type engines, using special head studs that required a 10 ft-lb installed preload. The stud tips featured female hex holes, allowing the use of a hex bit on a 3/8"-drive torque wrench.

Always read the stud maker's installation instructions, and follow them to the letter. Even if a specific type of head stud is designed to be installed with a slight preload, NEVER double-nut and severely tighten any stud, regardless of its design. In most cases, studs should be installed finger-tight, but as already noted, some specialty studs require a slight preload. Read the instructions!

Not all cylinder head studs will be easily visible or accessible when installed. On this Honda head, some studs are accessed through a hole in the cam bearing saddles. A hex wrench on an extension is required.

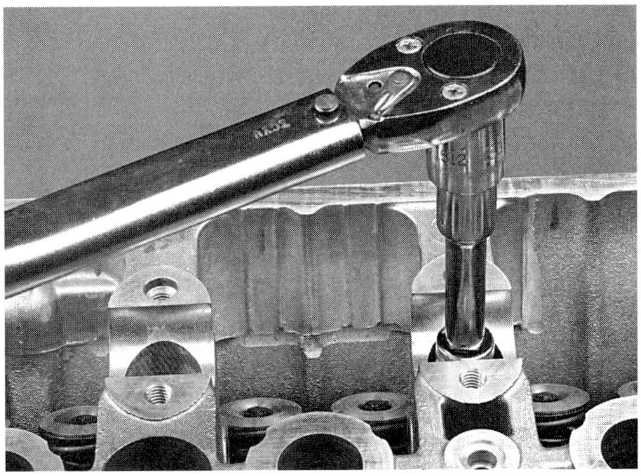

If a cylinder head stud requires a slight installed preload, slowly turn the stud, carefully approaching the final value. Don't yank and pull quickly, as this can easily result in overtorquing.

If head studs are tightened too far (beyond the maker's recommendation), abnormal stresses may result, which could cause the studs to "splay" at slight angles (no longer perpendicular to the deck surface). This will not only adversely affect achieving correct clamping load, but this can make cylinder head installation very difficult.

Always use a (supplied) hardened washer. Be sure to coat the washer with the correct lubricant. This provides the nut with a reduced-friction bearing surface for more accurate torque value and resulting clamping load. Here, because of difficult access, a pencil magnet is used to lower a washer onto the stud, which is located deep inside the cam saddle on a Honda engine.

On the same Honda race engine, a pencil magnet is used to position the nut onto the stud.

Studs stay with the block. During head service now or in the future, the threads in the parent aluminum won't be exposed to the wear and tear that would be generated with the use of head bolts.

A killer Honda race engine just before installation of the head and turbo. This powerplant was built to yank a whopping 900–1000 HP with a big turbo and direct-injected nitrous oxide. The use of studs for both main caps and cylinder head was mandatory.

This bullet-tipped stainless steel stud is used to mount a distributor clamp. With the stud in place, the clamp is dropped over the stud (making clamp positioning easy). The bullet tip allows the nut to be dropped onto the stud, without the hassle of trying to align threads at the very outset.

Here's a complete stud kit for a Honda head's intake manifold, complete with studs, washers and 12-point nuts.

These exhaust studs feature female hex drives to accommodate installation or removal. Note: whenever using a stainless steel fastener, it's advisable to apply a small dab of moly onto the threads to prevent galling.

Here we see stainless bullet-tipped studs being used to secure a carburetor. The studs provide easy drop-on alignment for both the base gasket and the carb, and the bullet tip provides hassle-free starting of the nut.

Here a set of ARP stainless steel studs, washers and nuts are used to secure the exhaust header. The stud layout allows you to position the gasket and the header, eliminating the need to hold the parts in place while trying to start the bolts. Plus, this stud setup looks way cool.

No need to use your third arm to hold the gasket in place while you grab the header. Studs offer superior clamping, provide easier installation and result in a clean and tidy appearance.

OTHER STUD ADVANTAGES

Studs, as opposed to bolts, also offer component installation advantages. Studs act as guide pins when aligning items such as gaskets, engine covers, etc. Studs are available (ARP for example) with "bullet" noses, where a slightly diminished diameter bare tip is featured. This greatly eases nut installation, allowing the nut to be dropped into position before thread engagement begins. Good examples of this include carburetor studs, distributor studs, exhaust header studs and oil pan studs. Instead of trying to hold a gasket or part in place while fumbling with bolts, trying to align the bolts to their holes, the gasket or part can be dropped over the studs (so alignment poses no further problems). With the part aligned in place, the washers and nuts can be installed without concern for the part being shifted out of alignment.

Another advantage of using studs is that you don't need to worry about insufficient clamping load because a too-long bolt bottomed out in its hole. The stud is stationary, and clamping load is achieved by tightening the nut.

Granted, studs can't be applied in all situations, due to component design. For instance, a V-block engine's intake manifold (depending on design) may place the fastener locations on opposing angles (where the bolt holes on each bank angle inboard). Obviously, studs can't be employed because you'd never be able to drop the intake manifold over the angled studs. Studs, where applicable, offer both installation convenience and superior clamping uniformity.

Again, regardless of stud diameter, install studs finger tight (just make sure you obtain full thread engagement) unless otherwise instructed by the manufacturer. Some studs feature a female hex drive on the tip, allowing the use of a hex bit to ease stud installation.

If you need to remove a stud from an OE component, keep in mind that the stud may have been locked in place with an anerobic thread-locking compound. If this is the case, applying a bit of heat to the stud (using a hand-held propane torch) will liquify the locking compound. The stud can then be double-nutted for removal.

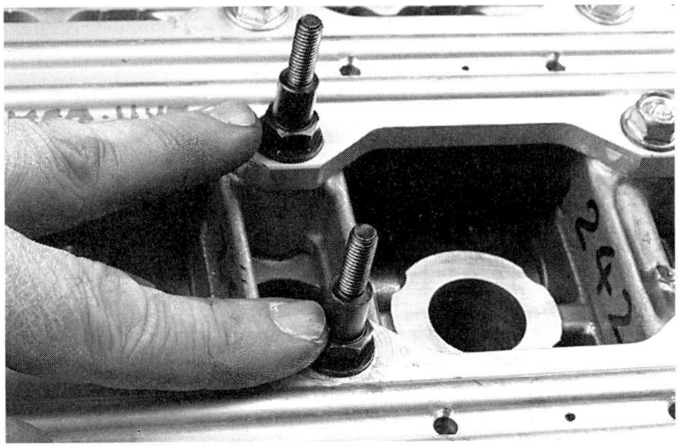

Specialty fasteners on some OE applications may feature a bolt/stud combination, as on the camshaft girdle plates on this Honda. The bottom half is a bolt, using the hex surface to torque the bolt and to secure the girdle. The exposed upper stud is then used to accept and fasten the cylinder head's cam/rocker cover.

This Honda head features a direct-drive distributor (driven by the intake cam). While Honda normally uses three 8mm bolts to secure the distributor, we installed a trio of ARP stainless 8mm studs.

The distributor studs were installed finger-tight, with a dab of Loctite 242 thread compound as insurance.

The studs really helped to align and guide the distributor into place.

Wherever they're practical to use, most serious engine builders almost always opt for studs instead of bolts.

Chapter 7

WHEEL FASTENERS

We're not going to refer to them as "lug nuts," as most people tend to do. Wheel fasteners exist in several styles, not all of which are "nuts." Wheel fasteners are available as nuts, bolts and "knock off" center-fastener styles.

Wheel fasteners—the combination of hub-mounted studs and removable nuts; or the use of wheel bolts that thread into hubs—are the single most important hardware items on any vehicle, and yet are often the least understood and appreciated. Here, we'll provide a basic overview of the often overlooked threaded wheel fastener—how to identify it and how to handle it. This includes information dealing with fastener thread size, seat styles and torque values. Our goal is to provide clear and accurate explanations of various wheel fastening systems that will help customers choose the correct type of fastener for a given application, and knowledge of safety-related issues.

Note: The common method of attaching a wheel to a hub involves a group of individual threaded wheel fasteners (wheel nuts or wheel bolts), involving a pattern of 4, 5, 6 or 8 fasteners per wheel. Another type of wheel clamping system involves attaching the wheel to the hub with a single spindle nut. This is most commonly referred to as a knock-off or pin-drive system. We'll discuss this at the end of this chapter.

WHEEL FASTENER TERMS AND STYLES

As I mentioned earlier, let's stop using the generic term "lug nuts" when referring to any and all wheel fasteners. Instead, let's make an attempt to accurately describe the different types of threaded fasteners that are used to clamp a wheel onto a hub. Remember that not all vehicles use nuts—for example, many German vehicles use wheel bolts instead, which engage into threaded holes in the hub face. So, constantly calling all wheel fasteners "lug nuts" simply isn't accurate.

The term "acorn" nut is a classic example of a term that's widely misused and misleading. If we refer to a wheel fastener's seat style, the term "acorn" would translate into a reference of a ball type or radius type seat. After all—take a look at an acorn. When viewed from the side, the belly of the acorn is somewhat radiused—which makes the term sensible when referring to a ball type or radiused seat style. However, some folks use the term "acorn" to refer to the head portion of a nut that features a ball-type rounded dome-head design. Yet others use the term to refer to a short-bodied nut that features a conical (straight taper) seat. So, when you hear the word "acorn" in reference to wheels, don't assume anything. In order to make sure that both you and the salesclerk are on the same page, clarify the term by pointing to a photo, diagram or better yet, showing him/her an actual sample to make sure that you're both talking about the same item. The only proper use of the term "acorn" is when referring to a nut with a rounded top dome. Just remember that the dome shape has nothing to do with the seat style.

Shown here are four examples of wheel fasteners. From left: spline-drive nut with a conical (tapered) seat; hex drive nut with conical seat and extension to accommodate thick alloy wheel; radiused (ball type) hex drive nut; and a wheel bolt that features a ball-type radiused seat.

Seat Styles

The "seat" refers to the contact area between the base of the fastener head and the wheel's fastener pocket. Although variations exist, three basic seat styles are in common use today. These include conical, radius and flat-seat types.

Conical—Conical wheel nuts (or wheel bolts) feature a straight taper at the seat area. Conical fasteners are also called "tapered" or "cone" types. All three names refer to the same style. As viewed from the side, you can see that the seat is "chamfered" on each side of the view. This type of taper is created most commonly in a 60-degree angle, although some light truck applications use a 90-degree angle. The angle indicates the degree of separation between the two walls, not the true vertical of the fastener.

Radius seats are also called "ball" or "rounded" seats. As the name implies, the seat features a radiused "ball" shape that nestles into a ball-shaped pocket in the wheel. Many European vehicles such as Porsche and Mercedes use radius seat styles, in either nut or bolt applications.

Flat—The flat seat style is most often called a "mag" style (the term "mag" doesn't imply anything in terms of geometric shape—it's simply a slang carryover from the early days of the American performance scene when race wheels were cast magnesium, and commonly used a flat seat because it was easier and cheaper to produce a round hole in a flat hub center area. As we all know, the unfortunate misnomer "mag" has stuck with us as a catch-call to denote any custom wheel, whether it's made of magnesium, aluminum or other alloys). Mag style (flat seat) nuts generally feature a smooth shank extension under the head, used to center the wheel and to provide needed thread engagement depth.

The golden rule, when discussing seat styles, is extremely simple: NEVER mix them! If a wheel is designed for conical 60-degree seats, the ONLY fastener seat style that can be used is a 60-degree conical type. The same rule applies for radius and flat seat styles. ONLY use the correct seat style of fastener for the seat style of the given wheel. Using the incorrect seat style can, and likely WILL, result in fastener loosening, wheel damage, and potential tragedy due to loss of vehicle control when the wheel wobbles or separates from the vehicle. It's simply impossible to overemphasize this point. The wheel fasteners are the ONLY connection between the vehicle and the wheels. Without proper fastening, you're flirting with disaster.

The three most important elements of wheel fastening:

1. Proper seat style
2. Correct thread size (diameter, pitch, length)
3. Proper clamping value

Thread Engagement

Aside from selecting the proper seat style for the wheels at hand, another consideration is thread engagement. In other words, how far the nut travels over the stud, or how far the wheel bolt travels inside the hub's threaded hole. The rule of thumb whenever discussing threaded fastener engagement is to achieve a minimum engagement length that is equal to the diameter of the fastener's threads. For instance, if the stud or bolt shank is 1/2" in diameter, the nut (or bolt) must engage its threads by at least 1/2". This provides the minimal theoretical engagement for proper clamping load. However, the industry usually takes advantage of greater thread engagement depth, for added safety (the deeper the threads engage, the longer it takes to actually lose the fastener; and in some cases, longer studs are used simply to aid in wheel installation onto the hub).

Thread Size—Thread sizing is not difficult to understand. The problem is that most people won't take the time to learn this very basic theory. Whether inch or metric, we need to know the diameter, the thread pitch and the thread length. For example, when dealing in an inch format, a nut identified as a 1/2 x 20 x 1 indicates that the thread diameter (the diameter of the stud it will accept) is 1/2-inch. The number "20" indicates thread pitch. That refers to the number of threads that exist per inch of length. In this case, the thread count, or "pitch" is 20 threads per inch. The final number refers to the length of the thread engagement. The number "1" means that the nut features 1 inch of

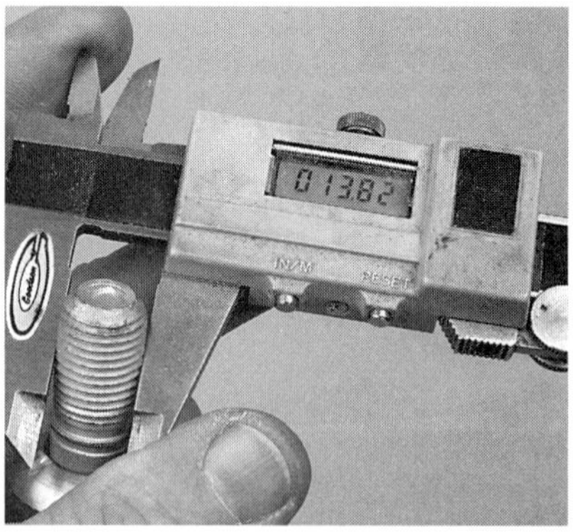

When measuring thread diameter with a caliper, you likely won't achieve the exact labeled size, due to variances in tolerance, allowance for thread clearance and possible tool calibration variances. This Porsche wheel bolt is a 14mm diameter size, but thread peaks measure a tad under when using a precision measuring tool.

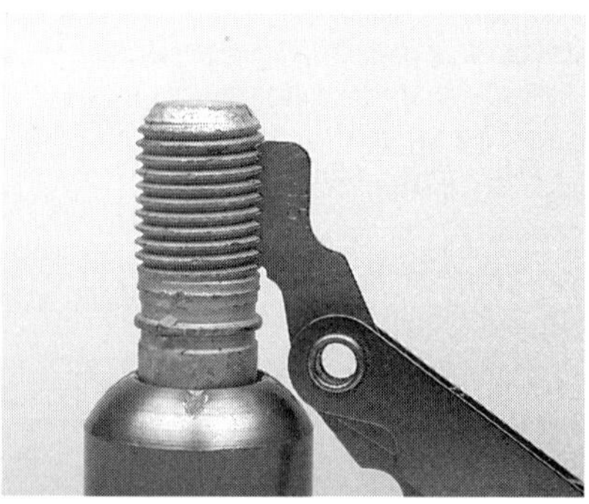

Make sure that the threads are clean before attempting to measure thread pitch.

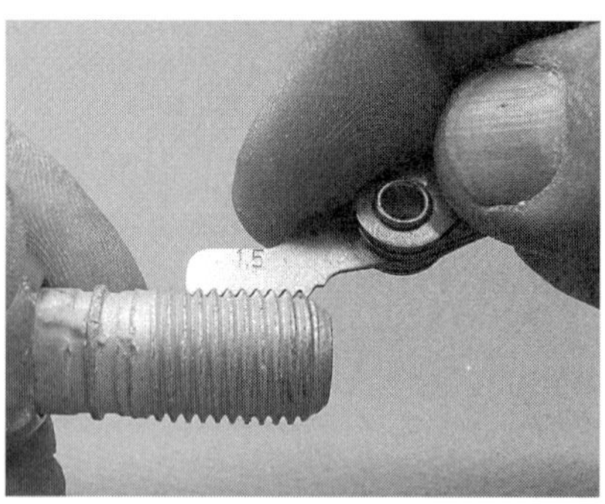

A thread pitch gauge can easily be used to check thread pitch on installed wheel studs, as shown here. It must mate perfectly when determining thread pitch.

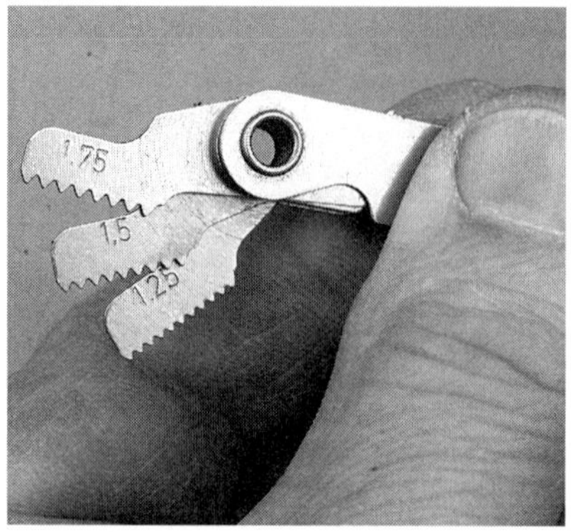

The most common thread pitches you'll encounter when dealing with metric wheel fasteners include 1.75, 1.5 and 1.25. Remember that metric pitch numbers represent the distance between threads. As you can see here, the 1.25 pitch (this means 1.25mm between threads) is the "finer" of the three shown here.

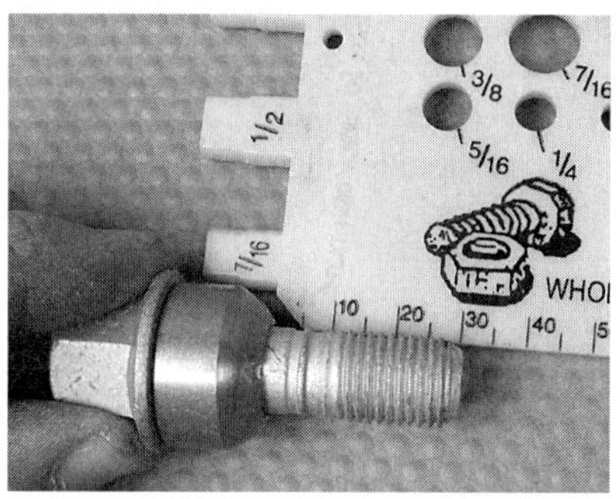

A wheel bolt's shank length is measured from the bottom of the seat to the tip of the threads. This Porsche wheel bolt would be labeled as 30mm in length.

This Porsche wheel bolt features a 1.5 pitch, which means that the distance from one thread to the next is 1.5mm.

internal thread engagement length. For more on thread sizing, see Chapter 1.

In order to accurately identify the thread diameter, pitch and length of any wheel fastener, you can handle this by obtaining a total of three readily-available tools: a combination bolt/nut sizing card (these are hard plastic and are available to cover both inch and metric sizes) and two thread pitch gauges (one inch and one metric). That's all you need to quickly and accurately read any wheel bolt or nut without any guesswork.

Head/Drive Styles

The "head" of a wheel bolt or nut features a specific drive design and appearance approach. The most conventional drive style involves a hex (6-sided) that requires the use of a 6-point wrench. However, other styles do exist, although not as commonly encountered. These include female hex drive "tuner" nuts which feature a relatively small and smooth outer body diameter (to fit into small-diameter fastener holes in some wheels). Instead of an external (male) hex head, a female hex drive hole is featured instead. Yet another drive style is the splined head which features a symmetric pattern of straight-cut external splines instead of hex walls. Naturally, the tuner hex drive nuts require the use of a hex bit as a driver; and the splined fasteners require a dedicated female spline socket.

In terms of appearance, nuts may be either open (the threaded hole runs all the way through the nut) or closed (threads terminate inside the nut). Closed-head styles are varied, and include flush tops, raised conical domes, raised flat-top domes, extended hex walls (for easier reach and handling and greater wrench contact area), radiused domes (these look like acorn bellies, featuring a nicely rounded, ball-type dome).

Body Length

All of the wheel nut and wheel bolt styles that we've mentioned are available in a wide variety of lengths, in terms of drive area or thread engagement area. Drive area height may be dictated by the thickness of the wheel's center hub, or by the design of the wheel (maybe the nut or bolt head needs to project out further for easier wrench access). The necessary thread engagement area will affect length as well, involving either inboard thread reach or outboard projection based on stud length. Often with lug centric wheels, a tapered (conical) seat style nut may be used that also features a short straight shank. The shank may both increase thread engagement length as well as aiding wheel centering on the hub.

This alloy wheel is designed to center itself onto the hub in a "hub centric" fashion, which means that the wheel relies on centering by mating its center hole onto the raised flange lip on the center of the hub. Many aftermarket alloy wheels are cast and machined with a larger center hole to accommodate a range of hub center diameters. This is easier and cheaper for the wheel maker, as opposed to making a variety of wheels with various hub center hole diameters. Depending on the vehicle application, a hub centric spacer may be needed. These spacers are made from plastics or aluminum. Either works fine, since they only serve to center the wheel during wheel installation. Here a plastic hub centric spacer is being installed into an aftermarket wheel's center hole recess.

Here a hub centric spacer has been installed onto the hub. It's always a good idea to apply a light coat of anti-seize paste onto the hub flange in order to prevent electrolysis (where dissimilar metals can corrode together). In order to prevent an alloy wheel from sticking onto a steel hub, apply a very thin coat of anti-seize on the hub face and around the base of each stud. It's also a good idea to coat the hub centric spacer as well, even if it's plastic, to prevent it from being "captured" onto the hub or wheel during disassembly. Don't apply too much paste, to avoid contaminating the rotor and pads.

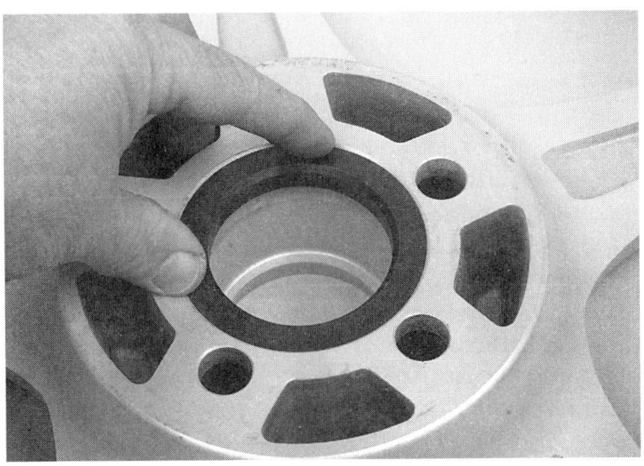

A hub centric spacer installed onto a wheel. The spacer may be installed onto the wheel or onto the hub prior to wheel installation onto the vehicle.

Hub Centric vs. Lug Centric

You may or may not be familiar with these terms, but they're real, and we're not making this up. These two terms are common and proper labels that indicate how the wheel centers onto the hub. The wheel must be attached to the axle hub as precisely as possible to avoid runout. In order to locate the wheel in a centered position, some method of locating must be used. If the wheel centers onto the hub via the fastener locations only, we refer to this as a "lug centric" method. This means that the location of the wheel studs (or the hub's bolt holes) must be accurately placed on the hub during vehicle manufacturing. If the wheel is centered by means of its center opening fit to the hub's raised center flange, it's a "hub centric" design. This means that the centering approach relies on the center hole of the wheel to the center locating dowel flange of the hub.

It was brought to our attention last year that some installer shops had never heard of these terms, which was surprising. Well, we didn't invent these labels on our own. They are common and accepted terms used to indicate the manner in which primary centering takes place.

Bolt Circle

The "bolt circle" refers to the spacing of the wheel's fastener holes, using a diameter, or circle, as the measurement. If we draw a circular line that runs through the center of all of the bolt holes, this circular diameter measurement indicates the bolt circle. For instance, if a 4.5" circle can be positioned at the center of the bolt hole group, this is a "4.5" bolt circle." If the circle measures 100mm, it's a 100mm bolt circle, etc.

The bolt pattern refers to both the number of fastener locations and the diameter of the hole pattern. A 5 x 100 bolt pattern refers to a wheel

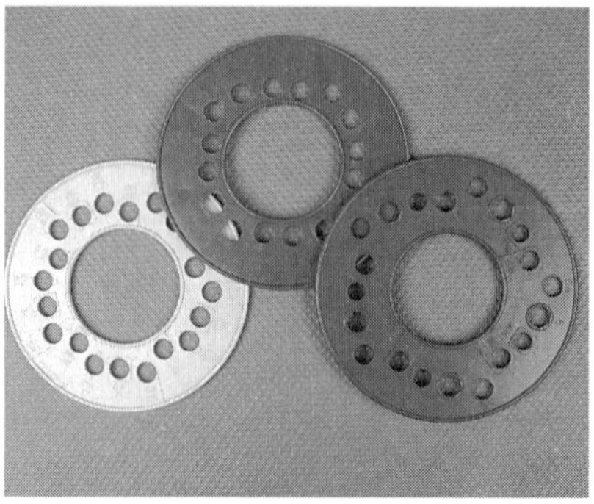

Bolt circle sizing cards offer a handy (and easy) method of determining the bolt circle size of a wheel or hub. These are available in versions that cover all known bolt circle dimensions, in both fractional and metric formats.

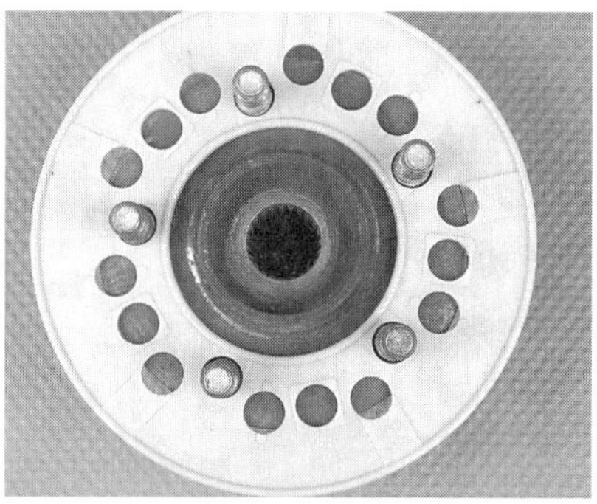

Here a fitting card is dropped over all five wheel studs on a hub. Simply look at the number that corresponds to each of these holes.

Bolt circle sizing cards offer a range of holes (to cover a range of bolt circles), with each hole clearly marked for bolt circle dimension. For example, if you test-fit this card onto a hub and find that all five studs match up to a set of five holes on the card, and the holes that match are labeled 4 1/2, you know that the bolt circle is 4 1/2 inches.

A sizing card can also be used to check a wheel's bolt circle dimension. Rotate the card until you find a perfect alignment of all of the wheel's holes. This particular example shows that this wheel features a 100mm bolt circle (we marked each hole for this illustration).

Bolt circle can be easily determined with a tape measure or ruler, as long as you understand where to measure. Remember: you're trying to measure an imaginary circle that runs through the center of all of the holes. If the wheel features 4 bolt holes, simply measure from the center of one hole, across the wheel center, to the center of the opposing hole. If the wheel features five holes, measure from the center of one hole to the outside edge of one of the holes that's furthest away. Here we measure 4 inches. In actuality, this is a 100mm bolt circle (if our tape measure was more accurate, this would show a measurement of 3.937 inches, which equals 100mm).

that features five fastener holes, laid out in a 100mm diameter path. A 5 x 4.75 pattern designation refers to a pattern with five holes running along a 4.75" diameter. In the "old" days, we could generically refer to a 5 x 4.5" pattern as a "Ford" pattern; or a 5 x 4.75" as a "Chevy" pattern. Aside from dealing with vintage domestics, those days are gone.

In order to measure a bolt circle using a ruler, remember that we're simply trying to find the diameter of an imaginary circle that runs through the center of each fastener hole. If the wheel or hub features directly opposing holes (as found on a 4-bolt, 6-bolt or 8-bolt wheel or hub), simply measure from the center of one hole to the center of the hole directly across the pattern (or from the outside of one hole to the inside of the opposite hole).

If the wheel or hub features a 5-bolt pattern, measure from the center of one hole to the outside edge of the hole furthest away.

If you don't wish to be bothered by measuring with a ruler, tape measure or dial caliper, you can cheat by using a specialty measuring device. Round plastic bolt circle template discs are available that feature multiple hole patterns (with each hole identified for bolt circle). Simply place the disc against a wheel hub face or onto the vehicle hub until all holes align. The holes that align to your wheel or hub indicate the bolt circle (again, the holes are labeled).

Another specialty tool that allows quick measuring of a wheel's bolt circle is in the form of a sliding caliper. This tool features two tapered dowels that slip into the wheel's bolt holes. With this specific tool, insert the dowels into two adjacent holes (two holes closest together). The graduated gauge face of the tool indicates the bolt circle, based on the number of bolt holes (the gauge features multiple pointers, for 4-bolt, 5-bolt, 6-bolt or 8-bolt).

While these tools are available from a number of sources, we'll name two as examples. The round bolt circle template discs are offered by Made For You Products at 760-868-6962. The sliding caliper tool is offered by Excalibur Wheel Accessories at 800-227-0504 (we sourced ours through American Tire Distributors at 800-222-1167).

Some common fractional bolt circles include

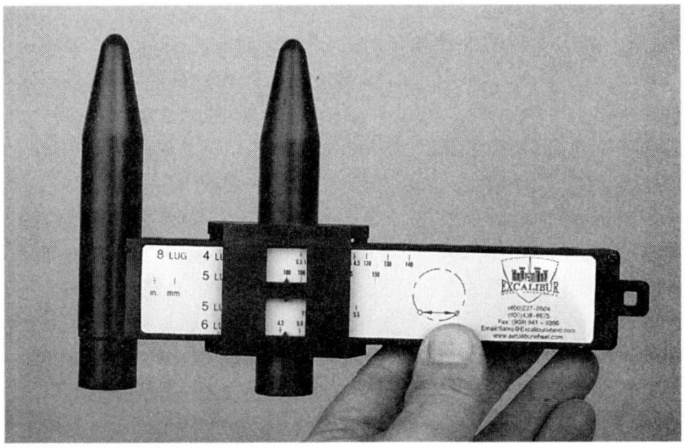

Here's a nifty tool for measuring wheel bolt circle. This sliding caliper tool is labeled for 4-lug, 5-lug, 6-lug and 8-lug wheels, and features incremented rulers for bolt circle dimensions.

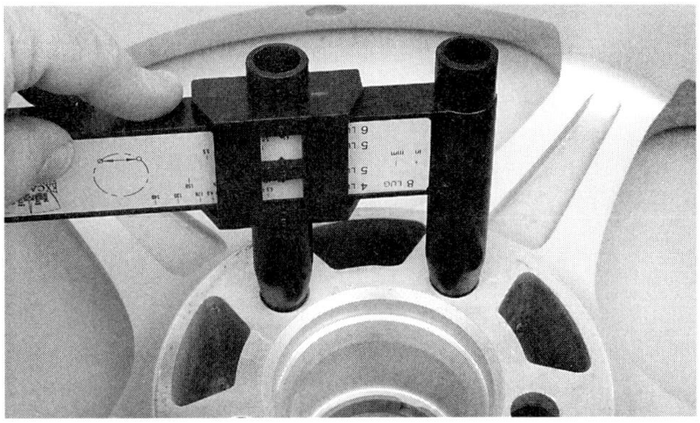

Simply insert the tool's pilots into two adjacent wheel bolt holes.

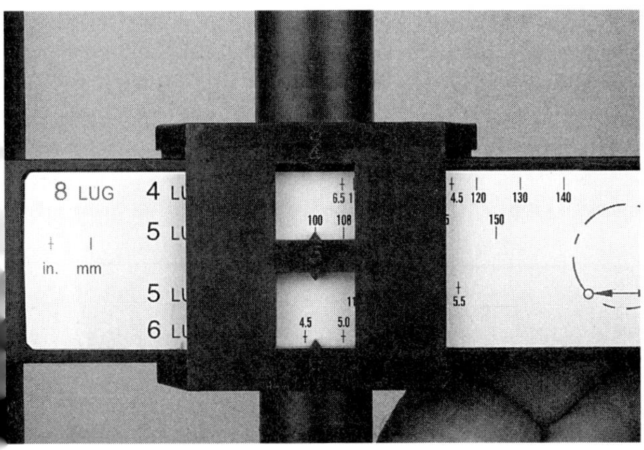

In the example of the wheel measured earlier, we examine the gauge window. To the right of the "5 lug" row, we see that the gauge pointer aligns with 100, so we know that this wheel features a 100mm bolt circle.

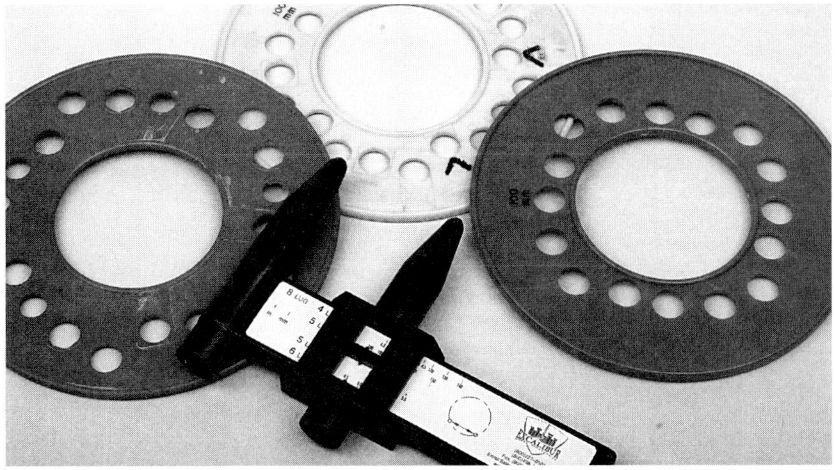

Bolt circle sizing aids are readily available from tire specialty tool suppliers. If you plan to visit a swap meet (looking for a set of vintage alloys), carry one of these in your pocket for easy and accurate identification. It beats the heck out of guessing.

4.00", 4.25", 4.50", 4.75" (and larger on some trucks and SUVs).

Some common metric bolt circles include 100mm, 110mm, 114mm, 115mm, 120mm, 130mm and 140mm. Other circle sizes may also be used in some applications.

WHEEL FASTENER TIGHTENING

Since this book focuses primarily on performance applications, we can assume that when we discuss wheels, we're referring primarily to lightweight custom alloy wheels. As a result, when we discuss installing and clamping the wheel onto the hub, we're interested in both form and function. In terms of form, we don't want to scratch the wheel. In terms of function, we want to secure the wheel properly in the pursuit of both safety and long-term performance.

Impact Guns

NEVER use an impact gun to remove or install threaded fasteners when dealing with custom alloy wheels. I know perfectly well that many who read this will ignore this advice and will get slam-happy with a 1/2"-drive gun anyway, but the advice stands. Unless you're in an race that requires wheel changing speed, there is no reason to ever approach an aftermarket alloy wheel with a pneumatic tool. I know it takes more time and energy to deal with removal or installation manually, but it's simply the most proper method to tighten a wheel fastener.

The main reason to avoid using an impact gun during removal is simply to avoid scratching the fastener pockets of the wheel. If you insist on removal with a gun, use only a clean thinwall socket and run the gun at a slow speed. It's easy to scratch

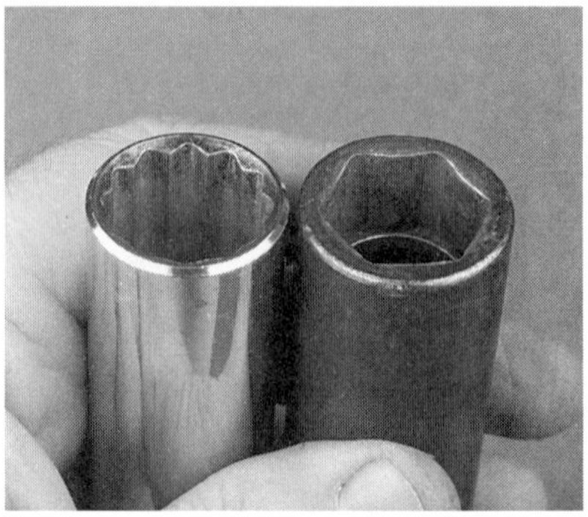

A thin-wall socket wrench (left) and a thick-wall "impact" grade socket wrench. While an impact socket wrench is designed to be used with an impact gun, when hand-tightening wheel fasteners, this tough-grade socket isn't needed. A thin-wall socket will provide more clearance in the wheel's fastener holes.

Don't allow the vehicle to fully rest on the tires before tightening the fasteners. Either ask a helper to step on the brake pedal (to hold the wheel stationary) while the vehicle is raised, or tighten all fasteners as far as you can by hand while the vehicle is raised, then gently lower the vehicle so that the tires are touching the floor, to obtain just enough friction to prevent wheel rotation during tightening.

Make sure that your socket wrench fits the hole, and that the wrench is clean and free of dirt and burrs.

A speed wrench (crank handle style wrench shown here) will aid in starting the fasteners, especially when the fasteners are recessed. It's always best to hand-snug the fasteners before using your torque wrench, this will seat the fasteners and will save time

When tightening a wheel fastener, secure the wrench head with one hand to stabilize the tool, and maintain socket wrench alignment. This will aid in achieving accurate tightening.

Before installing or removing an alloy wheel, take the time to verify socket wrench clearance in the hole recess. If the socket fits too tightly, you can easily gouge the inner walls and possibly chip off paint or clearcoat around the exposed hole perimeter. Also, make sure that the socket wrench is clean.

the pockets with the socket or the exiting nut or bolt, even while trying to maintain control of the gun.

The reason to avoid using an impact gun to install the fasteners is even more simple: it's just plain wrong.

Whether we're dealing with a bargain-basement, impurity-riddled aluminum casting or the finest and strongest billet- or forged-cast alloy wheel, proper clamping force is critical for both wheel care and performance. For one thing, over-tightening can gall or deform the wheel's fastener pocket seats. Over-tightening, in its extreme, can create a fracture in the alloy, which can lead to an eventual wheel failure. Excessive over- or unequal tightening can distort both the wheel center section and the hub. Considering the lightweight rotor hats featured on many of today's vehicles, that's an open invitation to disc brake warping and brake pedal bounce—which is annoying as hell, and perfectly avoidable.

When taper or round seat wheel fasteners are tightened, an interference fit is experienced. The male seat of the fastener contacts the female seat of the wheel and creates a small wedge contact when tightened, creating a pressure point that helps to lock the fastener in place. If either type of fastener is under-tightened, they can loosen during operation. If over-tightened, the fastener can become fatigued and can deform the material in the wheel's female seat pocket, which can result in fastener loosening. The shape of the radiused seat reduces the effect of over-tightening since contact pressure is more evenly distributed than with a taper seat style.

Flat-seat style is used almost exclusively with alloy wheels, since in the early days of alloy aftermarket wheels, the alloy material may not have been strong enough to handle the frictional forces created by tapered or radiused seats.

Some micrometer style (click type) torque wrenches are designed for right-hand use only (tightening clockwise), while others are designed for left-hand use only (counterclockwise for left-hand thread fasteners), and others are designed for both right- and left-hand use. If a torque wrench is labeled for right-hand use only, do not attempt to use it on left-hand threaded applications. You won't obtain accurate results, and you'll ruin the tool.

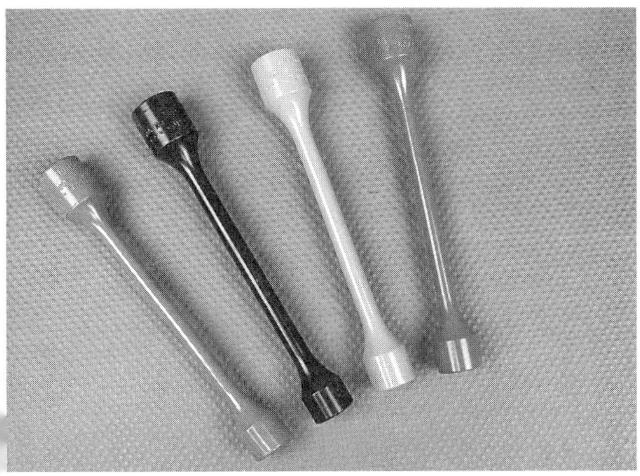

Torque sticks are intended for production-oriented shops where speed is important. Instead of taking the time to use a torque wrench, a torque stick may be used on an impact gun. The sticks are tempered steel and act as torsion bars. Attach one end to the impact gun and engage the other end onto the fastener and hit the trigger. The stick is designed to twist (relieving force) when a specific torque range is reached. These tools are color coded for both size and torque range. Personally, I don't like this approach. It isn't as accurate as a hand-held torque wrench, but on the other hand, it's better than simply guessing at fastener tightness. For a hobbyist, rod builder or racer, you don't need these things. Just buy a quality torque wrench and get into the habit of using it every single time you tighten a wheel fastener.

Over-tightening a flat-seat nut can deform the wheel, causing the aluminum under the washer to extrude, which displaces the aluminum, causing the nut to loosen.

Over-tightening can also stretch the wheel studs or wheel bolt shanks beyond their elastic range. All bolts or studs are designed to stretch a miniscule amount when optimal clamping load is achieved.

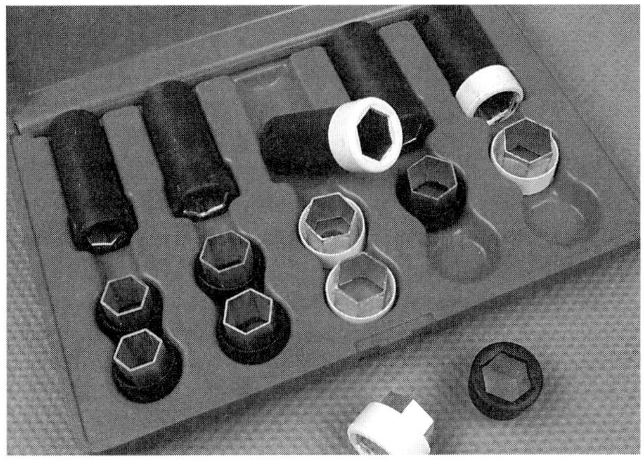

In order to prevent nicks and scratches on delicate chrome or anodized wheel fasteners, do your best to make sure that your socket wrench is perfectly clean and free of burrs. The set of sockets shown here features soft aluminum socket inserts with plastic shields. The aluminum (which is softer than the wheel nut or bolt head) won't gouge the fastener surface, and the plastic shield protects the fastener recess in the wheel.

This elasticity of the stud or bolt is what helps to secure the wheel on the hub. When torqued to specification, this is referred to as achieving the proper "clamping load." If the stud or bolt is excessively over-tightened, it's possible that it will stretch beyond its yield point, losing its "rubber band" effect. If stretched beyond the yield point, the stud or bolt becomes so weak that it cannot provide the clamping load needed. The result: the fastener loosens or the stud or bolt shank breaks. Far too many uninformed do-it-yourselfers approach wheel fastener tightening with the "tighter is always better" attitude.

Always follow the torque specifications listed by either the vehicle manufacturer or by the wheel maker. Don't guess. Actually take the time to pick up a calibrated torque wrench and tighten all of the wheel's fasteners, in the proper sequence, in several steps to achieve final (and equal) torque values.

As far as thread preparation is concerned, make sure the threads are clean and free of dirt, grease, grit, etc. As far as wheel fastening is concerned, specifications are generally listed based on dry (no lubricant) threads. Applying oil, grease or moly to the threads will result in inaccurate torque values (you'll end up over-tightening). Even if you use aluminum wheel nuts (which are popular in some racing situations, Porsche applications, etc.), the advice is the same. Simply make sure the threads are clean and dry. Aluminum wheel nuts are typically made from a very dense, strong 7075 alloy, and will function properly if handled correctly. While a race team will use speed guns for quick pit service, you have plenty of time to be careful in the shop, so install the fasteners clean, dry and with a torque wrench.

Retorque

Some will disagree with the need to retighten wheel fasteners, but my advice is to recheck the value of each fastener after about the first 50–100 miles of operation. Due to metal compression and

WHEEL FASTENER (NUT OR BOLT) TORQUE VALUES

While you should always refer to the vehicle manufacturer's recommendations for proper wheel fastener torque values, following is a generic guideline: (Courtesy The Tire Rack)

Hardware Size	Torque (ft-lbs)
10mm	45–55
12mm	70–80
14mm	85–90
7/16"	70–80
1/2"	75–85
9/16"	135–145

elongation and thermal stresses, the clamping loads may change during initial use (we're not saying they will change, but they might). When rechecking torque value, wait for the wheels to cool to ambient temperature (never torque a hot wheel). Loosen and retighten, to value, in sequence. Again, some will argue that this step is not necessary, but it's better to be safe than sorry, and it's better to catch a loose nut early as opposed to too-late.

Common Problems to Avoid

Make sure the fastener does not bottom-out. In other words, if using a closed-head nut, make sure the tip of the stud does not contact the inside of the nut roof. If the nut is a shanked "mag" type, make sure the shank is not too long. If the shank tip bottoms-out against the hub, clamping pressure at the seat will be lost.

Make sure that there is no grease, snow, ice, etc. lodged inside a closed nut. When tightened, this can create a hydraulic lock, causing the stud tip to bottom-out against this material. If the nut roof bottoms-out on its stud, the nut seat will not achieve enough clamping force.

Be careful not to mix inch and metric fasteners. Just because it CAN be threaded into place does not mean that correct thread engagement has been achieved. NEVER guess if you're using the correct nut or bolt.

As we mentioned earlier, NEVER mix seat styles. Proper clamping will only be achieved by using a conical seat fastener with a conical seat style wheel. A radiused fastener can only be used in a radiused seat pocket. A flat-seat wheel will only accept a flat-seat fastener. Again, just because you can thread the nut or bolt on by hand, don't think that you can "get by" by merely tightening it more. Right is right and wrong is wrong, and when it comes to wheels, there is no middle ground. This is your life we're talking about, so don't screw around.

Knock-Off/Pin-Drive System

Many race cars (as well as a few vintage street cars such as some old Corvettes, Ferraris, AC Cobras, etc.; and some new cars, such as Porsche's Carerra GT) feature a wheel fastening system that involves a single large center nut. Commonly called "knock-off" or "pin-drive" wheel, it features a large center hole in the wheel that slides over an extended male threaded spindle. A large center nut (sometimes hex sided, sometimes featuring two or three large "wings") secures the wheel to the hub. This system is often called a knock-off because (in the case of winged center nuts) a large hammer is sometimes used to loosen or tighten the center nut (where the nut is winged). Even if the nut is hexed, a large special wrench may feature extended leverage arms that allow the use of a lead hammer to knock the nut. The term "pin-drive" refers to the series of short stubby dowel pins located on the hub face that engage into holes on the rear of the wheel center. Locating the wheel onto these pins prevents the wheel from rotating in relation to the hub. Basically, the term knock-off and pin-drive are synonymous.

Most pin-drive knock-off setups will feature right-hand threads on the vehicle left side, and left-hand threads on the vehicle right side. This is done theoretically to compensate for rotating centrifugal force in an effort to prevent accidental nut loosening as the car is driven forward. It's important to be aware that a knock-off wheel system may feature left-hand threads on one side of the car before attempting to remove or install the wheel nuts. It's critical to recognize this before attempting to loosen a left-hand thread nut by trying to rotate it counterclockwise.

We'll show two examples of pin-drive wheels here. A vintage example is a 427 SC Cobra featuring Halibrand knock-off wheels. The knock-off nuts on the Cobra feature three "wings." These provide a leverage point for service. Using a heavy lead hammer, hit the wings in the appropriate direction to loosen or tighten the wing nuts. Lead hammers are employed to reduce damage to the aluminum wing nuts (the softer lead deforms instead of deforming the aluminum).

The center nuts on the 2005 Porsche Carrera GT feature a spline-type engagement, requiring a special dedicated Porsche tool for service. This socket tool is part of a "torque-multiplier" Porsche tool kit. The socket engages the nut, while a special torque multiplier gearbox attaches to the socket. The Porsche torque specification for these nuts is 407 ft-lbs. While that initially sounds like an impossibility (after all, how many folks have a 400+ ft-lb torque wrench lying in their toolbox?). However, that's where the torque multiplier comes in. Simply adjust your torque wrench to 128 ft-lbs and attach it to the multiplier tool. As you pull to a mere 128 ft-lbs, the multiplier gearbox converts this amount of force into an effective 407 ft-lbs. The tool isn't cheap (around a cool $2,000 at you local Porsche dealer), but it makes the job easy and precise.

Naturally, since only a single large nut secures the wheel in place, you want to make certain that the nut won't accidentally loosen. In addition to achieving the specified torque value when tightening the nut, a back-up system is often employed to maintain the nut's position. If the nut is the wing-nut style, stainless steel safety wire can be used to insure against loosening (wire is run through a small hole in one of the wings to a small hole in a wheel spoke). In the case of the Porsche Carrera GT, the alloy nuts feature a spring-loaded tab that engages in a spindle slot.

407 ft-lbs. of torque secure the alloy center nuts on the Carerra GT wheels. Here a GT is seen running down the front straight at Talladega at a cool 192 mph.

The Carerra GT's center nuts are color-coded red for left side and blue for right side (shown). The right side (blue) nuts feature left-hand thread.

Note the spline-drive design of the alloy nut. This requires the use of a special Porsche nut driver tool. If you look closely, this red nut (on the driver side of the car) is labeled with an arrow for direction to loosen. Each nut is so-marked to avoid confusion between thread directions. This red nut features right-hand threads, requiring counterclockwise rotation to loosen.

Porsche's specialty tool kit for removing and installing the center nuts on the Carerra GT. Included is the nut wrench, the torque multiplier and the stop-arm that rests against a wheel spoke. The Porsche tool engages onto the wheel nut. The torque multiplier tool is driven by a 1/2" drive torque wrench. By pulling a mere 128 ft-lbs on the torque wrench, the torque multiplier achieves a final at-the-nut torque value of 407 ft-lbs.

This cutaway front section of a Carerra GT (seen here at Porsche's training center in Atlanta, GA) shows the massive center spindle and the unique alloy center nut.

The venerable AC Cobra (the example shown here is an ERA replica) features "knock off" wheels, secured by a single center nut that features three wings. A lead hammer is used to "knock" the wing nuts on or off.

Knock-off wheels and hubs are often referred to as "pin-drive," because of the series of fat dowel pins on the hub. These pins engage into holes on the backside of the wheel to register the wheel onto the hub.

The knock-off spinner nut is made of aluminum. In order to tighten or loosen the spinner nut, a heavy lead hammer is used to beat the spinner wings. The lead deforms easily, but that's intentional. The lead (softer than the aluminum wings) is sacrificed, to prevent damage.

Chapter 8

ROD ENDS

Samples of rod ends that I borrowed from Aurora Bearing. Material selections include carbon steel, alloy steel, stainless steel and aluminum. A variety of coatings are available as well, to suit specific applications. Threads are available in a variety of diameters, including inch and metric formats.

Often called rod ends, spherical rod ends, rose joints or heim joints, these motion-transfer products allow connection between two points where pivoting movement is anticipated and required.

Rod ends are available in four basic formats, including male RH, male LH, female RH and female LH. Male rod ends feature a male threaded shank. Female rod end feature female (internal) threads. Either version is offered in right-hand thread or left-hand thread. Rod ends are also available with either fractional thread or metric thread.

Rod ends are constructed in several ways. Cheap rod ends are, well, cheap, and shouldn't be considered for any race car. The most inexpensive type that should be considered is the two-piece rod end that is fully swagged. In this case, the rod end body is formed around the bearing ball, and rides on the inner diameter of the rod end's housing (the housing interior serves as the race). The really cheap versions feature a low-grade steel housing and a press-in bearing race. Avoid these.

The better type—frankly, the only type you should consider for any steering or suspension application—is the three-piece type, where a separate bearing race is formed around the ball, and the race insert is staked into the body. The race itself may be made of mild, alloy or stainless steel. Rod end bodies may be made from mild steel, aluminum, stainless steel, or in some cases (where weight is everything and money is no object), titanium. High-quality aluminum rod ends offer plenty of strength. If the rod end is made from, say, 7075-T6 aluminum, its tensile strength is a bit greater than that of mild steel. The only downside of high tensile strength aluminum is that it's not as conformable as mild steel…it doesn't like to bend or stretch like steel. Other good choices are rods made from high-grade (17-4PH) stainless steel or 4130 or 4340 alloy steel.

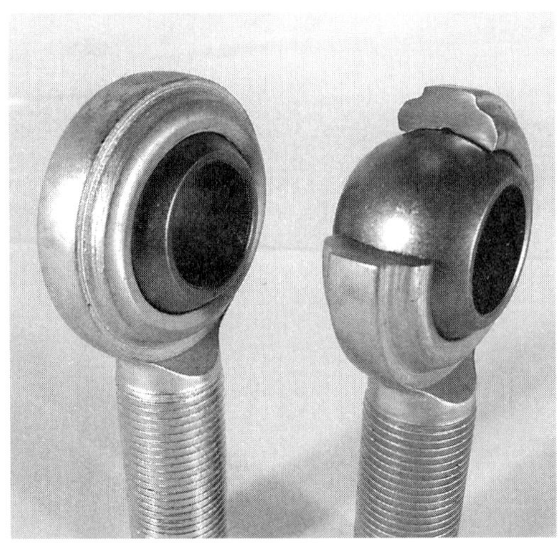

This is an example of a high quality 2-piece rod end. (Courtesy Aurora Bearing)

LINED VS. UNLINED BEARINGS

As far as the bearing is concerned, both metal-to-metal and Teflon-lined versions are available. Metal-to-metal bearings feature no liner between the bearing and its bearing bore housing. These bearings are designed with a slight clearance to allow bearing movement. Teflon-lined bearings are designed with zero clearance, with the liner acting as both a spacer and a lubricant.

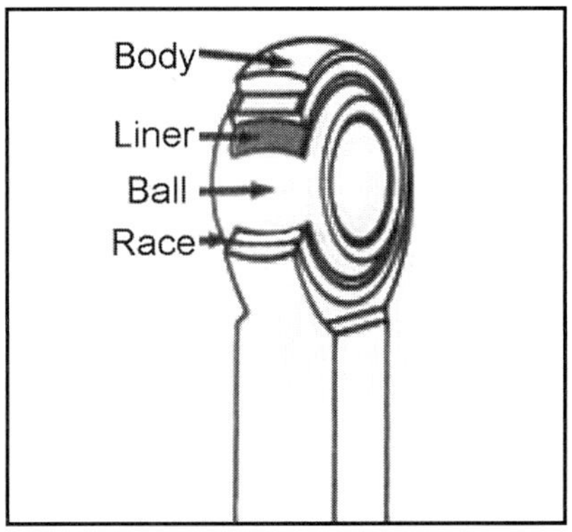

The PTFE liner is bonded to the inside wall of the race, causing the bearing ball to ride directly on the liner. As the ball moves (rotates), teflon is transferred from the composite teflon liner to the ball surface, providing needed lubrication (the liners are actually made of a carrier fabric impregnated with Teflon to make it durable). By the way, Teflon liners are available on both swaged two-piece rod ends as well as on three-piece rod ends.

There's no decent way to lubricate a metal-to-metal (unlined) bearing. However, some makers offer rod ends with a grease fitting. However, these probably shouldn't be your choice for a high-load application, since the tapped hole in the housing provides a potential stress-riser and results in a potentially weaker rod end.

As far as applications are concerned, metal-to-metal bearing rod ends are suitable for applications where minimal load and light duty cycles are involved, such as throttle linkage. For any application that involves high loads, and/or high duty cycles, Teflon/PTFE lined bearings are the way to go. They won't seize or gall, and they're quieter. Actually, for light-duty applications such as throttle linkage, it might be a disadvantage to use PTFE lined rod ends, since (due to the zero-clearance construction) a bit of initial effort is required to move the bearing, which could be annoying for light-sensitive throttle movement requirements. PTFE lined bearing rod ends are ideal for any suspension application, or where axial and/or radial loads are high. Just remember:

Heavy loads and/or high-duty cycles: PTFE lined bearings
Light loads and/or light-duty cycles: Metal-to-metal (unlined) bearings

In addition to providing lubricity, the Teflon liner eliminates clearance between the ball and race, providing that zero-clearance tight fit. Beware of cheap rod ends that may feature a Teflon-coated plastic liner, which may squish out during high load use. A high-quality Teflon liner will provide a compressive strength of 40,000 to 60,000 psi, while a cheap plastic/Teflon liner might only withstand 15,000 psi of compression.

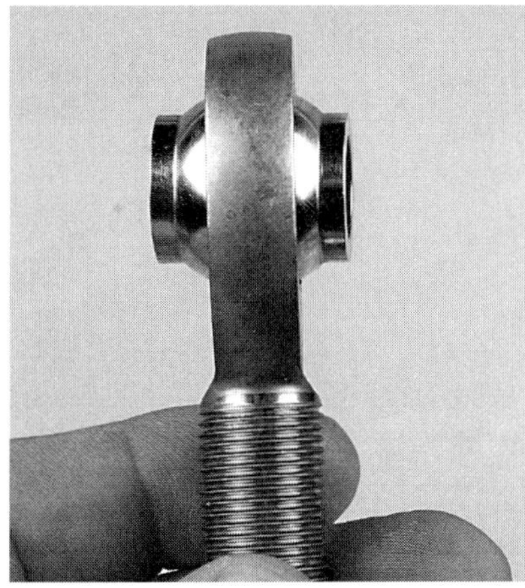

Example of a high-misalignment rod end. The extended shoulders on the bearing allow for greater ball angle, enabling the rod end to be used in a misaligned application. This is Aurora'a HXAM high-strength steel rod end with PTFE Teflon liner.

Replacement bearings are available from some manufacturers whose units may be serviced using a hydraulic press. Once replaced, the new bearing race should be staked in place. Follow the bearing maker's procedures for this.

ROD END SIZING

This is very basic. Both the I.D. of the bearing ball (the hole in the bearing through which you'll pass a bolt) and the shank length are included in the size, with the ball hole size listed first. For example, a rod end that features a ball with a 1/2" hole and a 1/2" diameter thread is referred to as 1/2" x 1/2". If the hole in the ball is 1/2" and the thread diameter is 5/8", the size is referred to as 1/2" x 5/8". In order to identify the rod end fully, add to this the configuration (male or female thread) and the thread direction (right-hand or left-hand). A rod end that is identified as 1/2" x 1/2" RH male features a 1/2" hole in the ball, and a male right-hand 1/2"-diameter thread. A rod end identified as 1/2" x 1/2" LH female features a 1/2" hole in the ball, and a female left-hand 1/2"-diameter thread. It's an easy format to understand.

Rod ends are offered in a wide range of sizes. In the inch-format family, bore sizes (the hole diameter in the bearing ball) are offered in 1/16" increments, similar to the format used in –AN plumbing sizes. A size 3 features a 3/16" bore, a size 4 features a 1/4" bore, a size 5 features a 5/16" bore, etc. You can order rod ends based on either the numeric code or by the actual bore dimension.

LEFT-HAND THREAD ROD ENDS

Why is left-hand thread necessary? If you have a threaded link with a rod end at each end, having a right-hand threaded rod end at one end and a left-hand threaded rod end at the opposite end, you're able to adjust the length of the assembly without

INCH FORMAT ROD ENDS

Code Size	Bore Size	Thread Size
3	3/16"	10-32
4	1/4"	1/4-28
5	5/16"	5/16-24
6	3/8"	3/8-24
7	7/16"	7/16-20
8	1/2"	1/2-20
10	5/8"	5/8-18
12	3/4"	3/4-16
16	1"	1 1/4-12

Note: Oversized shank rod ends are also available in certain sizes. For example, a rod end featuring a 1/2" bore with a 5/8" diameter thread, where a beefier threaded connection is desired.

METRIC FORMAT ROD ENDS (COMMONLY AVAILABLE SIZES)

Bore Size	Thread Size
8mm	8mm x 1.0
8mm	8mm x 1.25
10mm	10mm x 1.25
10mm	10mm x 1.50
12mm	12mm x 1.25
12mm	12mm x 1.75
14mm	14mm x 1.50
14mm	14mm x 2.0
16mm	16mm x 2.0

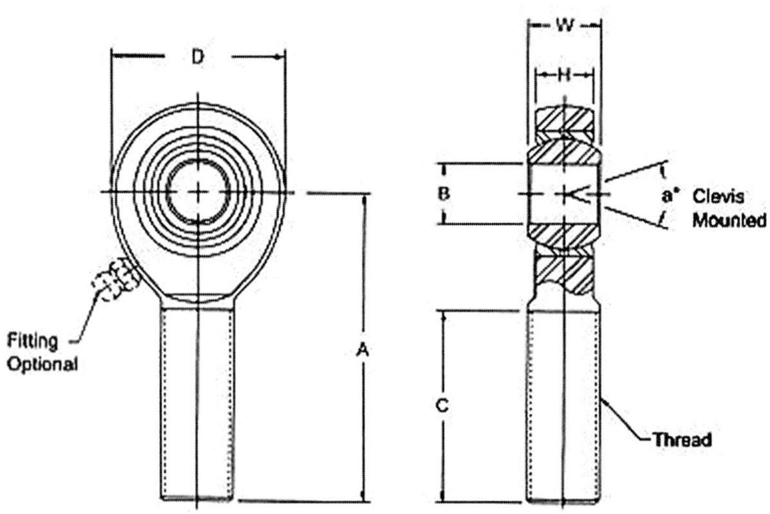

Rod end dimensions differ based on size. A is the rod end length, from center of the bearing to the end of the threads. B is the inside diameter of the ball. C is thread length. D is the width of the rod end head. H is rod end head housing thickness and W is the width of the ball. (Illustration courtesy Aurora Bearing)

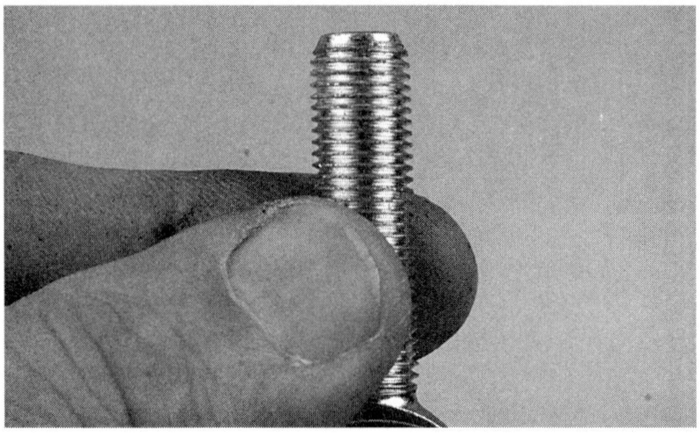

This thread is left-handed. Notice the thread spirals upwards at the right side. Left-hand thread inserts by rotating counterclockwise and backs out by rotating clockwise.

the need to disconnect one of the rod ends. Simply loosen the jam nuts that secure each rod end in place (always use a jam nut with a rod end) and rotate the link. As you rotate the link in one direction, both rod ends will thread deeper into the link (making the link shorter). When you rotate the link in the opposite direction, both rod ends will thread outboard from the link (making the assembly longer).

Here's a good example: Let's say you have a rack and pinion steering assembly in the car. Let's say this rack's tie rods feature 1/2" x 20 male RH threads. If the tie rod length is appropriate, you could install female RH 1/2" x 20 rods ends directly onto the tie rods. In order to adjust toe, you would then loosen or remove the rack boot clamps (to allow tie rod rotation) and rotate the entire tie rod in order to fine-tune length for proper toe angle.

If however, the rack's tie rods are too short, you could buy or fabricate threaded links that feature female 1/2" x 20 RH thread at each end. This would function, but when you need to adjust toe angle, you'd need to disconnect the rod end from the steering arm, rotate the link, re-attach the rod end, re-measure toe, etc. That's just plain dumb.

A better way to go would be to buy or make a link that features female RH thread at one end and LH thread at the opposite end. When you need to adjust toe angle, simply loosen the jam nuts, rotate the link to adjust toe, and retighten the jam nuts.

This RH/LH adjustable link approach is in common use in a variety of applications, including support/adjustment mounts for alternators, power steering pumps, A/C compressors (on street rods), etc. This offers an easy, stable and precise way to adjust an accessory drive unit's belt tension. And, by using rod ends, the assembly does not need to be perfectly 90-degrees to the attachment points,

It's difficult to visually determine if the bearing is lined or unlined. Chances are, when a new rod end features a bearing ball that rotates easily with your fingers, it's unlined. If the ball resists finger pressure, it's likely lined (since lined bearings intentionally have zero clearance when the Teflon liner is present). This rod end is unlined.

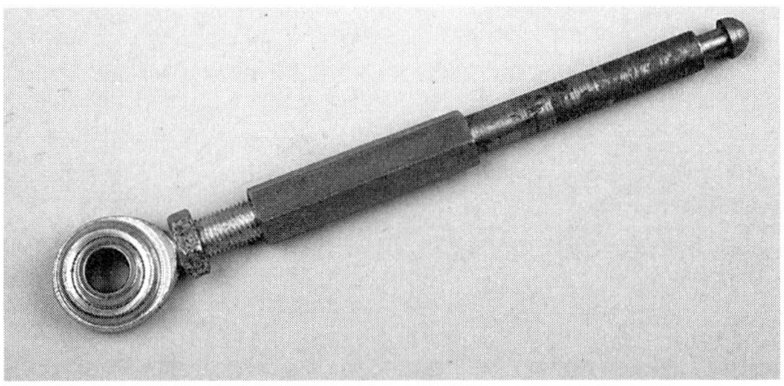

Plunger rod from a brake master cylinder, taking advantage of the solid connection, length-adjustment and pivoting angle compensation provided by a rod end.

This view shows the maximum clevis angle for this particular rod end, which we measured at 30 degrees from the straight position. In order to take full advantage of the ball's sweep, don't place a hex bolt head directly against the ball. Space the bolt head out with a thick washer.

because the rod end ball is free to rotate slightly to compensate for misalignment.

Note: For some applications, primarily those that require the smaller-sized 3/16" and 1/4" rod ends, special rod ends are available that feature a captive clevis (threaded stud) attached inside and

Safety Retaining Washer Dimensions (in inches)

Inside Dia.	Outside Dia.	Thickness
0.250	0.682	0.142
0.312	0.807	0.168
0.375	0.932	0.184
0.437	1.120	0.220
0.500	1.182	0.228
0.625	1.432	0.270

High misalignment ball. Notice the extended "shoulder" of the ball. This extends the clevis angle, allowing for greater angle and misalignment.

Standard with bearing ball. The face is flat-cut across the ball radius.

protruding from the ball bore. These are especially useful for throttle linkage hookups.

SAFETY RETAINING WASHERS

Certain race sanctioning bodies require the use of safety retaining washers for rod end bearings. These serve as a fail-safe device to keep the bearing in place, in case the ball or race tries to separate from the housing. These special washers are beveled to allow full freedom of movement for the rod end. The outside diameter of these washers is greater than the bearing race diameter for most of the available rod ends. Always check to make sure that the washer you choose does indeed exceed the O.D. of the race. These washers are designed as a back-up in case of a rod end bearing or race failure. It's cheap insurance that could prevent a real catastrophe. Think of it as similar to a driveshaft safety loop or a blowproof bellhousing. Nice to have, just in case.

Safety retaining washer for rod end bearings. This is designed to retain the bearing ball in case the bearing or race separates from the housing. Sometimes required by various race sanctioning bodies to conform to their tech rules. These cost anywhere from about $2 to $4 each. Available I.D. runs from 0.250" to 0.625", with O.D. running from 0.682" to 1.432".

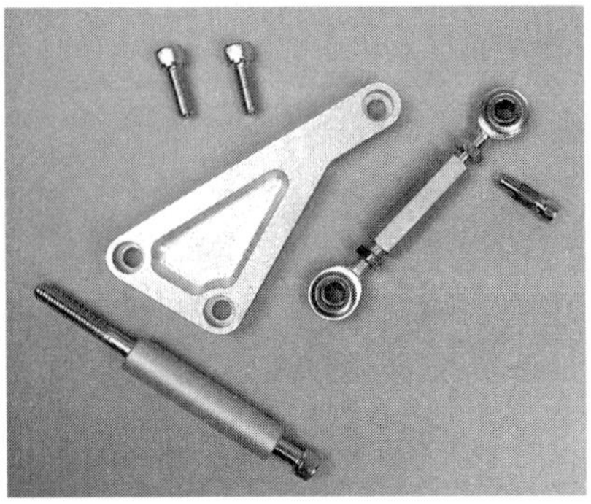

Here's a common store-bought kit for the mounting of a small-block Chevy alternator. The cast aluminum bracket and long bolt with tube spacer mounts the alternator, while the heim joints and remaining bolts attach the bracket.

Since the mounted position of our alternator was a tad high, the hex link was ultra-close to the alternator at the top end. In order to enhance eye appeal, we simply machined the hex area off of the link on the lathe. To make it match and look balanced, we turned down both ends, leaving a center hex for wrench access. We had the link red anodized at the same time as the lower bracket.

This alternator adjuster features 3/8" x 3/8" male rod ends, one with RH thread and one with LH thread.

The radius rods on the rear suspension of this Cobra replica features rod ends. These provide the pivot point for rod movement during suspension travel, and serve to provide a much more precise adjustment/location as opposed to eyelet ends with rubber or urethane bushings.

Here's an end-link mounted to the lower control arm of a 1973 Plymouth Duster that we fitted with a 3-piece anti-sway bar. We made a short tubular spacer to locate the rod end in the desired location.

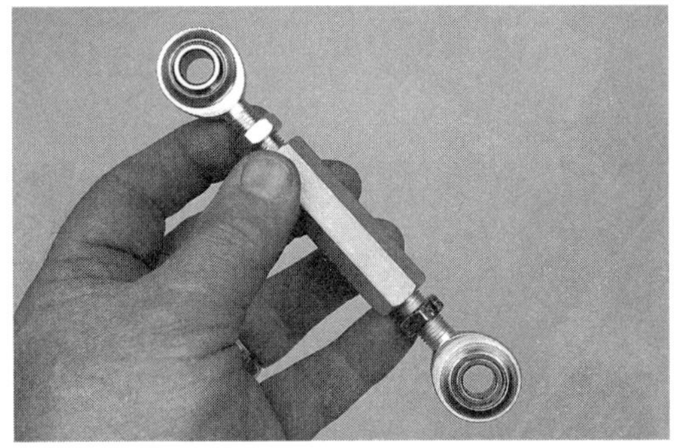

Remember: Always install jam nuts at each rod end. Once length is adjusted, the jam nuts will secure the assembly, preventing the link from rotating and accidentally changing length.

This Cobra also features tubular lower control arms, pivoting via rod ends.

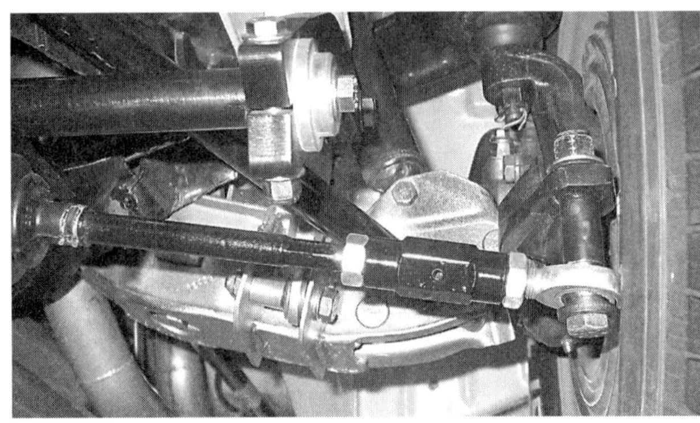

Here's the same Hemi Duster's rack and pinion steering hookup. Using a new Flaming River Pinto-style rack, we removed the original tie rods and replaced them with Dodge Neon tie rods that featured a beefier 14mm tie rod. Using 1" hex steel bar stock, we made a pair of link adjusters, drilling/tapping the tie rod end for 14mm x 1.5 RH thread, and drilling/tapping the steering arm end for 5/8" x 18 LH. We then installed a 1/2" x 5/8" male LH rod end to connect to the steering arm. Just to cut down on a bit of bulk and to dress it up a bit, we turned the ends of the link round on our lathe. We also drilled a small sight hole in the center of the hex link, just to provide a monitoring hole to check for thread travel of both the tie rod and the rod end. In order to adjust toe angle, we simply loosen both jam nuts and rotate the hex link using a 1" open-end wrench.

Chapter 9

RIVETS & CLECOS

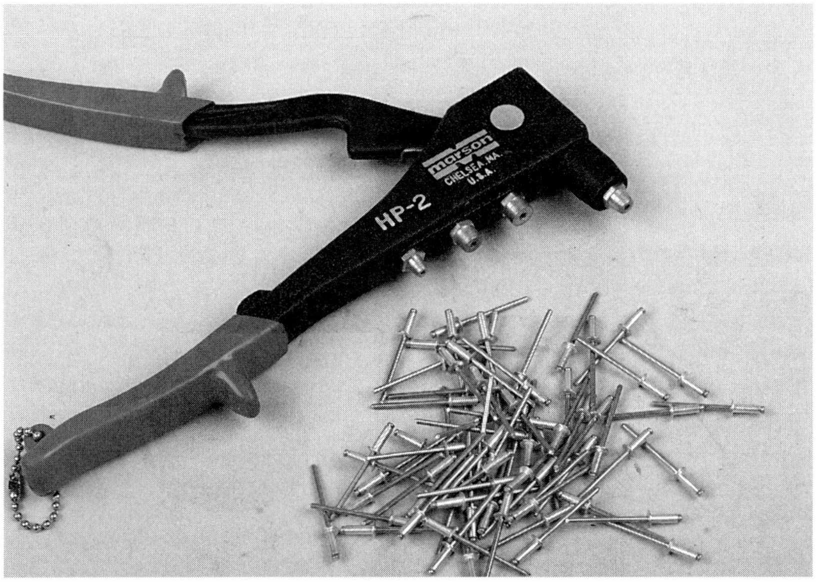

Rivets are easily installed, and are ideal for applications where access is limited to only one side of a workpiece. Rivets consist of two pieces, including the rivet body and the mandrel. Once the rivet body has been inserted into a hole, the rivet gun is operated to pull the mandrel back to expand the rivet body, effectively capturing the rivet body in the workpiece. As a result of the body expansion, the rivet creates a clamping force that pulls the two mating panels together.

TYPES OF RIVETS

Rivets are available in a variety of body diameters and grip depths (to accommodate different thickness of panel assemblies). In addition, rivets are available in various styles and materials, to suit a range of strength, clamping area and corrosion issues.

Steel body/steel mandrel—High strength. Use with steel materials.

Aluminum body/aluminum mandrel—Lightweight. Use with aluminum materials and in areas where rust prevention is a concern.

Aluminum body/steel mandrel—Light, but with increased clamp strength than all-aluminum.

Stainless steel body/stainless steel mandrel—Very high strength. Use with steel or stainless steel materials. Prevents rust/corrosion.

Stainless steel body/steel mandrel—Very high strength. Use where corrosion resistance is not a major factor.

Copper body/copper mandrel—Use where electrical conductivity is required.

Here's a dome head open-end rivet. The bulge you see at the right is part of the mandrel. When the rivet is installed in place and the rivet tool begins to draw the mandrel up through the rivet body, this mandrel bulge will expand the rivet body, creating a tight interference fit and clamping load.

Note: A few specialty performance aftermarket sources offer alloy rivets that have been anodized (clear or a variety of colors) to suit appearance needs).

Rivet Styles

Open-end rivets feature an opening at each end of the rivet body. This is the most common type that you'll encounter. Once installed, the exposed face will feature a hole (where the mandrel was originally located), and the backside will be open, revealing the remainder of the mandrel bulge that forced the backside of the rivet body to expand.

Closed-end rivets are the same as open-end rivets, except for a cup-shaped enclosed backside that hides the remaining mandrel bulge. This closed end seals the rivet backside, preventing moisture from entering the rear of the rivet.

T-rivets are designed for high-stress applications. The steel mandrel splits the rear of the rivet body into three long

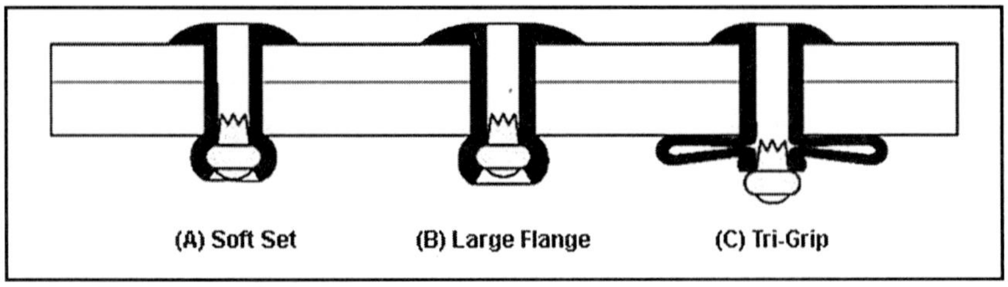

From left: standard dome head rivet, large-flange rivet and tri-grip rivet. (Courtesy Hanson Rivet)

"legs" for a larger footprint and higher clamping strength.

Rivet head styles vary as well. Typical head styles include the dome head, large-flange dome head and the countersunk head. The dome head is the most common. This provides a low profile and neat appearance. The head diameter is typically twice the diameter of the rivet body. The large-flange dome head simply features a larger-diameter head, twice the diameter of a regular dome head. The larger diameter head provides a larger top-side footprint, ideal for fragile or brittle materials where cracking is a concern. The countersunk rivet head is, as the name implies, countersunk into the face of the workpiece, with the installed rivet head flat and flush to the outer surface.

RIVET INSTALLATION

Once you've selected the rivet size and style, drill a hole through both panels that are to be attached.

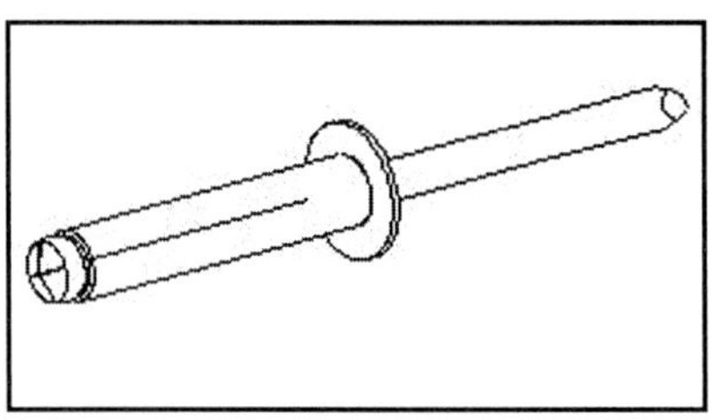

The T-rivet style features a split in the rivet body starting from the rear tip as seen here. (Courtesy Hanson Rivet)

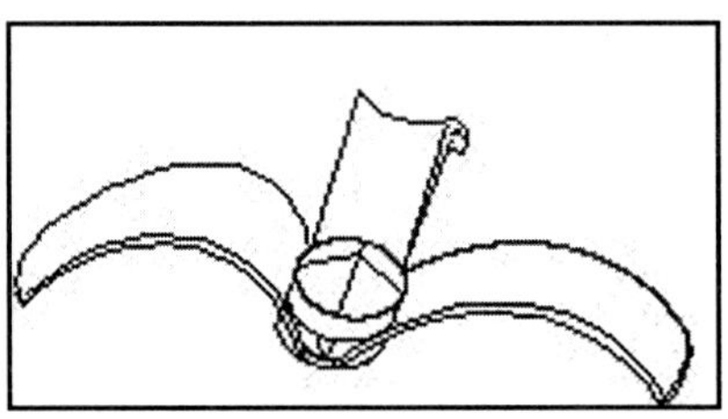

The T-rivet installed results in a tri-leg clamping area as seen here, suitable for structural applications. (Courtesy Hanson Rivet)

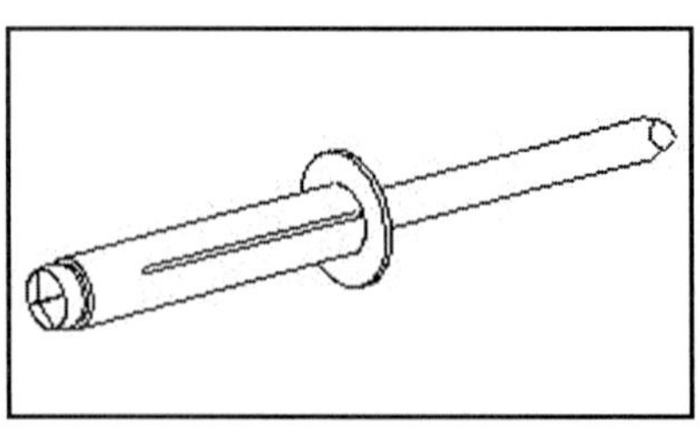

The Tri-grip rivet body features slits that start from under the rivet head, as seen here. (Courtesy Hanson Rivet)

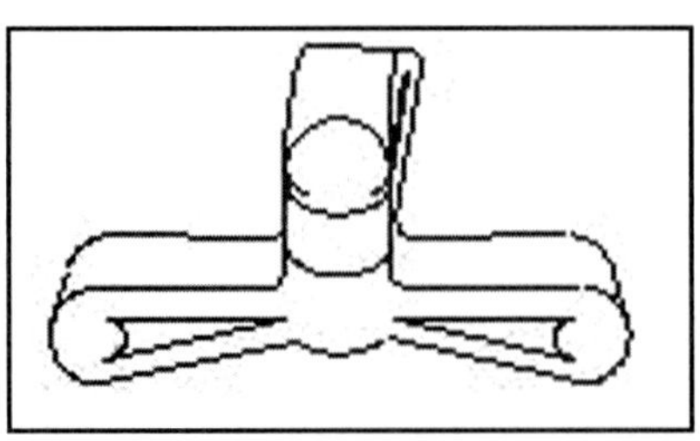

The Tri-grip rivet expands in a tri-leg pattern, but with the legs wrap-formed as see here. Suitable for soft or brittle materials such as plastics. (Courtesy Hanson Rivet)

RIVET SIZES

These "pop-style" rivets are commonly available in body diameters ranging from 3/32" to 1/4", with a variety of grip ranges available in each diameter size.

Listed here are typically available diameters and grip ranges (using open-end rivets as the examples here). All figures are in inches.

Body Dia.	Head Dia.	Grip Range
3/32	3/16	up to 1/8
3/32	3/16	1/8 to 1/4
1/8	1/4	1/32 to 1/16
1/8	1/4	1/16 to 1/8
1/8	0.220	1/16 to 1/8
1/8	3/8	1/16 to 1/8
1/8	1/4	1/8 to 3/16
1/8	0.220	1/8 to 3/16
1/8	1/4	3/16 to 1/4
1/8	0.220	3/16 to 1/4
1/8	3/8	3/16 to 1/4
1/8	1/4	1/4 to 5/16
1/8	1/4	5/16 to 3/8
1/8	1/4	3/8 to 1/2
5/32	5/16	1/16 to 1/8
5/32	5/16	1/8 to 3/16
5/32	5/16	3/16 to 1/4
5/32	5/16	1/4 to 3/8
5/32	5/16	3/8 to 1/2
3/16	3/8	1/16 to 1/8
3/16	3/8	1/8 to 1/4
3/16	5/8	1/8 to 1/4
3/16	3/8	1/4 to 3/8
3/16	5/8	1/4 to 3/8
3/16	3/8	3/8 to 1/2
3/16	5/8	3/8 to 1/2
3/16	3/8	1/2 to 5/8
3/16	5/8	1/2 to 5/8
3/16	3/8	5/8 to 3/4
3/16	5/8	5/8 to 3/4
1/4	1/2	1/16 to 1/4
1/4	1/2	1/4 to 3/8
1/4	1/2	3/8 to 1/2
1/4	1/2	5/8 to 3/4

In order to select the proper rivet for your application, select the rivet type according to the installation material and strength needed; select the rivet style that best suits your application; measure the assembled joint thickness in order to determine the needed grip range; select the rivet body diameter based on hole size and strength needed (the larger the rivet body diameter, the more strength).

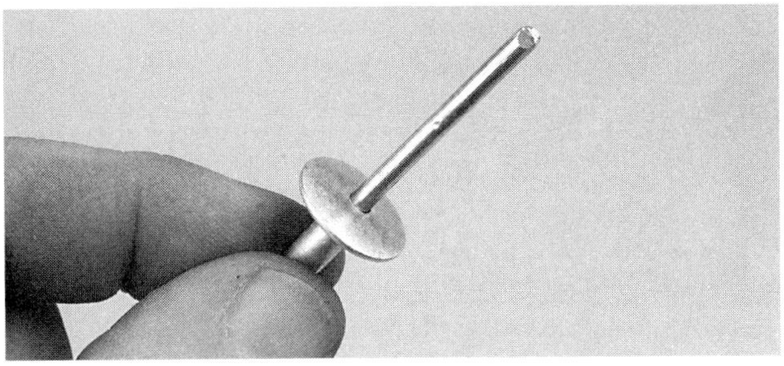

Large dome head rivet. The larger diameter head provides a greater footprint for spreading the clamping load over a larger area. Good for thin or fragile materials that might crack or warp with a small-diameter rivet head.

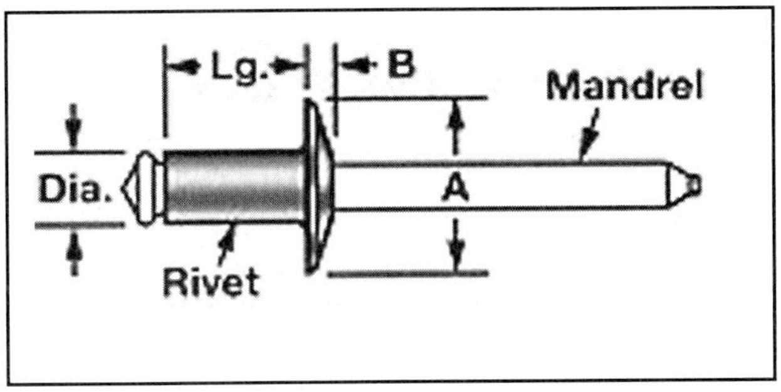

Rivet advertised diameter refers to the body diameter. The grip length or range refers to the length of the rivet body from under the head to the body tip. The mandrel is the driver pin that features a bulge end. (Courtesy McMaster-Carr)

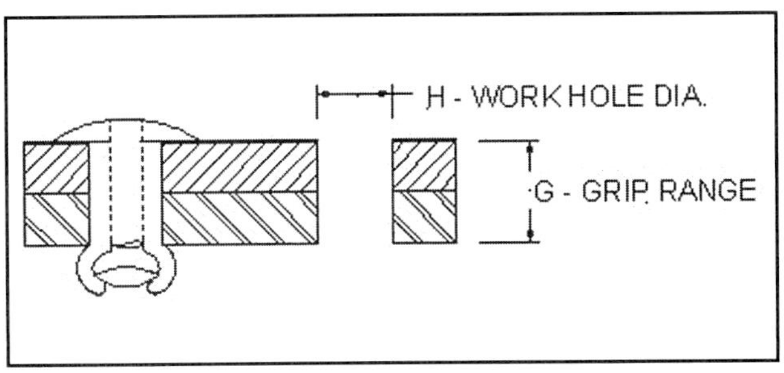

A cross-sectional view illustrating the rivet's grip range. (Courtesy Hanson Rivet)

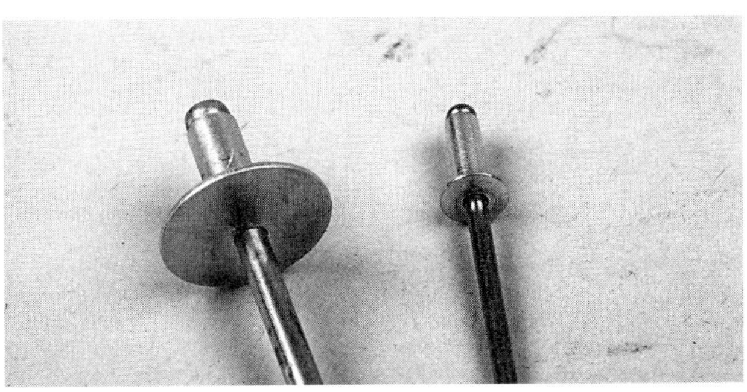

Two examples of river head sizes. On the left is a large-head 3/16" rivet compared to a standard head 1/8" rivet.

Refer to the rivet instructions for the hole size needed.

Also, based on the diameter of the rivet's mandrel, install the appropriate size nosepiece and thread the nosepiece onto the gun tip.

With the manual rivet gun's grips open, insert the rivet mandrel fully into the nosepiece.

Place the rivet body through the drilled hole (make sure the two panels are held together so that the rivet body fully inserts through both panels). Squeeze the rivet tool handles together until the mandrel snaps off. Depending on material thickness, you may need to squeeze the tool once, relax the grips, insert the tool deeper onto the mandrel, and squeeze the tool again.

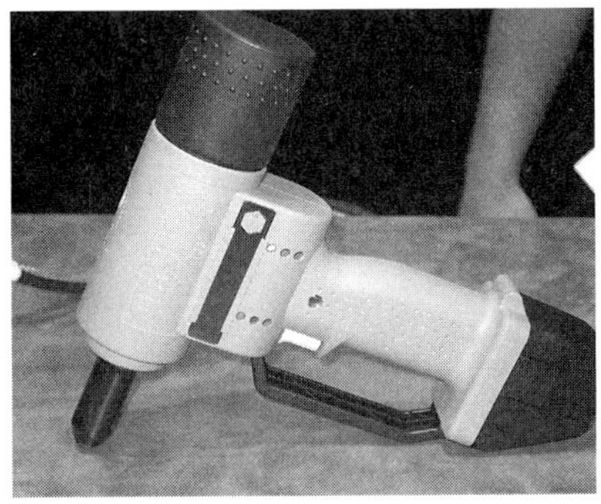

Cordless electric rivet guns are now available. Shown here is a Gesipa-brand gun (AccuBird model) that I saw at a recent racing trade show. Rivet tools are available in a variety of styles, including hand-operated, electric cordless, pneumatic and pneumatic/hydraulic. Unless you're performing repetitive production-line riveting, a hand-operated rivet tool will suffice just fine.

Regardless of the rivet tool, make sure you install the correct-size nosepiece to accommodate the diameter of the rivet mandrel. This tool features four different-diameter nosepieces. The spares screw into the tool for convenient storage.

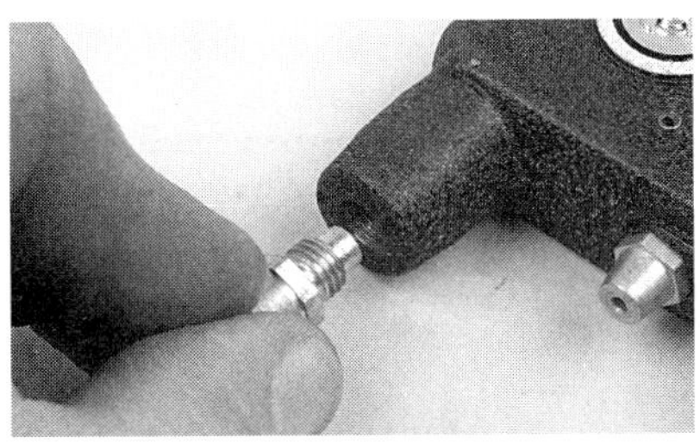

In order to change rivet tool noses (to accommodate various mandrel diameters, unthread the existing nosepiece and thread in the correct size. Note: in order to change nosepieces, compress the tool grips together to retract the internal mandrel clamp in the nose of the tool. You can do this by securing the grip-storage chain/strap to bring the grips together.

Snug the nosepiece using a wrench. This Marson rivet tool requires a 3/8" wrench.

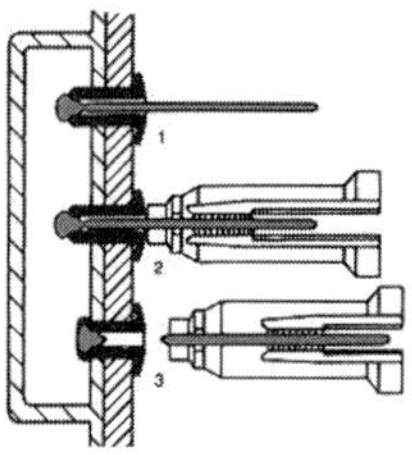

The rivet is inserted into the drilled hole. The rivet tool then grasps the exposed mandrel pin. As the tool is used to pull the mandrel, the bulge at the rear tip of the mandrel is drawn into the rivet body, expanding the rivet body. (Illustration courtesy Hanson Rivet)

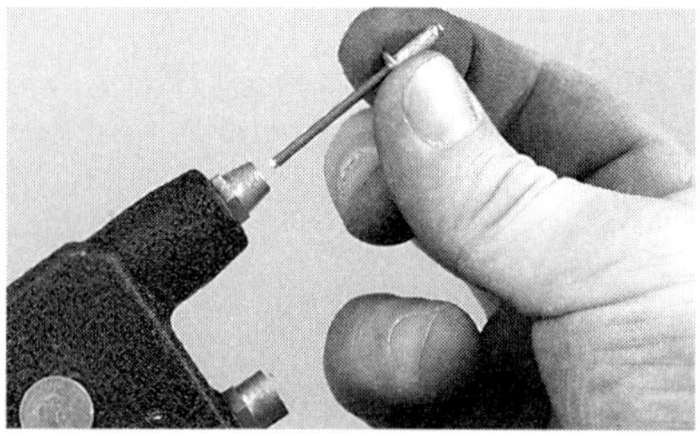

Insert the rivet mandrel into the tool nosepiece. The tool grips should be relaxed at this point. Do not squeeze the tool grips until the rivet is in place in the workpiece and you're ready to expand the rivet.

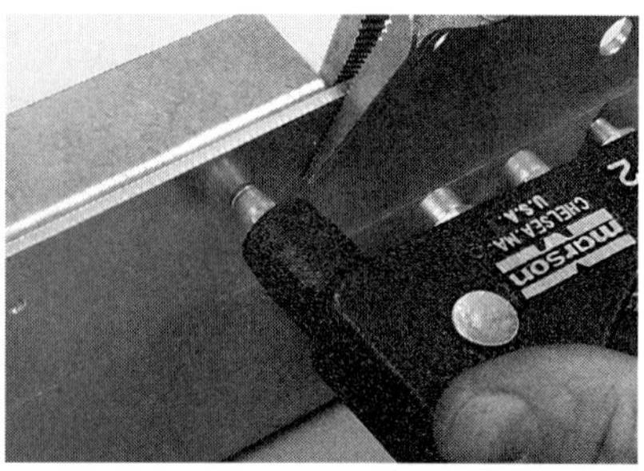

With the rivet body fully inserted, squeeze the tool handles together to begin drawing the mandrel through the rivet body.

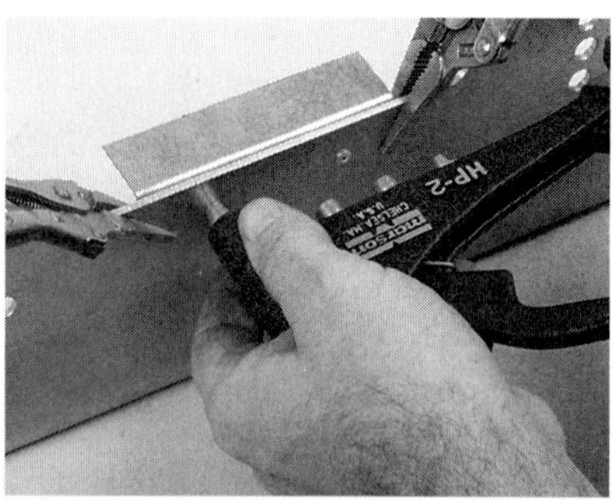

Once the second rivet is installed, we can remove our vise grips.

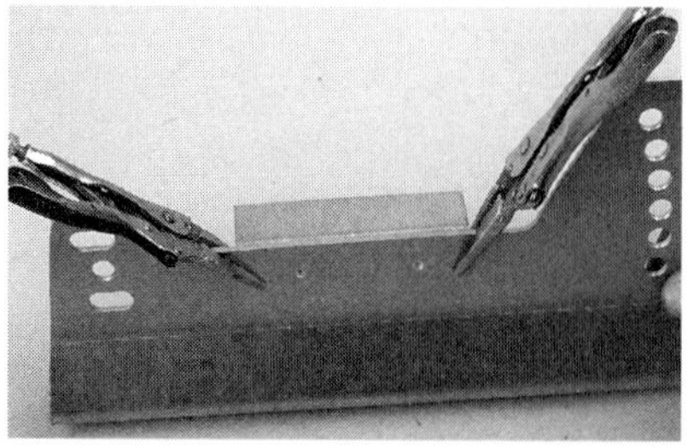

Sandwich the two panels together to achieve the desired installed position, then drill a hole through both panels (to accommodate the rivet body diameter).

Make sure that the rivet is fully inserted into the tool nosepiece. The rivet head should be flush against the nosepiece.

Whenever you squeeze the tool grips, always keep the tool nosepiece flush against the work surface.

Two dome head rivets installed.

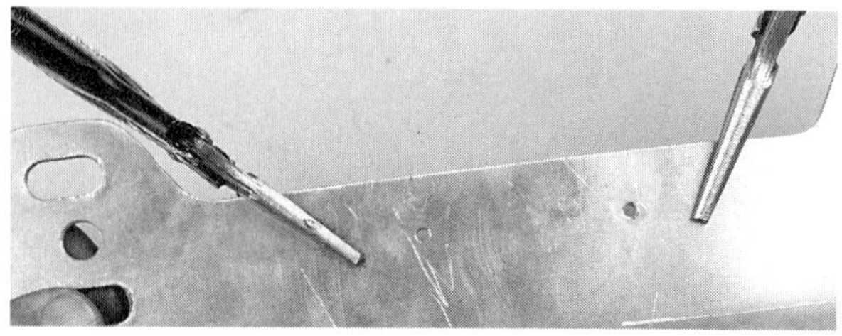

Clamp the two panel pieces together, securing them to maintain hole registry. Here we're using a pair of vise grips, but Clecos could be used as well.

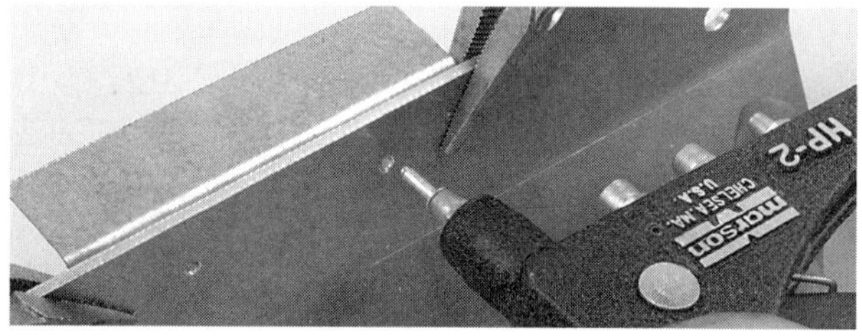

Insert the rivet body into the hole. Do not begin to squeeze the tool grips until the rivet is fully inserted in the workpiece.

Once the mandrel reaches the end of its travel and the rivet body has fully expanded, the mandrel breaks off. Be sure to discard the used mandrels in the trash instead of leaving them on the garage floor.

This rear view shows the expanded rivet bodies that were bulged by the mandrel.

Closeup of an installed dome head, open-end rivet.

Rear-side view of four fully-installed rivets. From left: 3/16" rivet with backing washer, 1/8" rivet with backing washer, 3/16" rivet without backing washer, and 1/8" rivet without backing washer.

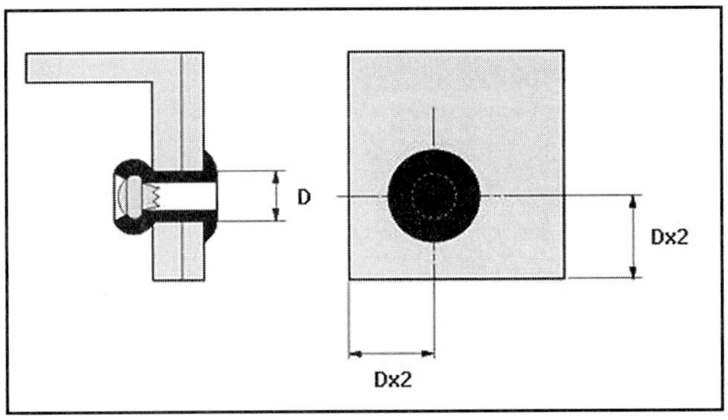

When installing a rivet near an edge, it's recommended to place the rivet at a distance from an edge of twice the rivet head diameter. If the rivet head is 1/4" in diameter, locate the center of the rivet at least 1/2" from an edge. (Courtesy Hanson Rivet)

Three rivets shown here. At top left, a 3/16" rivet inserted in the material, with a backing washer slipped over the rivet body. At center, a 1/8" rivet already fully installed with a backing washer. At bottom right, a 1/8" rivet fully installed with no washer.

A 3/16" rivet fully installed with a backing washer in place. You can see how the washer creates a larger surface area contact for spreading the clamping load, and that the bulged rivet end has secured against the washer, sandwiching the washer between the workpiece and the rivet bulge.

CLECOS

The name may sound funny, but these are way-cool fabrication aids.

If you're not familiar with Clecos, you need to be, especially if you perform any sheet metal or panel fabrication. I honestly don't know where the name originated (maybe Joe Cleco invented them), but it really doesn't matter. These are spring-loaded securing fasteners that hold two panels together during fabrication work. Let's say that you're making an access panel for the body of a drag car, to allow access to wiring or plumbing, etc. After you've cut your hole in the sheet metal (or fiberglass), you make a template for the panel cover. Now you need to attach it, but maybe you want to fit it and then remove it for painting or plating, etc. You can begin by drilling a 1/8" hole in one corner or edge of the panel. Hold the panel up to the installation area, align it into position, and use the panel hole as a guide for drilling a 1/8" hole into the body metal. Now, instead of installing a screw or a rivet to hold the panel in place while you drill the remaining holes, you can install a Cleco. Then you move to another location, drill a 1/8" hole through both layers, install another Cleco, and so on. Now, let's say that you want a finished hole size of 1/4". One by one, remove a Cleco and drill that hole to your final size. The Clecos secure the two panels together, holding everything in register while you continue your work. Or, you can use Clecos over and over, using them to temporarily install the panel while you perform other work that requires keeping the panel in exact register.

Another example: let's say that you're restoring a vehicle, and you want to replace a firewall. After you remove the original firewall, you can fit the new firewall into place. Then begin to drill 1/8" holes (where the firewall flange meets the cowling area) through both layers of material, installing

Clecos are very handy tools for fabrication/fitting of sheet material.

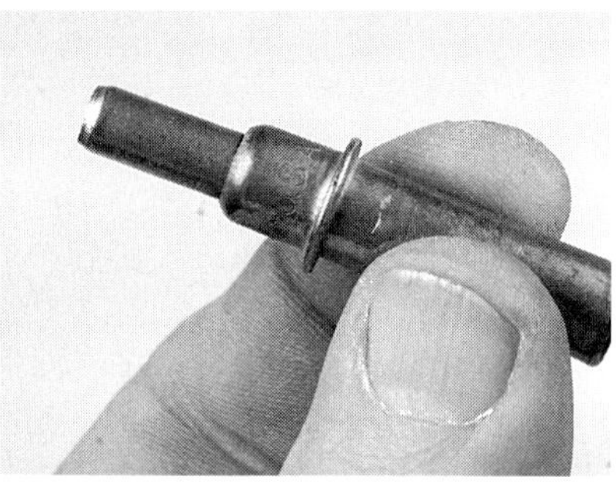

Each Cleco consists of a copper body that houses a spring-loaded plunger and needle set.

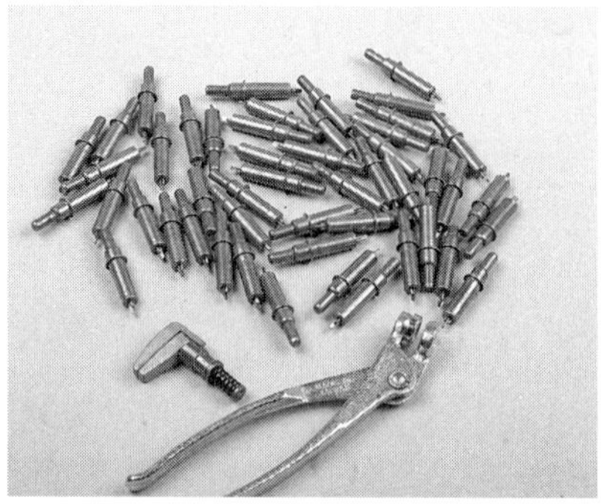

A Cleco kit can be purchased for about $30 or so, which should include pliers, a side clamp and about 25 Clecos. Individual Clecos run about 36 cents each or you can buy them in 10 packs of 25 (250) for about $12. The Cleco pliers run anywhere from about $4 to $10 and the Cleco clamp costs about $1.25. This isn't expensive stuff. Just keep in mind that everybody sells the same products. The only thing that will change is the price, depending on the seller.

The pinch clamp is operated by the Cleco pliers in the same manner as it handles the Cleco. By squeezing the handles, the jaw spreads and can be placed on the workpiece to clamp two sheets together in preparation of drilling the hole and inserting the Cleco.

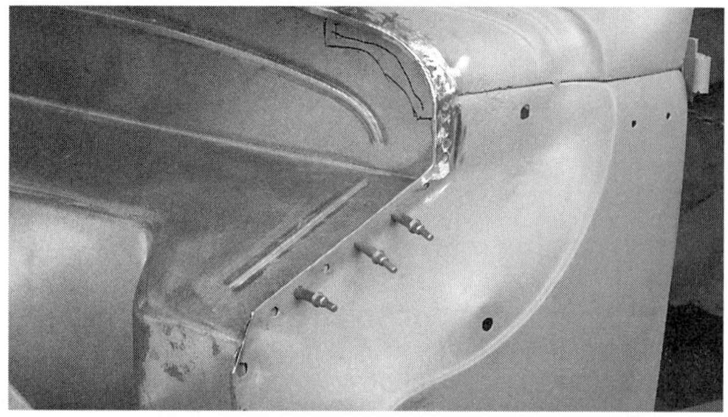

Clecos were used to position this new custom firewall being installed onto a 1939 Ford during a rod restoration. Here we've installed three Clecos to start with. The Clecos act as mini clamps, not only keeping holes in register, but also pulling the two layers of sheet metal together. This avoids the use of vise grips, or the time and trouble of installing temporary bolts and nuts, etc.

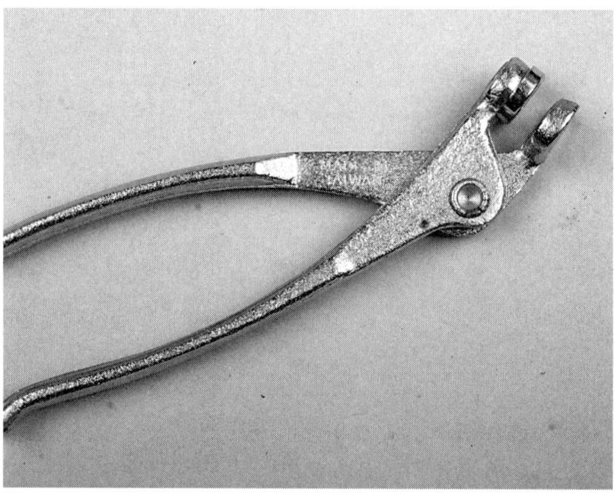

Cleco pliers are simple tool. One jaw features a closed concave surface that accept the Cleco plunger, while the other jaw features an open eyelet that captures the Cleco body's flange.

The pinch clamp (available separately or as part of a kit) features a spring-loaded vise jaw.

Clecos as you go. The Clecos will hold the firewall in place, allowing you to then begin your welding, removing one Cleco at a time as you progress. There are a multitude of uses for Clecos. Think of them as temporary and resusable rivets.

Clecos feature a copper body with an internal spring. At one end is featured a plunger button, and at the other end are a pair of barbed needles, separated by a center sliding pin. As the plunger is depressed, this center divider retracts, allowing the two needles to come together to create a 1/8" diameter needle pack. While the needles are together, the Cleco can be easily inserted into a 1/8" hole. Once the Cleco is fully seated, the plunger is released, causing the center pin to extend, spreading the two needles apart. This larger-diameter needle pack, along with their outside barbs, holds the Cleco in place and grips the two panels together. To remove the Cleco,

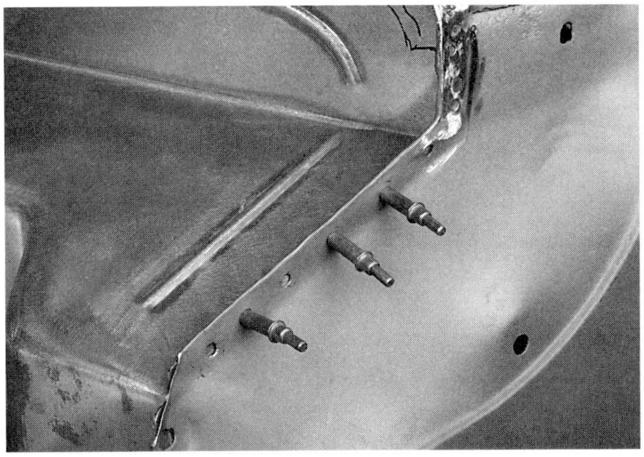

Once all of the Clecos have been installed, they can be removed one at a time for welding the sheet steel together. Having the entire firewall held in place and registered eases fabrication, since you know the two pieces aren't going to move in relation to each other. Think of them as temporary, reusable rivets. But instead of rivets, they don't need to be drilled out. Simply compress the plunger with the Cleco pliers and pull the Cleco needles out of the hole.

simply depress the plunger and remove the Cleco.

A special Cleco tool (pliers) is required, which is included in a Cleco kit. This is a simple pair of pliers that features an enclosed cup jaw at the top (to secure and push on the plunger) and a bottom jaw that features an open-end eyelet. The body of the Cleco slips into that open-end jaw. An exterior flange on the Cleco body rests into a recessed are in the lower plier jaw. It's a very simple tool to use.

Clecos are available in a variety of sizes and styles. For automotive use, the standard is the 1/8" (#30). This requires drilling a 1/8" hole. The needle pack provides a grip length of 0 to 1/4" (panel thickness).

Other common sizes include 3/32" with a 0–1/4" grip; 5/32" with a 0–1/4" grip; 3/16" with a 0–1/4" grip; and 1/4" with a 0–1/4" grip.

Still others are available with a 1/2" grip range (called extended series Clecos), mini Clecos that offer short bodies for tight locations, and even finger-operated Clecos that feature big wing-type finger grips that don't require Cleco pliers.

Once you buy a Cleco set, you'll have tons of fun as you discover new applications for its use. This is the same temporary rivet that the aircraft industry has used for decades. Clecos are commonly used to fit aluminum body panels during the manufacture of fuselages, wings, etc. Clecos are available from any aircraft parts supply shop, or from any of a number of racer suppliers such as Pegasus, Racer Wholesale, etc. Product quality won't vary. From my experience, it's all the same stuff, no matter where you buy the pieces.

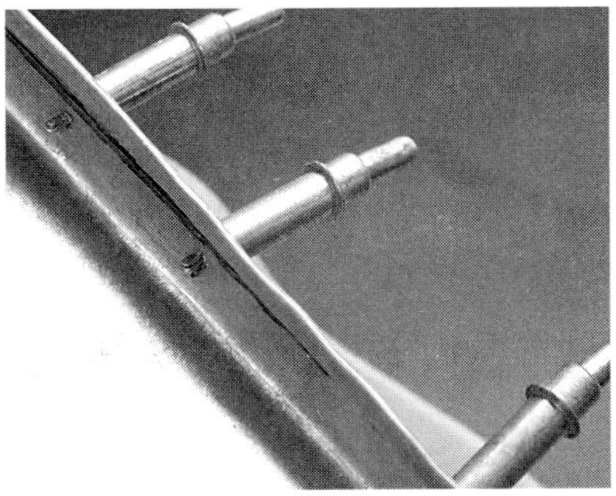

1. This backside view shows the expanded barbed Cleco needles locking the sheet metal together.

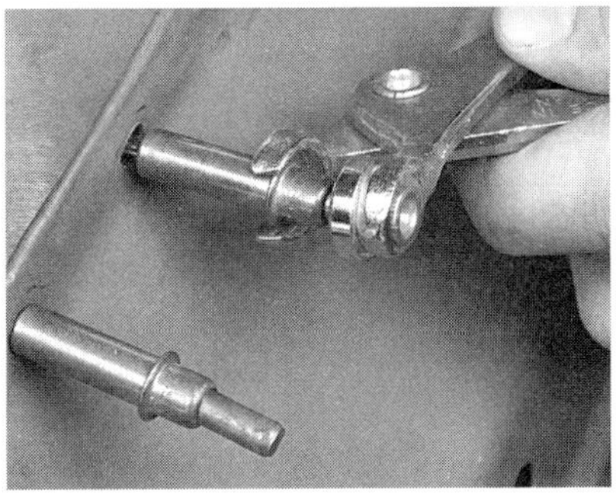

2. With the needles compacted together, slip the needle tip into the hole.

3. Once fully engaged, release and remove the pliers. This is easy stuff.

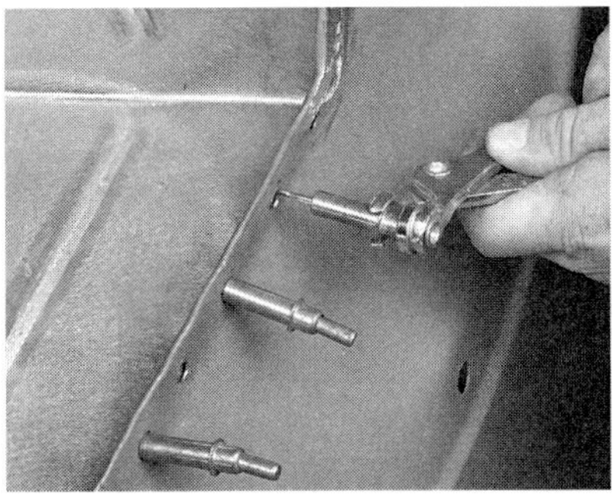

4. With the Cleco secured in the pliers, squeeze the pliers to retract the Cleco needles' center pin, which allows the needles to extend and compact together.

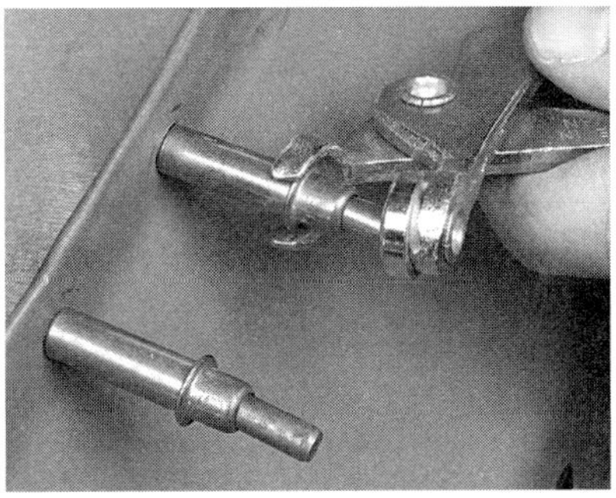

5. Once the needles are fully inserted, begin to release the pliers while still pushing the Cleco against the workpiece. This extends the center divider, forcing the needles to expand on the backside of the workpiece, similar to a rivet expanding.

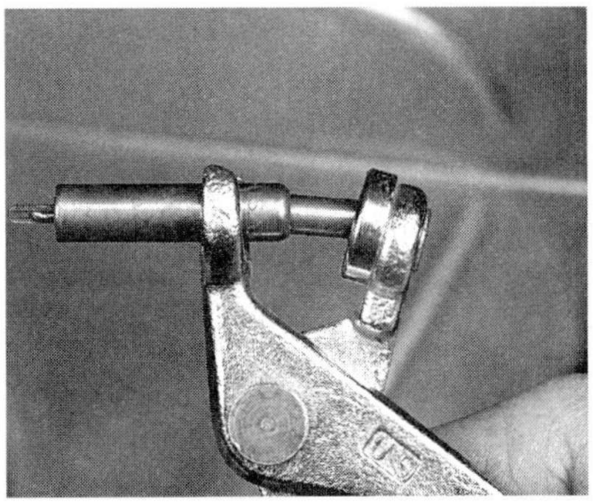

6. A Cleco positioned in the pliers. The Cleco body's flange lip rests against the lower plier jaw recessed seat.

7. With the Cleco fully compressed, the center driver pin (located between the needles) retracts, allowing the two barbed needles to spring together, making the needle package O.D. smaller.

Chapter 10
CLAMPS

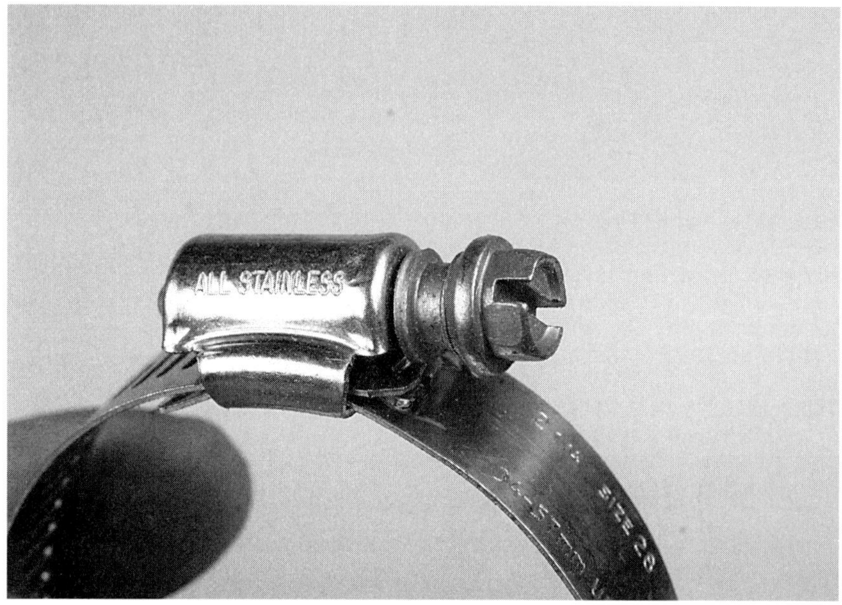

Worm-drive clamps derive their name due to the spiral-threaded screw that engages a series of slots in the band. The majority of worm-drive clamps can be driven with either a hex wrench or a screwdriver.

WORM-DRIVE CLAMPS

Among the many methods of securing a slip-on hose to a fixed male tube, the worm-drive clamp is probably the most common. Worm-drive clamps work similar to a rack and pinion assembly. A toothed, or serrated band runs across an angle-geared pinion bolt. As the bolt is rotated, the clamp is adjusted larger or smaller as needed. Worm-drive clamp styles vary, with specialty designs offered beyond the conventional types that you commonly find at auto parts or hardware stores. Some offer flat inner surfaces, some have double-bead inner surfaces, some are convex to accommodate flange-to-flange mating, etc.

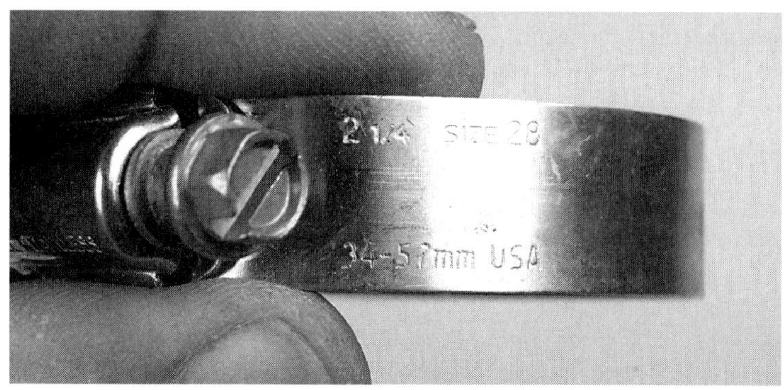

Any worm drive hose clamp selected for automotive use should always be of stainless steel construction. Size range (hose O.D.) is marked on the band. This one is ideal for 2 1/2" hose, but will function within a range of 34 – 57mm, or about 1.5" to 2.5".

CLAMSHELL TUBE COUPLING

This is a way-cool flexible tube connector from Adel-Wiggins, their W991 coupling. This series of clamps (available in a range of sizes) is a clamshell-type clamp that provides one-handed installation, easy on/off operation with secure flip-type latches. The internal design, requiring flanged tube adapters on the two mating tubes, locks the two tubes together and seals the connection with O-ring seals. This is a very popular tube coupling found on the most sophisticated drag, oval track and road race cars. It offers a burst pressure rating of 375 psi.

T-BOLT CLAMPS

T-bolt clamps are so-named due to their design. Instead of a spiral-toothed drive that engages on the band, T-bolt clamps feature a smooth, solid band throughout, with the adjustment handled by a positive-positioned specially-shape bolt. The bolt is T-shaped at one end and features a threaded section at the opposite end. The T-bar end is locked inside a rolled loop, allowing the T bolt to pivot inside that loop, but anchors the T-bar in position. The bolt shank runs through a hollow guide on the opposite side of the band travel, and is secured with a locking nut. This provides a very strong, sturdy and finely-adjustable method of clamping which is much hardier than a worm-drive clamp. The only downside is that T-bolt clamps do not offer the wide adjustment latitude provided by worm-drive clamps. When you order a T-bolt clamp, you must order according to the specific diameter for the application. You'll have a bit of leeway, but not much.

Dual-bead worm drive clamp. The inner shield provides excellent sealing capabilities and protects soft hose surfaces. (Courtesy Murray)

Embossed-style worm drive clamp. The band is non-perforated, providing added protection for soft hose surfaces. Popular in some European OE applications. (Courtesy Murray)

Here's one of my shop's recent restorations. The massive air ducting that feeds the cold air box on this 1964 Ford Fairlane Thunderbolt drag car is secured to the cast aluminum air box snorkels with "T-bolt" clamps. This provides not only a more positive attachment, but a cleaner appearance as compared to traditional worm-drive clamps. Besides, we used T-bolt clamps on a T-Bolt car. I mean, c'mon...how cool is that?

Constant-tension turbo seal worm drive clamps feature a convoluted heat treated band that provides a spring-like tension for thermal cycling hose applications. This example (Murray) features a unique dual bead shield that concentrates band sealing pressure. Flanged edges protect soft-compound hoses. (Courtesy Murray)

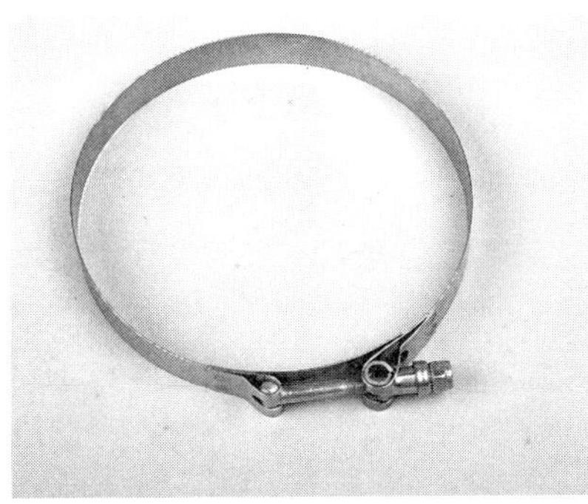

T-bolt clamps must be ordered for the specific size required, as they are not as widely adjustable as worm drive clamps. T-bolt clamps do offer a range, but the range is very limited. This is a more-exacting-fit type of band clamp.

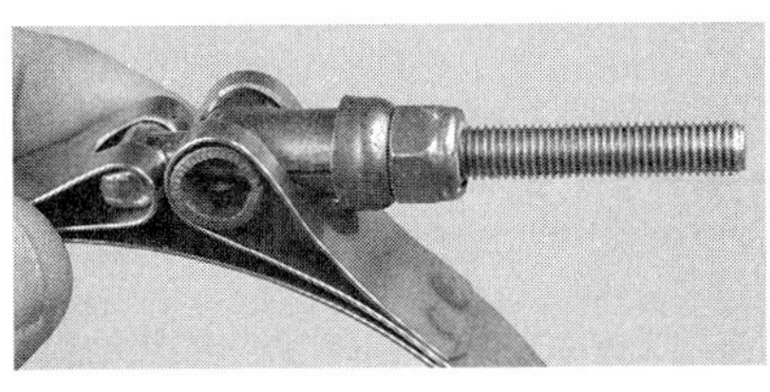

The same T-bolt clamp, tightened to its minor diameter limit. Never tighten a T-bolt clamp to the point where the bolt retaining loops touch, as this creates a dead-stop and will not allow clamping force. T-bolt clamps are marked for a specific recommended size (this one is marked as a 5.5"), and although some adjustment range is possible (maybe as much as 1/2" to 3/4", depending on size), you should always order T-bolt clamps for the specific diameter at hand.

Note the Adel-Wiggins clamshell couplings on the ends of the two hoses dangling from this race engine. The quick-release feature is very handy.

1964 Ford Thunderbolt, restored to original glory. All hose clamps used on this vehicle were T-bolt style.

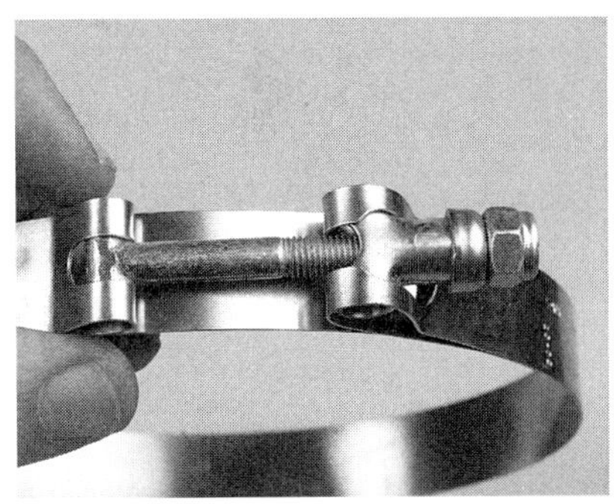

T-bolt clamps are so-named due to the retaining and adjusting design. A T-shaped bolt (seen here) is anchored at the "T" end. The threaded shank passes through a hollow guide, and is secured with a locking-style nut, usually a nylon-locking nut. T-bolt clamps provide superior clamping and adjustment control. This 5.75" T-bolt clamp is shown adjusted at its largest diameter. This clamp will achieve a 5.75" diameter at about the mid-point of the bolt's threads.

T-bolt clamps feature positive hose connections. The edges of the bands are rounded to protect the hose outer surface. T-bolt clamps are an excellent choice for securing turbo air intake plumbing. Choose wide-band clamps for a large footprint.

T-bolt clamp with spring, designed to compensate for expansion and contraction of the hose due to temperature and/or cold hose flow. (Courtesy Murray)

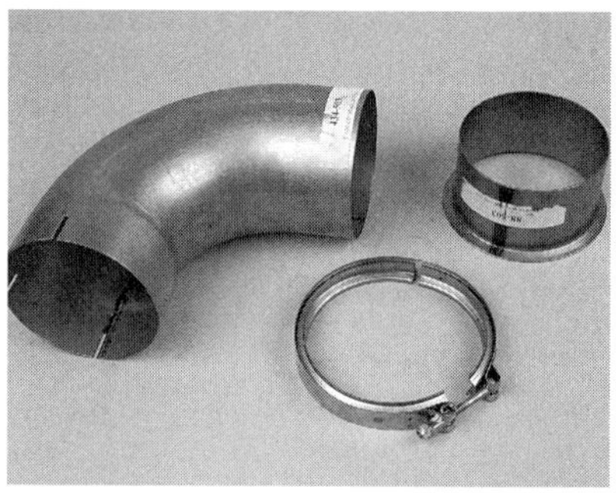

This turbo exhaust pipe will be welded to a flange pipe (upper right). The flange pipe will then be mated to the turbocharger's outlet flange with the V-band T-bolt clamp. Note the female channel on the clamp's inside diameter.

In addition to smooth-bore T-bolt clamps, where the band surface is flat, T-bolt clamps are also available that feature a convex band (large relief groove on the inside band diameter). These are called "V-band" clamps, and are designed to secure two flanges together. For instance, let's say that you have a turbocharger exhaust flange (flat outer face with a tapered outer perimeter) that must mate to a flanged pipe. The V-band clamp will capture both flanges, drawing them together as the band is tightened.

PINCH BAND CLAMPS

Pinch band clamps require a specialty tool to squeeze the clamp and to lock the clamp.

WIRE GRIP CLAMPS

Wire grip clamps (popular with many import cars) feature tempered wire instead of a steel strap. These clamps are often non-reusable.

ADEL CLAMPS

The adel clamp has been used as a generic label for many designs of clamps that are intended to secure hose, fluid lines or electrical harnesses to an adjacent surface, such as bulkheads, firewalls, inner fenders, body bellies, etc. In fact, the name Adel

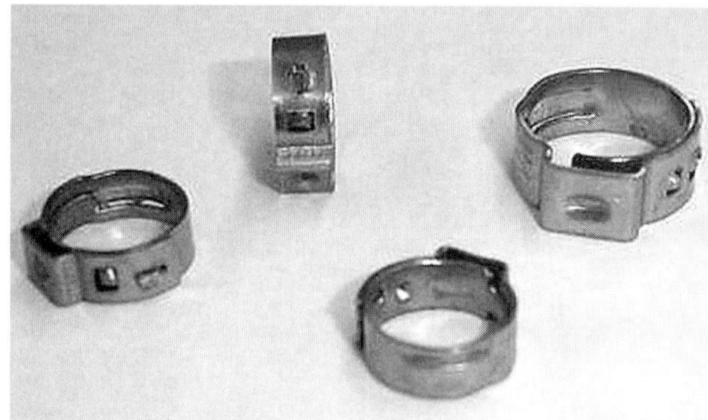

Seamless pinch band clamps suitable for a variety of hose sealing applications. (Courtesy Murray)

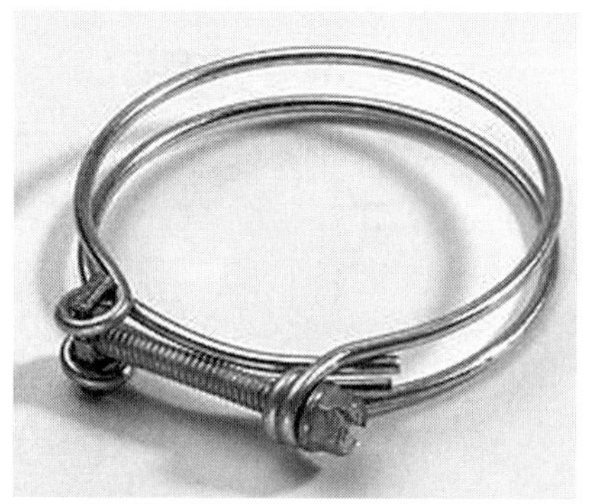

Typical wire grip hose clamp. Features tensioned steel wires that create double-contact sealing pressure points. A machine screw (threaded through a captive rectangular nut) adjusts clamp pressure. (Courtesy Murray)

actually refers to the Adel-Wiggins Group, the primary innovators of these types of clamps (of course, plenty of other makers have copies this basic loop-bracket design).

"Adel" clamps are available in a variety of shapes, materials and sizes.

Types of clamps include loop clamps (the type you're probably most familiar with and which are commonly available at hardware stores). Other styles include locking type (this style simply features an additional extended mounting bracket), block clamps (where the loop is not rounded, but rather is formed in a more oval shape to accommodate non-concentric objects) and saddle clamps (U-shaped, with mounting tabs on each leg).

Adel-Wiggins also produces a wide range of specialty, aerospace, aircraft and marine designs.

This type of clamp is available with or without insulation. Insulation material (wrapped around the band) protects hoses or wires from chaffing and cuts. The bands are constructed of aluminum, corrosion-resistant steel or low carbon steel, depending on the application.

Clamp band coatings (cushions) are diverse. These include ethylene propylene (usually purple), which is resistant to phosphate-ester hydraulic fluid, but is not resistant to petroleum products. Nitrile coatings are resistant to fuels but not resistant to

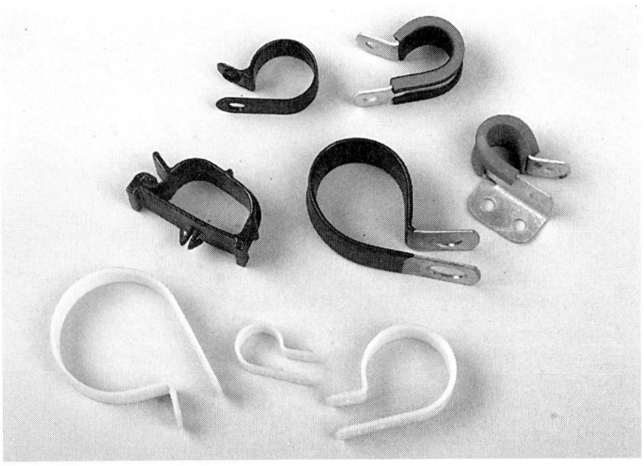

An assortment of common adel, or "loop" type clamps. This style of clamp is available in plastics (often ABS or nylon), as well as steel and aluminum.

phosphate-ester-based fluids. Color is usually yellow. Chloroprene is a general-purpose coating for use with petroleum-based hydraulic fluids. Color is usually black with a blue stripe. Neither nitrile or chloroprene should be used with titanium tubing. Silicone coatings are intended for phosphate-ester based fluids and other synthetic fluids, but not for petroleum based fluids. Color is usually natural white or red. Fluorosilicone coating is designed for use in elevated temperatures in petroleum based fluid environments, but are not resistant to phosphate-ester based fluids. Usually blue in color.

With that said, most of the adel-type clamps you'll find are insulated with black rubber, the exact composition of which normally isn't identified (if you ask your local hardware store or speed shop if the insulator is silicone or chloroprene, they'll probably look at you as if you were from Mars). Unless you plan to soak the clamps in various fluids, don't worry about it. Most of the time, you'll simply use these clamps under the hood, in the trunk or in the interior to secure runs of wires or hoses. If something catastrophic happens that results in the clamps being soaked in oil or fuel, clamp insulator damage will be the least of your worries.

HOSE SEPARATORS AND LINE CLAMPS

Hose separators offer exactly that—a clean method of organizing and separating two hoses that run in parallel. The custom aftermarket offers these in machined aluminum from a variety of sources. These are ordered based on the hole diameters. They're offered with two identical-sized holes, as well as in hole-size combinations, where two dissimilar-diameter hoses must run next to each other. These separators can also be used to clamp a hose or harness to a round bar, etc. Just use your imagination.

This aluminum loop clamp features a wrap-around rubber cushion to protect tubes, hoses or wires.

Plastic loop clamps are commonly available in both black and white. On rare occasions, I've seen them in other colors. Note that the black clamp shown here features a short locating shoulder on the bottom of the upper tab. This helps to guide the two holes together for a more exacting fit. Unfortunately, this style is not commonly found at the local hardware store. Plastic loop clamps are available in a wide range of sizes, from 1/8" all the way up to 2.5" (larger sizes may be available as well). It's a good idea to buy a kit assortment.

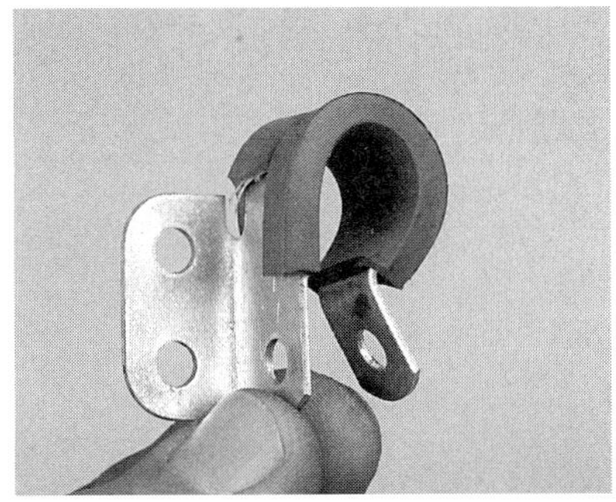

This 90-degree loop clamp offers options regarding mounting locations, such as the need to run wires or tubes across and 90-degrees to square tubing or at the edge of a flat wall.

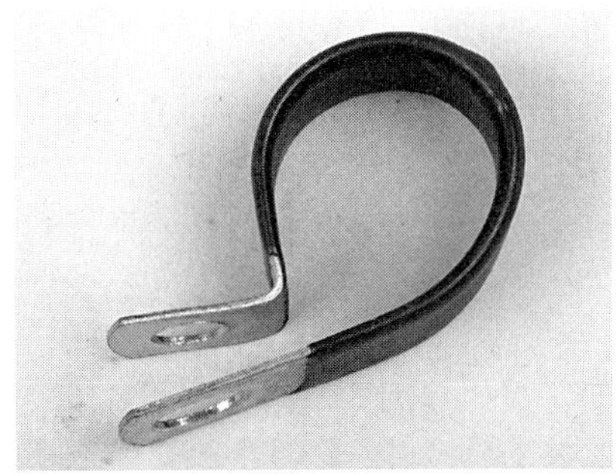

This is a steel loop clamp, dip-coated with plastic. These are very inexpensive, but they don't offer the cushion or abrasion resistance of other styles.

This is an aluminum line clamp that features the basic adel, or loop clamp profile. It's really slick, featuring a flat mounting surface and sturdy clamshell construction. Available in a variety of sizes. These are great for securing runs of brake line tubing along a frame or along flat wall surfaces.

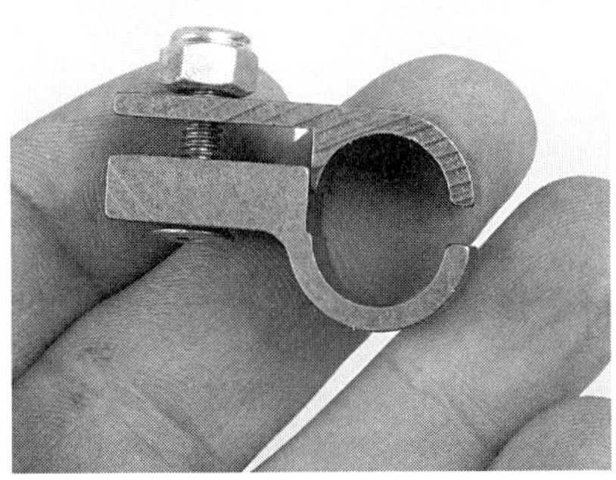

Construction of these line clamps is wonderful. You can see here how the two pieces register together for a positive assembly. If you mount this to a flat wall (sheet metal, fiberglass, etc.), you can secure the machine screw with the nut. If you're attaching the clamp to a frame, where you don't have ready access to the backside of the frame wall, simply tap the frame hole for the size of the machine screw and thread the screw into the frame instead of messing with the nut. Machine screw sizes for these clamps will vary depending on clamp size, but common sizes include 8x32 and 10x32.

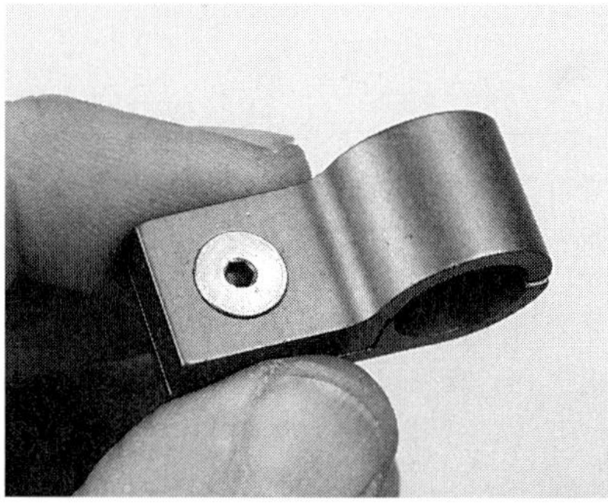

I just love these little guys. They do the job and they look clean and tidy. Here, I'm holding a 3/8" clamp, designed to accommodate a 3/8" O.D. tube. Note: If you're using a hard clamp like this for tubing, always use the correct size to match the tube. You don't want to pinch a tube, nor do you want the tube to rattle around in the clamp. For wiring runs, the clamp can be slightly larger than the wire diameter.

Here an aluminum hose separator is used on a motorcycle handlebar to secure brake hose and wire harness. The larger 1" hole clamps to the bar, while the smaller 3/4" hole accommodates the hose and wire.

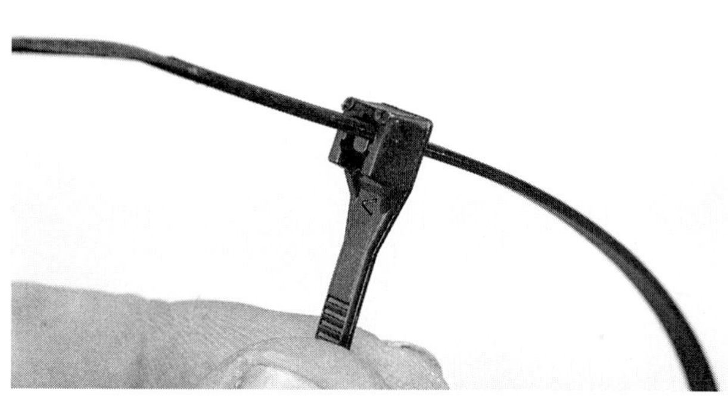

The locking tab is tapered at the entrance and "toothed" at the exit, which provides a secure lock of the strap position. You can pull it tighter, but it won't back off. However, be aware that not all wire ties are created equal. Some cheap imports are so poorly made that they won't positively engage and lock. Only buy known-brand wire ties to avoid this problem. We once did a job for Porsche, and while in the field, we bought a dozen bags of generic wire ties at a retail chain store, only to find that they were all junk and would not lock. Don't assume that they're all of good quality.

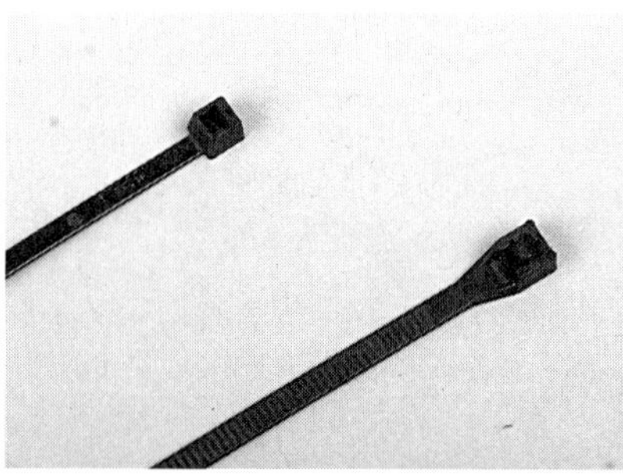

We're all familiar with nylon tie straps (often called wire ties).These locking straps are useful in a variety of applications, and are offered in lengths and widths to suit any need. While the most common colors are white and black, they're also available in just about every color of the rainbow.

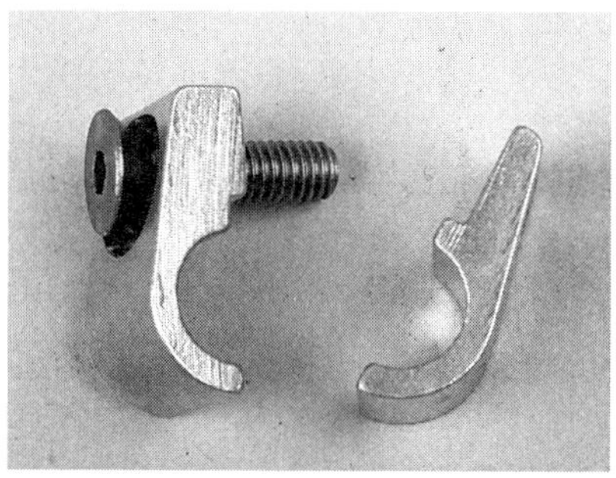

Aluminum line clamp separated. The two halves are secured together, and to a mounting surface, with a stainless steel flat-top machine screw. When assembled, this style features a tapered, or teardrop side profile.

This firewall sealing grommet system from Seals-It is very cool. This allows you to very neatly pass a hose assembly through a bulkhead while sealing it at the same time. And the really neat feature is that the entire hose assembly (with hose end attached) will pass through. A variety of firewall sealing grommet assemblies are offered. (Courtesy Seals-It)

"Line clamp" is the generic term used to describe non-adjustable solid-material clamps that are surface-mounted and used to secure runs of hard lines (fuel lines, brake lines, etc.) or wiring harnesses along a frame or along body material. For the automotive aftermarket, these are commonly available in plastic and in aluminum. The aluminum line clamps are popular among street rodders for their "billet" appearance. These line clamps are offered in two basic styles. One-piece clamps feature an open "hook" (once installed onto a flat surface, the hook secures the line or wire against the flat surface). The two-piece eyelet style is shaped somewhat like the letter "P" when assembled. This locates the eyelet (that carries the line or wire) to be offset from the mounting surface.

Note: The styles of clamps discussed here are the most commonly used in automotive applications, but by no means represent all styles. Clamp designs are incredibly varied. The more you look, the more odd and unique clamp styles you'll find, especially if you start looking at hardware sources for aircraft, marine and industrial applications.

FIREWALL SEALS

Specialty grommets are available where plumbing or wires run through a bulkhead. The rubber seals out moisture and dust, and insulates the plumbing or wires from sharp metal edges.

Chapter 11
PIPE THREADS

NPT thread is tapered. Thread diameter is smaller at the tip and increases as thread length approaches the top, or drive head.

PIPE THREAD

Pipe threads are designed to provide connections for the transfer of fluids or gasses. A variety of pipe thread designs exist, but for automotive purposes, we'll discuss only those that apply. These include NPT (National Pipe Thread), straight thread (sometimes referred to as NPSM, or National Standard Free-Fitting Straight Mechanical Pipe thread) and BSPT (British Standard Taper Pipe Thread).

Both NPT and BSPT threads are tapered, which theoretically aids in sealing. Straight pipe thread has no body taper. Regardless, when installing either type of pipe thread connections or fittings, always apply a Teflon-based thread sealer to assure a leak-proof connection. The exceptions: where a straight-thread line nut will secure a flared tube (as in brake lines or some nitrous lines). In those cases, the flare and its seat creates the seal, so no additional sealing compound is needed. In cases where a straight-thread pipe thread fitting features an O-ring, additional sealing compound may not be needed.

Pipe Size Does Not Reflect Thread Size

Pipe thread is sized based on the inside dimension of the tube or flow path, not on the thread dimension itself, unlike bolts or studs. Many people are easily confused when looking at pipe thread, and that's understandable. After all, when you look at a hole that looks about 7/16" in diameter, and somebody says that you need a 1/4" pipe fitting, it doesn't look right, because you can plainly see that the hole is larger than 1/4". Just remember that pipe thread sizes are based on the theoretical diameter of the flow path, not on the thread size itself.

NPT Size	Measured Thread O.D.
1/8"	0.400"
1/4"	0.520"
3/8"	0.670"
1/2"	0.825"
3/4"	1.030"

NPT Thread Diameter

Since tapered pipe thread is, well, tapered, measurements will vary depending on where you measure the thread diameter. As an example, I've measured five sizes of male NPT fittings, using a dial caliper, at about the halfway point on the thread length. This will give you a close idea of how to identify NPT size if you try to perform a measurement. I measured all of these at the major thread diameter (measuring across the thread peaks).

NPT (tapered) thread taps are readily available, but use caution when tapping tapered threads. If you drive the tap too deeply, you'll create a threaded hole that may be too deep (the head of the fitting might bottom out before the threads have a chance to seal). Run the tap about halfway, remove the tap and try your fitting. Tap deeper if needed.

BRITISH PIPE THREADS

In addition to our commonly found NPT sizes, other foreign formats exist that find their way onto U.S. shores, in many import vehicles. For instance, many folks who plan to hot-rod Honda engines run into odd pipe threads that they don't recognize. In many cases, they end up forcing NPT

Here are examples of NPT thread (at left is a 1/8" NPT plug, followed by AN to NPT adapters. All NPT thread is shown here at the top of each adapter). Starting at the left is 1/8" NPT, 1/4" NPT, 3/8" NPT, 1/2" NPT and 3/4" NPT

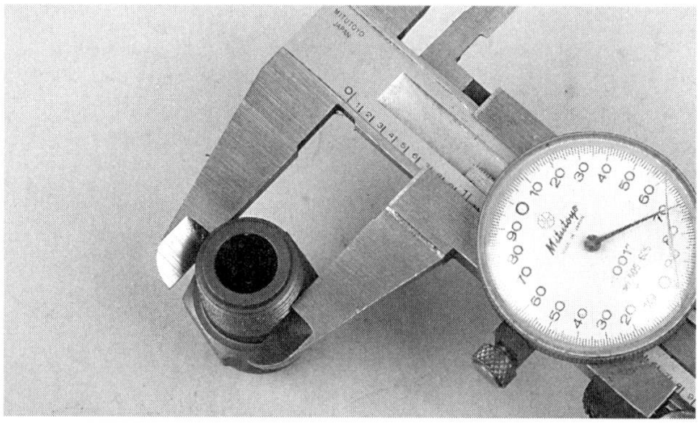

In order to provide examples (see the accompanying short chart of thread diameters), I measured each size with dial calipers at about the midpoint of thread length.

fittings into threaded holes, where they sometimes function without leaks; and sometimes they don't work and leak like crazy. It's best to identify what you're dealing with so that you don't mess up the OE female threaded holes.

BSP threads (British Standard Pipe) are available in two types, including BSPT (British Standard Pipe Taper) and BSPP (British Standard Pipe Parallel). These are two completely different types and are not interchangeable.

BSPT threads are tapered, similar to NPT. However, thread pitch differs from that on NPT.

NPT THREAD SIZES

(Note fitting I.D. and –AN size compatibility for flow)

Pipe Thread Size	Threads Per Inch	Theoretical Fitting I.D.	Closest –AN Fitting
1/16"	27	1/16"	
1/8"	27	1/8"	4
1/4"	18	1/4"	6
3/8"	18	3/8"	8
1/2"	14	1/2"	10
3/4"	14	3/4"	12
1"	11 1/2	1"	16

(When you need to match flow size, consider the NPT fitting I.D. when using –AN to NPT adapters. For example, when adapting a –10 AN hose end to a AN-to-NPT adapter, try to use an adapter that features a 1/2" NPT thread. Using a smaller NPT, say, 3/8" NPT, will begin to restrict flow when used with a –10 AN hose assembly.)

NPT THREAD SIZE AND DRILL SIZE (FOR TAPPING)

NPT Size (inches)	Threads/Inch	Drill Size (inches)
1/16	27	C (0.2420)
1/8	27	Q (0.3320)
1/4	18	7/16
3/8	18	37/64
1/2	14	45/64
3/4	14	29/32
1	11	1 9/32

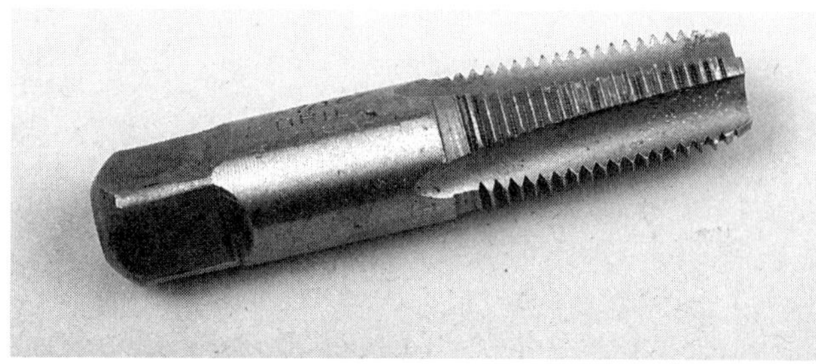

A tapered (NPT) tap. Don't run the tap to full depth right off the bat. Tap about halfway, then check the fit of your fitting. Tap further if needed.

BSPP threads are "straight" or parallel threads, and require an O-ring seal.

Note: Just because these are "foreign" threads, don't assume that they're metric, because they're not. Thread pitch is measured in the number of threads per inch, just like we see in inch format threads.

SEALANTS: A WORD TO THE WISE

When selecting a thread sealant, don't grab just any sealant that may be handy. Certain applications call for certain types of sealant. Teflon pipe sealant is available in both a paste and in tape form. While the tape may be more convenient (no mess), you should not use it in certain jobs, such as fuel or nitrous line connections. The reason: when using Teflon tape, if you're not extremely careful, small pieces of excess tape (maybe a piece hanging off the end of a thread) can dislodge and plug a tiny orifice (jet, transfer port, etc.). When dealing with fuel, whether it's gas, diesel, nitrous oxide, etc., use Teflon paste instead of tape.

In addition to aiding in sealing, the use of pipe thread sealant offers other benefits, including reducing the chance of thread galling, permitting easier joint tightening, and making it easier to

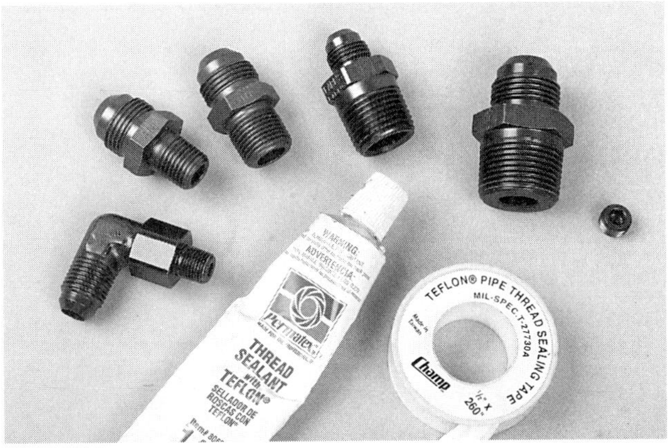

Sealing NPT threads can be handled with Teflon thread sealant paste or with Teflon tape.

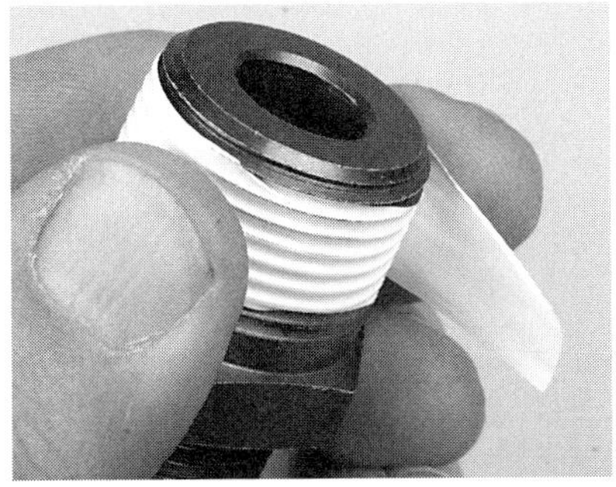

This wrap is almost done. As you can see, since the fitting features right-hand thread, as the fitting is threaded into its hole, the end of the tape wrap will face away from the direction of installation. If you wrap the tape following thread direction, the exposed end may try to unravel. Generally, two layers of wrap should suffice.

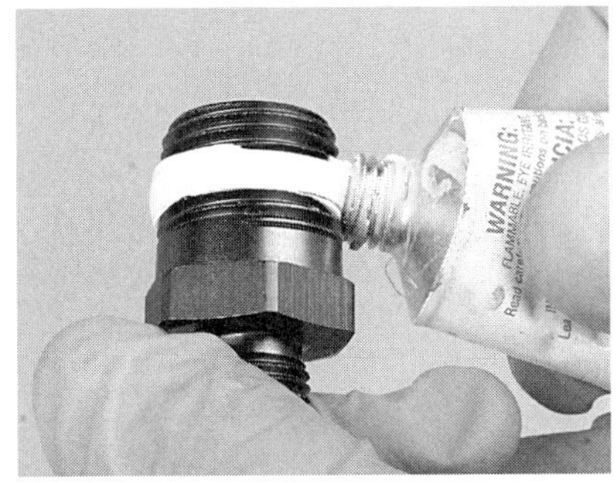

Applying Teflon sealant paste is easy but it can be messy. A little bit goes a long way. Rotate the fitting while applying the paste in order to coat the entire circumference of the threads. Once the paste has been applied, work it into the threads and rotate the fitting to spread the paste around the circumference. Make sure to wipe any excess paste off of the tip area.

Don't tear Teflon tape. Instead, cut it cleanly with a razor or scissors. Wrap Teflon tape opposite thread direction. As you view the end (the end of the thread that will first enter the hole), wrap the tape in a clockwise direction. This will prevent the tape from unwrapping as you thread the fitting into the hole.

A pair of 1/8" NPT plugs on a fuel rail. Whenever sealing NPT threads that will be exposed to fuel or oil, it's best to use Teflon pipe paste instead of Teflon tape. The reason: if any excess sealer enters the fluid system (which you should never allow to happen, by the way), small crumbs of tape can easily clog small metering orifices. You can use either type, but just to be safe, don't use tape on fuel or oil applications.

Here we used NPT plugs on the water pump and intake manifold. We always seal all NPT threads with Teflon tape or paste, but for show cars, we're very careful not to let any show if we can help it. For a race engine, revealing a bit of thread sealant is a good thing, since if you see a bit around the attachment area, you don't need to wonder if you remembered to apply the sealer.

disassemble the joint or fitting in the future.

Galling occurs when two pieces of metal "weld" themselves together due to friction between the male and female threads. This is very common when dealing with alloys such as aluminum and stainless steel. Common instances where galling is an issue involves the installation of aluminum fittings or threaded plugs in an aluminum component, and when stainless steel fasteners are threaded into steel or aluminum. If you're using stainless bolts or nuts, it's always a good idea to apply a small bit of anti-seize on the threads first. If it's a plumbing connection (where fluids or gasses or air must be sealed), use a pipe thread sealant.

When applying Teflon tape to a fitting or adapter, wind the tape in the opposite clock direction as the threads. If you wind the tape in the same direction as the thread, you provide the tape with a great excuse to unravel as you thread the adapter into its hole.

By the way, when you're using Teflon tape, cut it to length using a clean, sharp blade or scissors. Granted, it tears apart easily, but when you stretch and tear it, you generate a sloppy end, with strands of Teflon hanging off the end. Not only is this sloppy, but if you're not careful to place them over the threads and squish them down with your finger, there's always a chance that these strands can break loose and contaminate the fluid supply.

PIPE THREAD

(British Standard Pipe Parallel, BSPP/BSFF)

PIPE THREAD

(Japanese Pipe Parallel, PF)

MARYLAND METRICS -- THREAD DATA CHART:									
British Pipe Thread (Whitworth Form) -- Parallel (BSPP/BSPF)									
Thread Designation BSPP/BSPF	Nominal Size BSPP/BSPF	Thread Form Type	Major Diameter mm d=D	Pitch mm p	Threads per inch tpi	Pitch Diameter mm d2=D2	Minor Diameter Male Thd. d3	Thread Height H1	Tap Drill Diameter mm
G 1/8	1/8"	BSPP/BSPF	9.728	0.907	28	9.147	8.566	0.581	8.7
G 1/4	1/4"	BSPP/BSPF	13.157	1.337	19	12.301	11.445	0.856	11.6
G 3/8	3/8"	BSPP/BSPF	16.662	1.337	19	15.806	14.95	0.856	15
G 1/2	1/2"	BSPP/BSPF	20.955	1.814	14	19.793	18.631	1.162	19
G 5/8	5/8"	BSPP/BSPF	22.911	1.814	14	21.749	20.587	1.162	20.75
G 3/4	3/4"	BSPP/BSPF	26.441	1.814	14	25.279	24.117	1.162	24.5
G 7/8	7/8"	BSPP/BSPF	30.201	1.814	14	29.039	27.877	1.162	28
G 1	1"	BSPP/BSPF	33.249	2.309	11	31.77	30.291	1.479	30.5
G 1 1/8	1 1/8"	BSPP/BSPF	37.897	2.309	11	36.418	34.939	1.479	35
G 1 1/4	1 1/4"	BSPP/BSPF	41.91	2.309	11	40.431	38.952	1.479	39.5
G 1 3/8	1 3/8"	BSPP/BSPF	44.323	2.309	11	42.844	41.365	1.479	41.5
G 1 1/2	1 1/2"	BSPP/BSPF	47.803	2.309	11	46.324	44.845	1.479	45
G 1 3/4	1 3/4"	BSPP/BSPF	53.746	2.309	11	52.267	50.788	1.479	51
G 2	2"	BSPP/BSPF	59.614	2.309	11	58.135	56.656	1.479	57
G 2 1/4	2 1/4"	BSPP/BSPF	65.71	2.309	11	64.231	62.752	1.479	63
G 2 1/2	2 1/2"	BSPP/BSPF	75.184	2.309	11	73.705	72.226	1.479	72.5
G 2 3/4	2 3/4"	BSPP/BSPF	81.534	2.309	11	80.055	78.576	1.479	79
G 3	3"	BSPP/BSPF	87.884	2.309	11	86.405	84.926	1.479	85.5
G 3 1/4	3 1/4"	BSPP/BSPF	93.98	2.309	11	92.501	91.022	1.479	91
G 3 1/2	3 1/2"	BSPP/BSPF	100.33	2.309	11	98.351	97.372	1.479	97.75
G 3 3/4	3 3/4"	BSPP/BSPF	106.68	2.309	11	105.201	103.722	1.479	104
G 4	4"	BSPP/BSPF	113.03	2.309	11	111.55	110.072	1.479	110.5
G 4 1/2	4 1/2"	BSPP/BSPF	125.73	2.309	11	124.251	122.772	1.479	123
G 5	5"	BSPP/BSPF	138.43	2.309	11	136.951	135.472	1.479	136
G 5 1/2	5 1/2"	BSPP/BSPF	151.13	2.309	11	149.651	148.172	1.479	148.5
G 6	6"	BSPP/BSPF	163.83	2.309	11	162.351	160.872	1.479	161.5

To make matters even more confusing, the British pipe threads are also referred to as Japanese pipe threads. BSPP is the same as JPP (Japanese Pipe Parallel, sometimes called PF); and BSPT is the same as Japanese Pipe Taper (JPT or PT).

PIPE THREAD

(British Standard Pipe Taper, BSPT)

PIPE THREAD

(Japanese Pipe Taper, PT)

MARYLAND METRICS -- THREAD DATA CHART:										
British Pipe Thread (Whitworth Form) -- Taper (BSPT)										
Thread Designation BSPT	Nominal Size BSPT	Thread Form Type	Major Diameter mm d=D	Pitch mm p	Threads per inch tpi	Pitch Diameter mm d2=D2	Minor Diameter Male Thd. d3	Thread Height H1	Rounding r	Tap Drill Diameter mm
R 1/16	1/16"	BSPT	7.723	0.907	28	7.142	6.561	0.581	0.125	6.4
R 1/8	1/8"	BSPT	9.728	0.907	28	9.147	8.566	0.581	0.125	8.4
R 1/4	1/4"	BSPT	13.157	1.337	19	12.301	11.445	0.856	0.184	11.2
R 3/8	3/8"	BSPT	16.662	1.337	19	15.806	14.950	0.856	0.184	14.75
R 1/2	1/2"	BSPT	20.955	1.814	14	19.793	18.631	1.162	0.249	18.25
R 3/4	3/4"	BSPT	26.441	1.814	14	25.279	24.117	1.162	0.249	23.75
R 1	1"	BSPT	33.249	2.309	11	31.77	30.291	1.479	0.317	30
R 1 1/4	1 1/4"	BSPT	41.910	2.309	11	40.431	38.952	1.479	0.317	38.5
R 1 1/2	1 1/2"	BSPT	47.803	2.309	11	46.324	44.845	1.479	0.317	44.5
R 2	2"	BSPT	59.614	2.309	11	58.135	56.656	1.479	0.317	56
R 2 1/2	2 1/2"	BSPT	75.184	2.309	11	73.705	72.226	1.479	0.317	71
R 3	3"	BSPT	87.884	2.309	11	86.405	84.926	1.479	0.317	85.5
R 4	4"	BSPT	113.030	2.309	11	111.551	110.072	1.479	0.317	110.5
R 5	5"	BSPT	138.430	2.309	11	136.951	135.472	1.479	0.317	136
R 6	6"	BSPT	163.830	2.309	11	162.351	160.872	1.479	0.317	161.5

To make matters even more confusing, the British pipe threads are also referred to as Japanese pipe threads. BSPP is the same as JPP (Japanese Pipe Parallel, sometimes called PF); and BSPT is the same as Japanese Pipe Taper (JPT or PT).

Chapter 12

FLUID PLUMBING

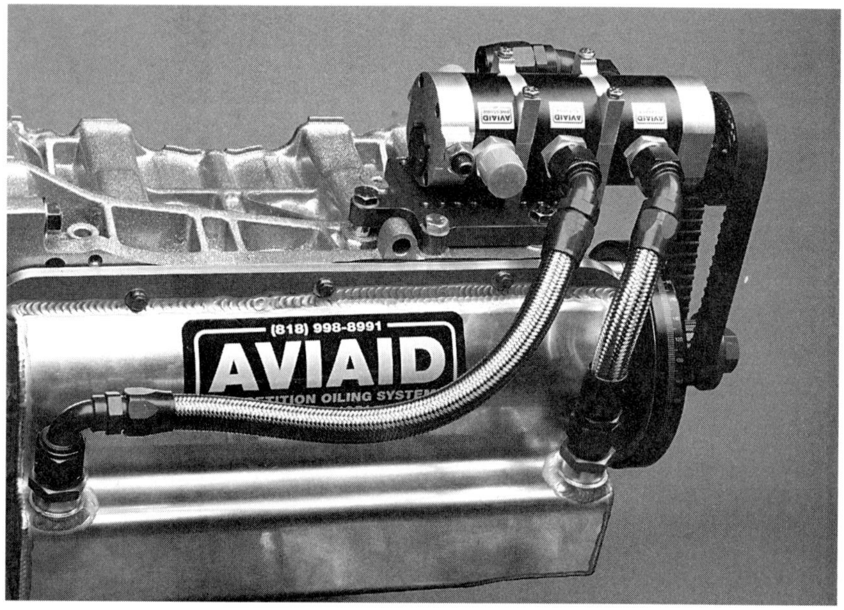

Here's a 3-stage dry sump pump fitted to a Honda race engine (the engine block was rotated on the stand to better show the pump and pan). A pair of –12 AN hoses connect the dry sump pan to the pump.

ENGINE OILING SYSTEMS

A street engine and many race engines utilize a wet sump system to deliver pressurized oil to the engine's internal rotating and reciprocating assemblies. Engine oil is stored in the oil pan sump (the big reservoir section of the pan). The oil is picked up by a mechanically driven oil pump, which obtains the sump's oil via a submerged oil pickup. As the pump rotates (the pump runs whenever the engine runs, since it's driven by the crank or indirectly via the distributor drive), oil is pressurized and distributed from this main source throughout the engine through the oil passages in the block, crank journals, main bearings, rod bearings, and eventually, to the valvetrain. Although this system does deliver oil to all required areas, the oil must be pushed through all of these passages to eventually route to the upper end of the engine. The delivered oil then is free to drain back to the sump, with delivery and drain-back serving as an ongoing cycle during engine operation.

Dry Sump Systems

A dry sump system takes a different, and much more direct approach. This uses an externally mounted (and usually belt-driven) oil pump, which takes advantage of external plumbing to directly deliver oil to the engine. The oil pan does not feature a sump reservoir. Instead, the dry sump pan serves only as a containment cover and as a location from which oil can be returned to the reservoir. The oil reservoir (the oil supply) is a remote container (usually mounted in the cockpit of the race vehicle). In very basic terms, the external pump pulls oil from the remote reservoir, and then pushes the oil (under pressure) to wherever directed by the plumbing. In this manner, oil can be directly sent to the main bearings through a block port, and oil can be delivered directly to the valvetrain with one or two plumbed hoses (depending on the application), oil can be directly delivered to a turbocharger, etc. Dry sump pumps are offered in stages, depending on how many direct-delivery and return (scavenge) routes are required. The advantages include immediate oil delivery to specific areas, and elimination of oil starvation caused by angle and centrifugal forces (when the race car accelerates, brakes, or turns, which may momentarily expose an in-the-pan oil pump pickup, allowing air (not oil) to travel throughout the block. In other words, the potential for oil aeration is eliminated. Since the scavenge sections of the dry sump pump pull oil from the dry sump pan, this creates a vacuum, pulling excess oil from the surfaces of the crank and rods, reducing parasitic drag (which means more efficient use of engine power).

Plumbing A Dry Sump System—When plumbing a dry sump system, size requirements may vary somewhat, but a general rule of thumb is to use –10, –12 and –16 hose, fitting and hose end sizes (for some turbo applications, a –6 hose size may be recommended). Generally, a larger size is required for the return hose that runs out from the pump to the remote oil reservoir (for instance, if –12 hoses are used for all feeds, a –16 hose would be used for oil return to the reservoir).

DRY SUMP ADVANTAGES

- More consistent oil pressure
- Oil isn't pushed away from the oil pickup during severe turns, acceleration or braking, as may occur with a wet sump system
- Adjustable oil pressure
- Increased oil system capacity
- Shorter profile pan, allowing engine to be mounted lower in chassis for lower center of gravity.
- Increase in engine power
- Positive oil delivery to critical areas of the engine
- Cooler oil supply (since the oil is quickly returned to the remote reservoir instead of being stored in the hot oil pan)

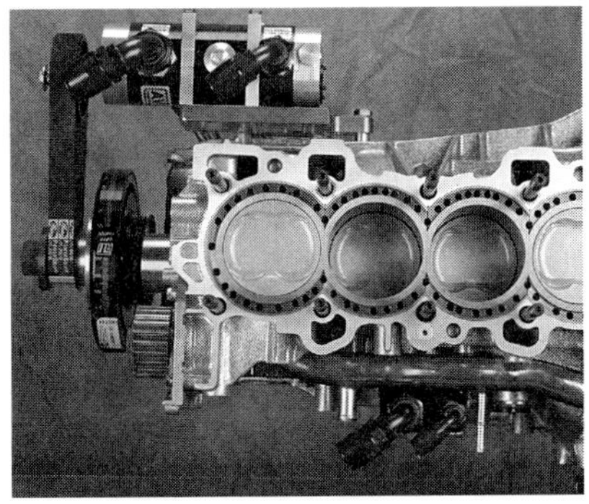

This overhead view shows the pump side-mounted to the block. The toothed Gilmer belt connects the small diameter toothed crank pulley to the pump pulley. The dry sump pump runs at approximately half of the crank speed. Notice the two hose ends on the top of the pump. The larger –16 will be plumbed to the upper port of the reservoir, serving to return oil from the pump's scavenge back to the tank. The –12 will be plumbed to the block's main oil port to feed the engine.

The –16 hose shown here plumbs scavenged oil from the pump to the remote reservoir. A fitting/hose connection at the bottom of the reservoir is the feed line from the reservoir to the pump. Note the oil delivery –12 hose to the oil filter adapter on the block. The reservoir shown here was mounted on our shop's engine stand for display purposes. The reservoir will be mounted in the race car cockpit area (obviously with longer hoses).

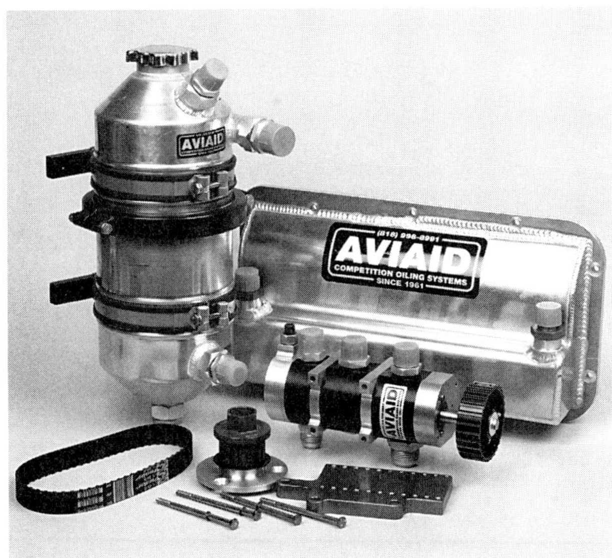

A dry sump "kit" including reservoir, pan, pump, belt, crank pulley and pump mounting bracket. The angled fitting at the top of the reservoir will accept a breather.

While a dry sump oil pan will feature threaded bungs to accept suction line fittings (where drainback oil is pulled by the pump), plumbing to the block or other areas may require drilling and tapping or the purchase of special adapters. Typically, if an adapter is needed to feed oil to the block's main galley (at the stock oil filter location, for example), aftermarket adapters are readily available for all popular blocks. By the way, if you're using the OE filter location for oil feed, you'll need to remotely locate an oil filter.

An externally mounted dry sump oil pump is most commonly driven by a notched belt (toothed belt often referred to as a Gilmer belt), directly from the crankshaft snout (a pulley is installed on the crank, with the belt connecting the crank pulley to the pump pulley). Dry sump pumps, via the proper pulley diameters, are generally driven at about half of crankshaft speed. Dry sump pumps are also available that are driven by the nose of the camshaft, with the pump mounted on the face of a special timing cover.

Dry Sump Components—A dry sump pump features a pressure section and a scavenge section. The pressure section delivers oil to the engine, while the scavenge section pulls "leftover" oil from the dry sump pan and sends it back to the remote oil reservoir.

Dry sump pumps are built in stages, with one pressure section and one or more scavenge sections. The additional scavenge sections (as few as one, as many as 6) allow oil to be scavenged more quickly and efficiently from specific areas of the engine, instead of waiting for the oil to be drawn down into the pan for scavenge pickup.

A typical dry sump oil system features a host of components including:

- Dry sump pump
- Dry sump pump pulley
- Toothed pump drive belt
- Pump fittings (male –AN 37-degree flare to accept –AN hose ends)
- Pump mounting bracket
- In-line screened oil filter(s)
- Remote in-line oil filter and filter mount
- Remote-mounted oil tank
- Breather (separate or on-tank)
- Oil filter block-off plate for engine block
- –AN hose assemblies as required

This Aviaid dry sump pump mounting bracket features multiple threaded holes, providing plenty of latitude when mounting the pump for optimum location (for pulley alignment).

The toothed pump drive belt slips onto the pump pulley by engaging the teeth. The pulley is also adjustable fore/aft to fine-tune belt alignment.

When routing your plumbing hoses, make sure they don't interfere or rub on any moving parts. Here the two hose ends are installed at a sufficient angle to prevent the hoses from touching the timing belt, and the hoses are secured together with an aluminum hose separator.

Here you can see the feed hose (at the bottom of the reservoir tank), where the pump draws oil from the tank to the pump.

A 5-stage dry sump pump. Note the four scavenge hose connections at the pan.

A dry sump system allows multiple oil feeds. Here a direct oil feed hose is plumbed to a turbocharger using a –6 AN hose.

This race engine features scavenge/vent plumbing from the valve cover to aid in reducing parasitic drag and to vent the engine within a contained system.

Quick-connect couplers allow quick connection/disconnection of oil hose assemblies without loss of oil. The Jiffy-Tite coupler shown here features valving that shuts off when the connection is uncoupled. Use of these quick-connect couplers allows faster engine removal with no oily mess.

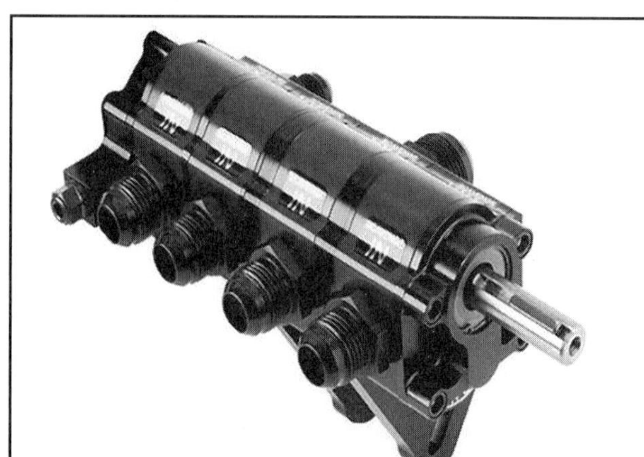

Dry sump pumps are built in stages. The ports marked IN are for scavenge hoses and oil feed from the reservoir. The ports marked "OUT" are for pressure output from the pump and for return to the reservoir. Pulleys are keyed and provide a positive retention onto the pump drive shaft. (Courtesy Barnes)

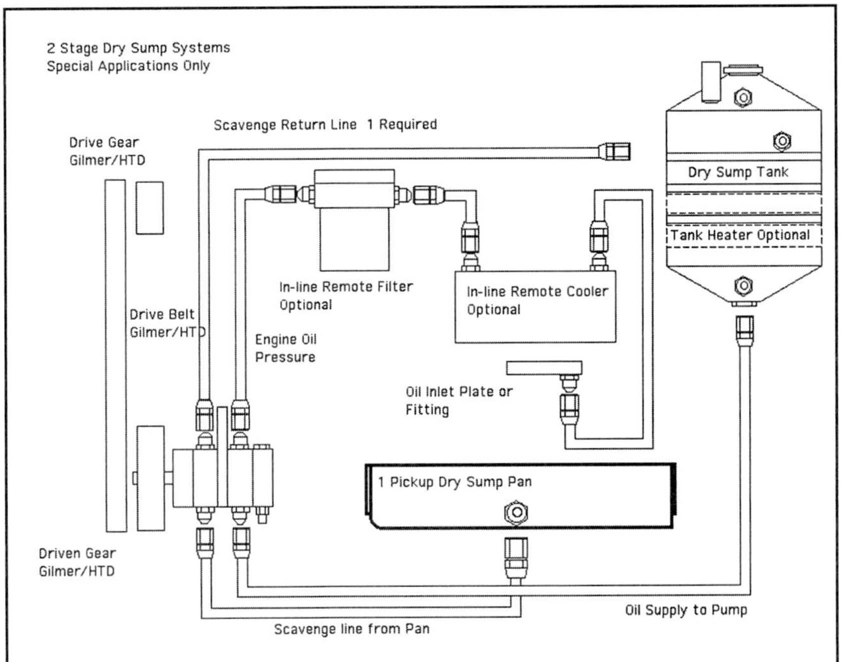

(Courtesy Aviaid)

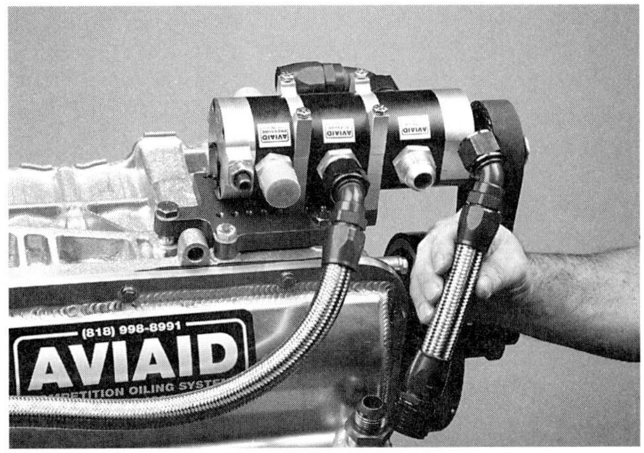

In this plumbing hookup, we installed –12 male AN flare adapters to the pan, to accept –12 hose ends. For the scavenge hose being installed in this photo, we chose a straight hose end for the pan connection and a 45-degree hose end at the pump. The capped fitting seen on the scavenge side of this pump will be used to connect to the reservoir's bottom port to pull oil from the tank to the pump. Due to the range of hose end shapes available, it's easy to create direct and tidy plumbing regardless of the engine configuration.

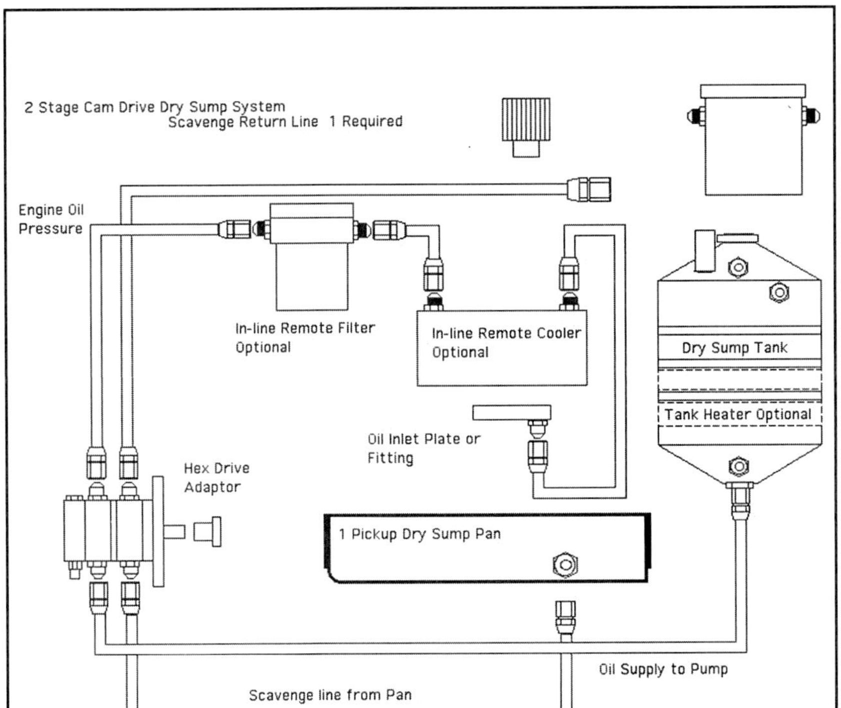

(Courtesy Aviaid)

Always use aluminum –AN wrenches to service the hose end connections.

Caution: If an engine fails (by that we refer to any problem that allows metal debris to enter the oiling system), don't neglect the dry sump system during the rebuild process. The external oil pump must be disassembled, cleaned and inspected for damage (in-line filters can be plumbed in-line in the oil return lines during system installation, which is always a good idea, but even if a filter is present, the pump should still be inspected). Naturally, the remote oil reservoir should also be cleaned. However, many people tend to overlook the hoses (and a remote oil cooler, if present). If

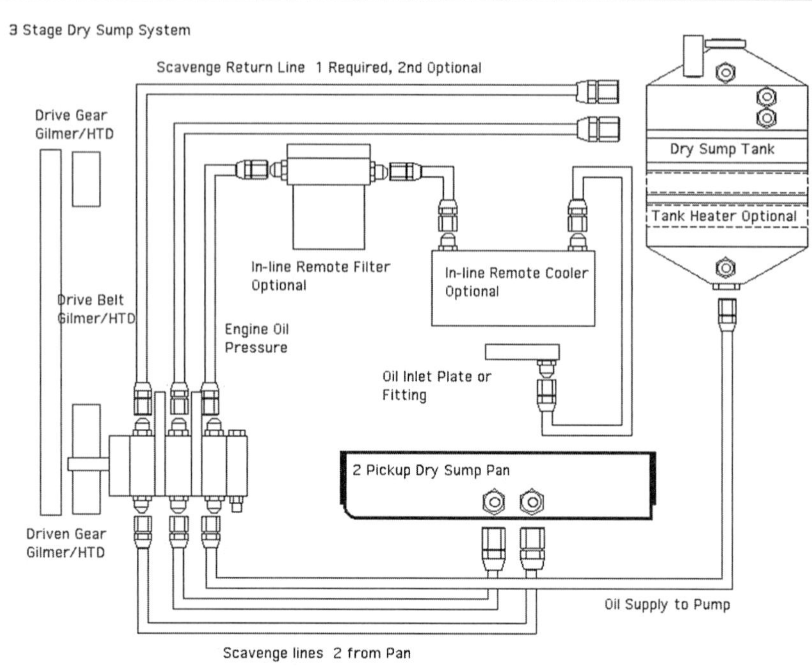

(Courtesy Aviaid)

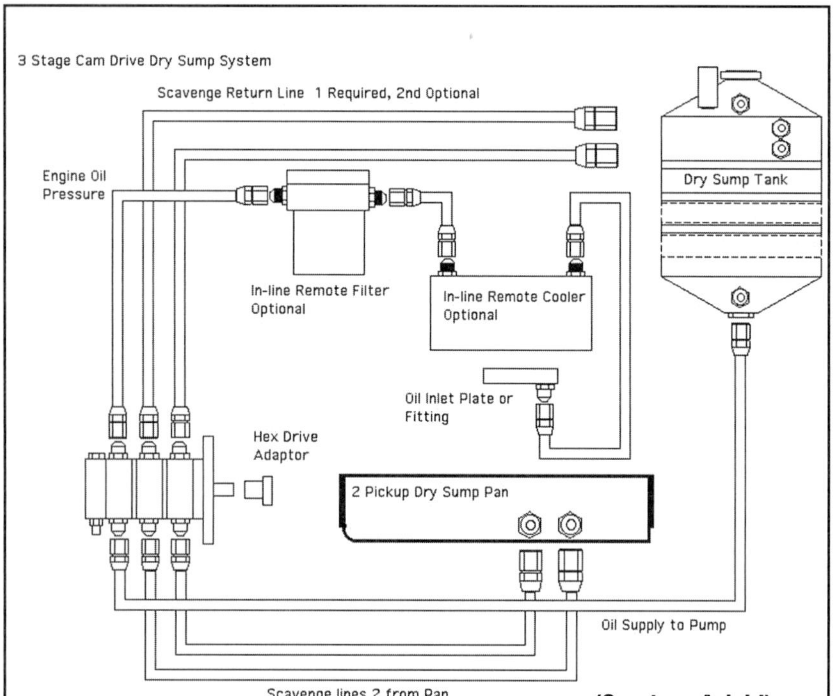

(Courtesy Aviaid)

you suspect metal contamination, although you may try to flush and clean all of the hoses, if you're dealing with hoses that you can't also visually inspect (long hose or hose fitted with angled hose ends), the safest route may be to replace the hoses. That's a call you'll need to make on your own, but be aware that random debris might be lodged in the hose inner wall. If you're certain that the hose is clean, by all means feel free to reuse it. If you have any doubts though, spend the money and replace it.

As far as external oil coolers are concerned (and this holds true for any remote cooler, whether it's for the engine, transmission or axle drive), if you suspect that metal debris has entered the cooler, replace the cooler. Unless you have access to an X-ray machine, you'll never be certain that the cooler tubes are clean. If you reuse a contaminated cooler, you're asking for trouble, since the debris can dislodge and run into the engine. Considering the cost of a cooler (even the most expensive type) versus the cost of the engine, it's good insurance to toss the suspect cooler and to buy another.

Plumbing Remote Oil Filters and Coolers

Remember: if you ever blow an engine, NEVER attempt to reuse the oil cooler, since you'll never be able to guarantee that you can flush the cooler of all debris. Regardless of how much you paid for the cooler, trash it or hang it on the garage wall as an offering to the God of speed and buy a new one. By the way, when plumbing an auxiliary cooler, you can use either fitting location on the cooler as IN or OUT. As long as fluid flows through the cooler, the cooler itself doesn't care which direction it flows.

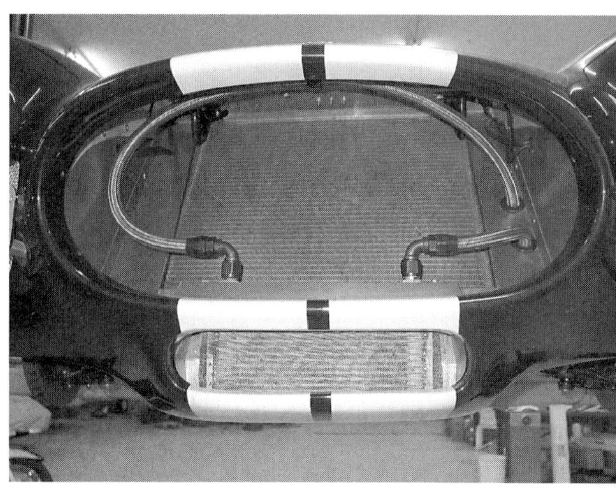

Engine oil cooler in a Cobra replica. The aluminum oil cooler is mounted behind the lower oval air scoop plumbed with –8 AN hose assemblies.

Installing a filter adapter to the block in place of the screw-on oil filter allows plumbing to/from an external oil cooler and a remote-mounted oil filter. Filter adapters are readily available for just about any engine block.

This cast aluminum remote oil filter mounting base features clearly-marked oil flow directions at each port. The two unused ports were plugged with NPT screw-in plugs. The plumbed port at the upper right is the feed from the engine block to the remote filter. The plumbed port at the lower left is the outlet flow from the filter, back to the engine block.

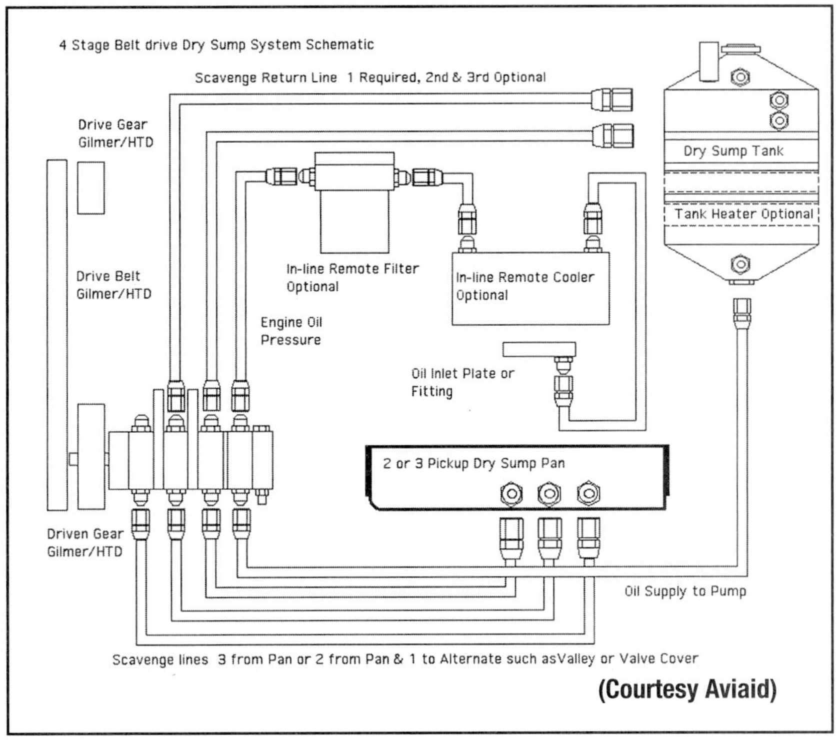

(Courtesy Aviaid)

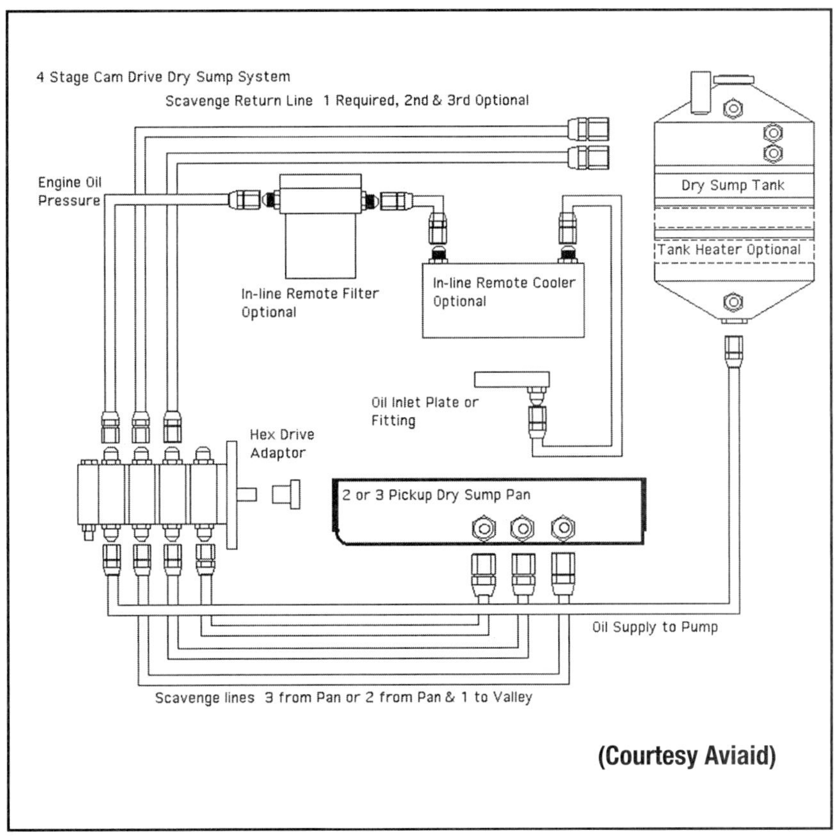

(Courtesy Aviaid)

Any remote oil filter base should be securely attached to eliminate any chance of movement. Be sure to use locking washers, locking nuts and/or thread locking compound on the mount fasteners.

If you're using a screw-on oil filter, make sure it's of a robust design with a heavy or double-wall construction, especially if you've bumped up the engine oil pressure. I've seen cheap filters burst open under as little as 50 psi oil pressure. The NAPA "racing" filter does a good job, as well as others on the market. Just stay away from "Brand X" cheapo filters. When we road raced Ford Mustangs in showroom stock classes, we once ran a few test laps with a fresh (and expensive) 302 engine, and made the mistake of using a cheap oil filter for our "break-in" laps. The sucker burst open when the engine pulled about 6,000 RPM, and while we shut the engine down before any damage occurred, it made an unholy mess. Pay more for your filters. It's worth it.

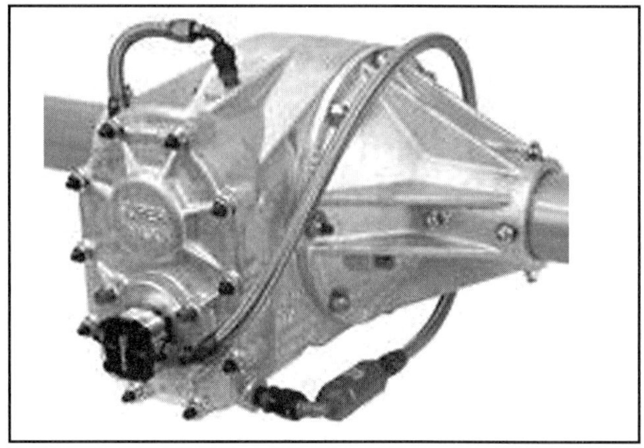

Differential lube via an externally-mounted direct-drive oil pump on a race car rear axle assembly. This allows gear lube to be delivered under pressure to the gears without relying on only "splash" lubrication. (Courtesy Speedway Engineering)

If you're confused as to which port feeds which way, the offset port (seen here at the right) is the "OUT" where oil is pushed out of the engine block, and the port that's somewhat centered is the "IN" where oil runs into the engine block.

A spin-on oil filter adapter. This screws onto the engine block's original oil filter location (using an O-ring seal).

Axle housing oil cooler plumbed here with –6 AN plumbing.

DIFFERENTIAL OILING

While not needed on a street vehicle, some race applications can benefit from pressurized oiling of the drive gear assembly. Normally, the gear assembly in an axle housing carries gear lubricant in the housing sump (at the base of the center housing), which is picked up by the ring gear and "splash" delivered as the gear assembly rotates (the thick lube is carried from the ring gear to the pinion gear, spider gears, etc.). In order to deliver a constant and more efficient lube delivery, an external oil pump may be installed, delivering lubricant directly to critical areas via external plumbing. Pumps are available in three styles: belt driven (driven from the pinion input), electric or direct-drive (mechanically driven by the gears). Plumbing can be handled with stainless braided hose, AN fittings and hose ends. Generally, –6 or –8 size hoses are used.

FUEL PLUMBING

Fuel lines will be plumbed using either 3/8" I.D. (-6 AN) or 1/2" I.D. (-8 AN) hose or tubing. Whether you're using a mechanical or electric fuel pump, a fuel pressure regulator may be installed between the pump and carburetor or injection rail. As far as a fuel filter is concerned, a filter should be installed between the pump and carb or injection, but a filter may also be installed between the tank and external pump.

Pressure Gauges

If you plan on installing a fuel pressure gauge (which is always a good idea), place it between the pump and carb or injection. To install a pressure gauge, this can be mounted to a T fitting and plumbed inline, or an adapter (straight, 45 degree, 90 degree etc.) can be purchased with a threaded port into which the gauge can be installed. For carburetor setups, fuel pressure will generally be in the 4 to 6 psi

Note the fuel feed on this blown pro street engine. The fuel feed from the remote pump runs from the firewall to the pressure regulator. Individual hoses run to T-adapters which branch off with stainless tubes to feed each carb's bowls.

All we can say is...wow. A Bank's 4-turbo drag engine plumbed with enough polished tubing to drain Niagara Falls.

range (applications vary, but that's a general range), while injectors use a higher pressure pump and generally require in the range of 30 to 40 psi (again, specific applications vary). Just keep this in mind when buying a fuel pressure regulator. An adjustable regulator is best, to help you dial in the desired pressure being delivered to the carb or injectors.

Routing Fuel Lines

One main rule when routing fuel lines is to avoid

This custom engine is fed with 3/8" tubing. A pair of carb bowl adapters featuring –6 AN male ends connect to –6 tube nut assemblies.

Extended-reach (long) stand-off pipes are readily available to space your fuel log further from the carb in cases where linkage or other obstructions are in the way.

as many sharp restrictive bends as possible. Another is to route your fuel plumbing away from high heat sources as much as possible for obvious reasons. If a section of fuel plumbing is close to extreme heat (such as exhaust, turbo, etc.), thermal insulating sleeve material can be installed onto the line.

When using threaded connections such as fittings with NPT thread where a thread sealer is needed, use Teflon thread sealant paste instead of Teflon tape. Teflon tape, if not installed very carefully, can cause fuel system contamination (if strands or small pieces of tape enter the fuel system), causing needle and seat malfunctions, orifice plugging, etc.

If you plan to use braided hose and –AN hose ends, in cases where hose end clock position (installed angle of angled hose ends) cause binds or twisting in the hose, making connections difficult, consider the use of swivel union fittings in the hose run (this is a straight two-piece fitting that allows swiveling at its center, giving you the freedom of rotating the hose sections to avoid binds). These unions feature –AN thread at each end, allowing connection with two additional straight hose ends (hose to union).

Avoid using cheap inline fuel filters that can create flow restrictions. If you're using –AN hose

You can purchase a ready-made carburetor fuel feed or you can assemble your own using hose, hose ends and adapters.

Here's a completely tubular fuel feed using 3/8" tubing and –6 AN tube nuts and adapters.

Carburetor bowl adapters are offered for all popular Holley, Carter/Edelbrock, Demon and other carbs, allowing easy connection if you choose to use –AN hose or tube assemblies.

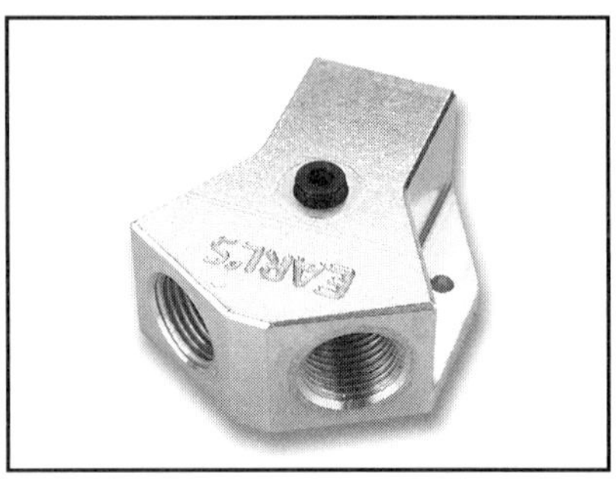

Fuel distribution block. These are offered in several shapes, and in several different thread sizes. This Y-style, for example, allows running the main fuel feed to the stem of this block, then running two individual fuel lines to each carb bowl. The block features mounting holes for locating it where you desire.

You can always plumb a fuel pressure gauge into the fuel feed by either buying a fuel feed assembly that features an open end at each end on the horizontal run, or by obtaining an in-line adapter fitting that features a 1/8" NPT port to accept the gauge.

Extended carb bowl adapters allow you to place the horizontal fuel feed further from the carb to clear any obstructions. Extended carb bowl adapters are available in a variety of lengths.

Always install a fuel filter in any fuel plumbing system. Locating this inline filter between the fuel pump and the carburetor or injector log offers easy service. Note the two sealing washers (one at each end). When filter service is performed, it's best to install new sealing washers instead of reusing the originals.

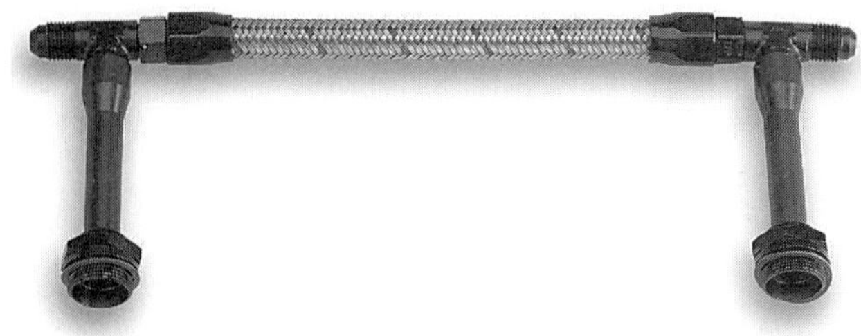

Carburetor fuel feed assemblies are available for specific carb makes and models, based on carb bowl inlet thread size and distance from bowl port to bowl port, so there's really no need to make your own. This feed assembly features a T adapter at each end of the horizontal run, providing a choice of which direction you want to route the feed from the pump, and allowing you to either cap the unused end or install a pressure gauge, etc.

This engine is plumbed with two fuel feeds...one to the carb (far left in this photo) and one for the extra fuel for the nitrous system. Each feed features its own fuel pressure regulator.

A shut-off valve can be installed virtually anywhere in the fuel plumbing between the tank and the carb or injector log. This can serve as a kill switch, theft deterrent, or as a safety shut-off during vehicle service to prevent fuel spillage.

The fuel inlet on this Edelbrock carb features a straight nipple, causing the fuel hose to interfere with the owner's air cleaner base.

A banjo fuel fitting may be used in place of the straight nipple, transforming the angle of entry to 90 degrees. The banjo set includes the banjo housing with nipple, banjo bolt and two sealing crush washers.

After installing sealing washers, slip the banjo housing over the bolt shank, followed by a second sealing washer. The set is now ready to be installed onto the carb.

Install the banjo onto the carb, rotating the nipple in the desired location, then tighten the banjo bolt. Don't overtighten.

There's no such thing as a radiator that's too big. Take advantage of the space that's available.

What a beautiful job. This custom setup hides just about everything. The only readily visible plumbing is the upper coolant tube. With the exception of the black insets on the valve covers and carb shroud, everything is painted with the same silver color. Hidden behind the upper radiator shroud is a short piece of silicone hose that attaches the coolant tube to the radiator, to serve as a vibration/movement isolator between the engine and radiator.

assemblies, "billet" aluminum filters are available in both –6 and –8 sizes, with male –AN threads at each end. These are serviceable (many use a sintered bronze filter that can be removed and cleaned or replaced).

COOLANT PLUMBING

Engine cooling efficiency is a critical issue in any application, and is especially troublesome in some street rods where cramped underhood spaces and sometimes inadequate airflow contributes to engine overheating. Pay attention to the basics first. Make sure that the coolant passages in the block are clean and free of rust. If the engine is being built from scratch, this is easy to check while the block is bare. For an engine that's already assembled and in the vehicle, you can still perform a flush by running water through the system (tying into coolant plumbing or removing the water pump and flushing through the water jacket ports). Before you flush an assembled engine, remove the thermostat so water is free to run through the entire system.

Avoid the use of RTV silicone if possible. When RTV is used (to help seal a water pump or

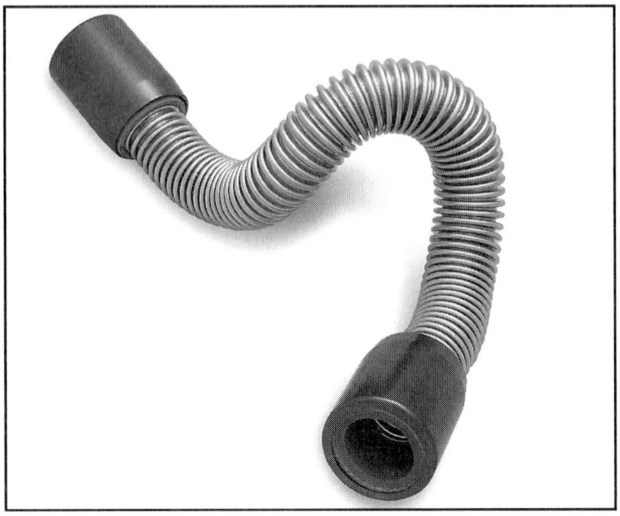

This flexible metal tubing system (Earl's Flex-A-Form shown here) offers a very clean and custom appearance for a streetrod application. A silicon rubber sealing sleeve is located inside the hose end. These systems are available in a variety of commonly required I.D. sizes. The tube can be hand-bent without kinking, and retains its shape.

thermostat gasket, etc.), use this sealant sparingly. Excess RTV can easily enter the cooling system and can plug or restrict water jackets. If it gets into the radiator core, this will cause major flow problems as well.

As far as hose is concerned, you have options. You can use commonly available reinforced rubber hose, high temp silicone hose, reinforced braided hose (stainless steel or Kevlar) or a combination of flexible hose and hard tubing. For upper and lower radiator plumbing, I.D. size will require a minimum of 1 1/4" (equivalent to a –20 AN), although some race car applications can get away with –16 (1" I.D.). If you opt for –AN plumbing, this is available to handle all external coolant plumbing sizes (radiator to water pump inlet, radiator to engine water outlet, heater hoses, etc.).

While some street rodders/custom car owners opt to use stainless steel braid hose sleeves and clamp-on fake hose ends simply for appearance, you can employ actual –AN hose assemblies for the radiator connections, but for street applications, this presents pros and cons. On a positive note, this will provide reliable coolant connections and it will make the assembly look "pro." On the downside, you'll need to be able to make the connections. This means either buying a radiator that already features –AN fittings, or modifying your existing radiator by welding –AN fittings in place of the slip-on style necks. You'll also need to buy a thermostat housing that features a male –AN thread, and if you want the water pump connected in the same manner, you'll need to weld an AN fitting onto the pump inlet neck. This leads to another downside. If the radiator or water pump requires replacement down the road, and the component failure occurs while you're on a trip, you're looking at some downtime, since you're not going to find a radiator or water pump with those fittings sitting on an auto parts store shelf in Anytown, USA. I'm not trying to talk you out of a true –AN radiator plumbing setup. Just be aware of the future potential pitfalls.

For custom/streetrod applications, a great option for radiator (and heater hose) plumbing is flexible metal tubing that's made specifically for this use. This is a ribbed tube that allows flexing to create the desired routing, but remains in the bent state (it won't flop around and lose it's shape). This is sold under various names. As an example Earl's offers their Form-A-Flex system. The hose ends feature aluminum housings with captive stainless clamps and high-grade silicon rubber insulators for a reliable sealing connection at the necks. The tubing (at least all of the versions I've seen) are bright chrome-plated or anodized. The hose ends (since they're aluminum) can be had in chrome or color-anodized finishes. These assemblies look good and work great.

Yet another option involves the combination of rubber hose and metal tubing. If the engine didn't move or vibrate in relation to the radiator, all-metal tubing connections would be fine. However, since the engine does move (rocking on its mounts or simply vibrating), flexible hose is required to accommodate this movement, to prevent damage to the radiator necks. So, as long as there's enough rubber hose to allow deflection, there's no reason that you can't use both tubing and hose, if for no other reason than to gain the appearance you want (after all, steel or alloy tubing can be bent to shape and can be painted, powder coated, plated or anodized for a desired finish appearance.

If metal tubing will be used as part of the hose assembly, the tubes should be beaded for superior hose retention (reducing hose slip-off). An easy-to-use tube bead forming tool can be used to create this retention bead. Earls' for example, offers a line of these tools, each dedicated to a specific tube I.D. size, called their EZ-Beader tool. The downside is that you must purchase an individual tool for each diameter, but since most builders use only one or two, or three tube sizes anyway, it's not that big of a deal. Universal tube bead tools are also available, but the good ones are big bucks. The tube beading tool we used here is available in sizes including 3/8", 1/2", 5/8", 3/4", 1", 1 1/4", 1 1/2", 1 3/4" and 2".

Especially for street use, always run a mixture of water and anti-freeze in the cooling system (generally, a 50/50 mix is common). The chemicals

Tube bead forming tools are easy to use. This is a two-piece tool. As the inner section rotates clockwise, the three bead forming balls expand into the tube.

Rotate the inner section (via the smaller hex) counterclockwise to retract the balls flush with the O.D. of the tool body, and insert the tool into the tube.

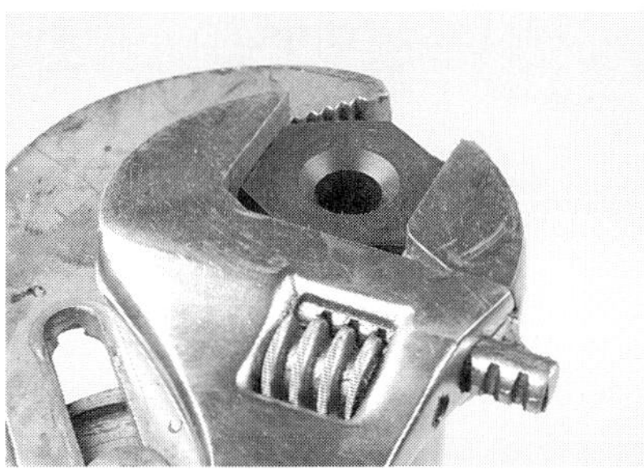

Holding the tool body (via the larger hex), rotate the inner section clockwise (tightening) to expand the balls.

As you continue to tighten the inner section, the balls expand further, pushing the tube outward at the areas of ball contact.

With the balls expanded, rotate the main tool body 360 degrees to form the bead. Tighten the inner section further, rotate the main body again, etc. until the entire bead is formed.

Once the bead has been formed, loosen the inner section to retract the balls and remove the tool from the tube.

The result: a perfectly formed bead around the entire circumference of the tube.

The bead will aid in hose retention. Slip the hose onto the tube and place a clamp behind the bead area (above the thumb seen here). This creates a much more reliable slip-on hose connection as opposed to a smooth, un-beaded tube.

Pay attention to the basics. Always use new hoses, and inspect the hoses periodically. Remember that coolant hoses tend to deteriorate from the inside, so feel and look for signs of sponginess and brittleness.

in automotive coolant/anti-freeze contain rust inhibitors, they help to transfer heat, and they help to lubricate the water pump seals. Running straight water is never a good idea for a street application (and in most race engines as well). As far as water is concerned, don't use tap water, whether it comes from a town's treatment center or if it comes from a residential well. In order to minimize the chances of rust, corrosion and calcification, use only distilled water whenever filling or topping any cooling system.

Here's a common problem. This guy used flex-hose for his coolant. The hose is bent so far that it has begun to collapse and kink at the bend. This will restrict coolant flow severely.

This water pump to radiator connection uses a short length of silicon rubber hose.

This expansion tank also features a small air bleed hose (from the radiator to the tank). Any air in the system will migrate to the tank.

The polished aluminum hose end on this flexible tube contains a silicon rubber insulator that provides the sealing for the connection.

This Cobra replica features an original type cooling system expansion tank. This provides a high water level container, necessary whenever the top of the radiator is positioned lower than the highest water level of the engine assembly. This provides the air bleed-off and fill point for the system. Note the upper radiator hose assembly that features a straight hose from the expansion tank to a 90-degree tube, and a 45-degree section of hose from the tube to the radiator neck.

This 1936 Chevy truck street rod uses a Form-A-Flex tube to connect the thermostat housing to the radiator inlet. Once this tube is hand-bent to the desired shape, the shape is retained.

Some small displacement engines, like this Honda B18C1, use an external cooling transfer tube (the horizontal black tube seen here). Each end features an O-ring and a formed bead stopper. This tube connects the water pump to the block and thermostat housing.

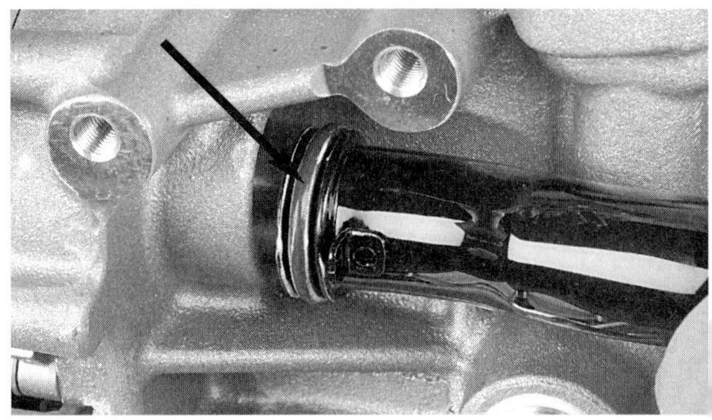

This external coolant transfer tube simply slips into the block port. The opposite end of the tube features a mounting bracket to secure the tube. Whenever servicing an external coolant transfer tube, always replace the O-rings. Apply lithium lube to make insertion easier.

BRAKE SYSTEM PLUMBING

Plumbing the brake system, although extremely critical, isn't all that complicated. Basically, the master cylinder (when the brake pedal is depressed) pushes fluid to the calipers/wheel cylinders, causing the disc brake caliper pistons to push the pads against the rotors; or causing the drum brake wheel cylinders to push the shoes outward against the inside surfaces of the drums. This is an enclosed (no air) hydraulic fluid system.

Additional items to consider are proportioning, or metering valves. Since you generally want the front brakes to apply more immediate force than the rear brakes (for vehicle control), a proportioning valve is plumbed in-line between the master cylinder and the rear brake line to establish a proportional hydraulic force between front and rear brakes. For performance/racing use, an adjustable proportioning valve allows fine-tuning of this front to-rear brake bias, again for handling and vehicle control issues under braking.

The brake fluid transfer plumbing should consist of hard tubing everywhere except where flexible hose is needed for suspension and steering movement (from the frame to the calipers and from the frame to the rear axle housing mount). If movement is expected from one point to another, a flexible hose must be used.

As far as hard-line tubing size is concerned, 3/16" or 1/4" I.D. can be used throughout, though the most common size is 3/16". For flexible hose locations, -3 AN or –4 AN (3/16" I.D. to 1/4" I.D.) can be used. If you're running 3/16" hard line, use –3 hose. If running 1/4" hard line, either –3 or –4 AN can be used.

For performance or racing use, it's always better to use a stainless steel braided reinforced, lined dedicated brake hose as opposed to OEM rubber hoses. Remember: a hydraulic braking system is just that…a hydraulic system. You can't compress fluid, so under pressure, it looks for places to escape. Under braking pressure, rubber hoses tend to expand under hydraulic pressure, so you're wasting a bit of the needed pressure by expanding the hoses instead of forcing all of the pressure to the wheel cylinders where it's needed. Reinforced braided hose (again, dedicated reinforced and lined brake hose specifically designed for brake system use) won't expand, so all of the pressure that you generate is delivered to the calipers and rear cylinders. The result: a firmer brake pedal and more efficient braking performance. These brake hoses are generally made with a hard-walled Teflon liner. This, coupled with the reinforcement of the braid, provides much better performance than any OEM-type synthetic rubber/fabric reinforced hose.

Racing brake hoses are too difficult for home assembly, so don't even think about making your own. Buy them already assembled, in the length you need and with the appropriate fittings/hose ends already installed. These hoses are also excellent choices for hydraulic clutch systems as well. Race-type brake hoses are available with any ends you desire, including for double flare, ISO bubble, British bubble, banjo and –AN, with straight or angled ends.

The inlet end of the brake hose must be anchored securely to the chassis. The hard line plumbing from the master cylinder will connect to the flexible hoses.

Adequate flex room is required (hose length) to allow suspension and steering travel without pulling the hose. Make sure the hose routing won't rub or pinch during travel.

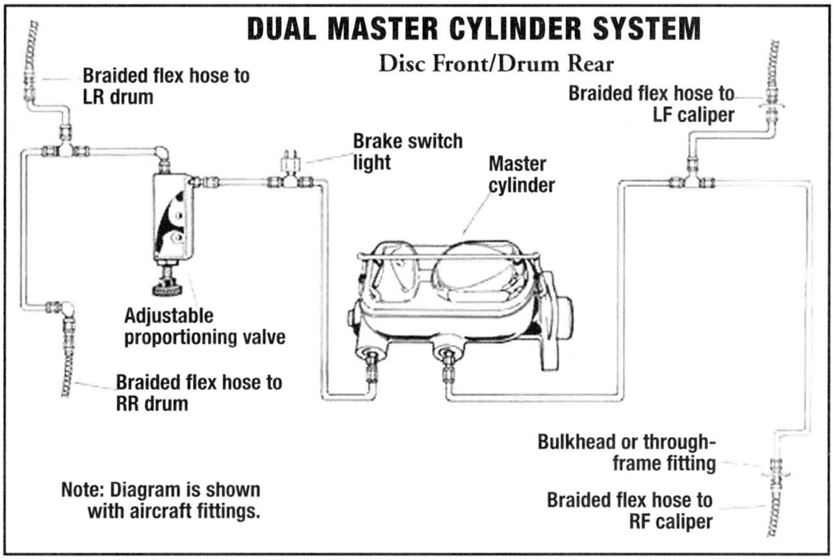

Typical disc/drum routing. Notice the adjustable proportioning valve. (Courtesy Wagner)

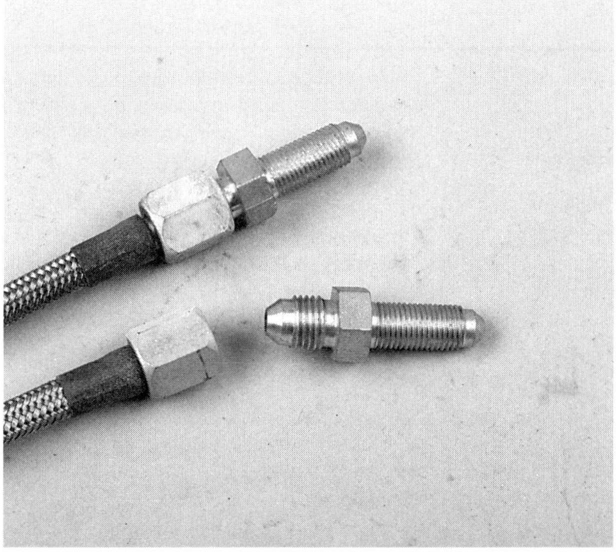

A pair of –3 braided brake hoses with female hose ends to accept –3AN adapters.

Race brake hose with straight female fittings.

Race brake hose with a banjo fitting on the left end.

Double flare tube.

British bubble flare.

ISO bubble flare.

If your calipers feature external fluid crossover tubes, inspect them for corrosion and damage. An insulation is often used between the line and the caliper body to reduce the chance of vibrational contact damage.

A popular move for street rods involves mounting the brake master cylinder (and power booster) on the vehicle frame. This cleans up the firewall area. Just remember that if the master cylinder is mounted lower than the calipers, residual valves will be needed. Also, if the master cylinder is mounted backwards and behind the pedal area (allowing brake pedal linkage to push the cylinder), remember that the master cylinder's outlet port closest to the mounting flange and pedal linkage is to be plumbed for the front brakes, while the outlet port closest to the master cylinder nose is for the rear brakes. Don't get confused just because the master cylinder is mounted 180 degrees from its normal orientation.

Note that the master cylinder and power booster are mounted backward in this in-frame installation. When plumbing, you need to remember that the rear reservoir line on a dual master cylinder (reservoir line closest to the booster) feeds the front brakes, and the front reservoir line (closest to the cylinder housing nose) feeds the rear brakes.

Calipers With External Transfer Tubes

Multi-piston calipers are designed with an external tube connecting the two caliper halves allowing the brake fluid to flow from the inboard housing to the outboard housing. This method of brake fluid transfer is popular with calipers which operate in the upper temperature ranges as the cooling process is aided by permitting cooler air to flow over the crossover tube.

Adjustable Proportioning Valves

The adjustable proportioning valve is basically a device which reduces pressure that is conducted through by approximately 50%. When the knob is screwed in, (closest to housing), the valve has no effect. With the knob completely out, the valve reduces 50 to 60% at pressures above 100 to 300 psi, depending on the brand being used. Adjustable proportioning valves are recommended for reducing the amount of rear brake pressure, preventing lock-up during hard braking on drag race and street vehicles.

Residual Pressure Valves

These are small in-line valves that prevent excessive brake fluid flow-back. If your master cylinder is mounted lower than the calipers, installing these anti-drainback valves prevent fluid from quickly draining from the calipers to the master cylinder reservoir during period of vehicle storage. If drainback occurs, this also creates a slight vacuum in the system, which can cause the caliper pistons and pads to retract slightly (possibly resulting in a low or no-pedal the first time you step on the pedal.

Tandem master cylinders are an option for race car applications, where one master cylinder is dedicated for front brake only and the other for rear brake only. Proportioning the brake pressure is partially handled by an adjustable twin-bar brake pedal assembly, allowing adjustment for balancing initial pressure between the two cylinders.

Some fluid reservoirs (OEM and performance aftermarket) are made of plastic, which work fine and dandy. Just be sure to avoid nicking them, and keep high heat sources away (of course, you shouldn't have high heat levels anywhere near your master cylinders anyway).

Install this valve in-line between the caliper and master cylinder reservoir. Close to the master cylinder is a good location.

Residual valves are commonly available in two pressure versions: 2-lb and 10-lb. For disc brakes, use a 2-lb valve. For drum brakes, use a 10-lb valve. This valve basically maintains about 2 or 10 (depending on which version you're using) pounds of pressure at all times, eliminating the drainback. Again, you only need these valves if the master cylinder reservoir is mounted lower than the calipers or wheel cylinders.

Caution: Do not install a 10-lb residual valve on a disc brake system, as this will result in severe brake drag.

Brake Fluid

For racing applications, always use a DOT 3 or 4 fluid. However, consider the dry and wet boiling points of the fluid, since the DOT numbers really don't provide the information you need. Use a fluid with a dry boiling point of at least 450 degrees F.

The two most important aspects of a racing brake fluid are compressibility and its rate of moisture assimilation. Compressibility refers to pedal feel...how firm the pedal feels when the brake pedal is depressed (also assuming the system is properly bled of air). Different brake fluids can dramatically alter pedal feel, with no other system changes.

Silicone DOT 5 brake fluid has its place, but not in racing and not on a system equipped with anti-lock brakes. It won't hurt paint, but that's about it's only real benefit. Granted, it's not a hygroscopic as glycol, but it can still carry moisture on its surface. I use it for collector cars that only get driven on occasion and in cases where I'm paranoid about an ultra-bucks paint job.

Glycol brake fluid is highly hygroscopic, which means that it loves to soak up moisture from the air. Depending on the chemistry of the brake fluid, it can have a high rate of moisture absorption or a low rate of moisture-soak. When a brake fluid is contaminated with as little as 2% moisture content, its boiling point may be reduced as much as 50%, which reduces its racing use to nil. Condensation that occurs within the brake system is probably the biggest contributor to moisture absorption. The ONLY cure for this is to change your brake fluid often!

We need to talk about DOT 5 silicone fluid in an effort to resolve this question once and for all. Some people love it and some hate it. In a nutshell, DON'T USE IT FOR RACING. The only advantages of silicone fluid is the fact that it won't harm a painted surface, and that the fluid itself is non-hygroscopic (it won't suck moisture from the air). That's why it's popular with restoration, show car and street rod folks. However, in terms of performance, this stuff can cause real problems.

Let me preface these comments by saying that I've used silicone fluid in my own street rods and kit cars and restorations, simply because of the harmless effect it has on a painted fender or firewall if accidentally spilled, so I'm as guilty as anyone else who uses this stuff. The problems, however, can far outweigh the advantages under certain conditions. Silicone fluid can become very unstable at temperature, resulting in a low or ever-changing pedal feel. On the race track, this is obviously not a good thing. It's also affected by atmospheric pressure. At higher temperatures, it expands, requiring more pedal pressure and more pedal travel. I'll probably continue to use it in a car that is stored for long periods of time, or in rarely driven restored

BRAKE FLUID SPECIFICATIONS

Brand	Dry Boil Point	Wet Boil Point
Brake Man Hi Temp 577 Fluid	577 deg. F	300 deg. F
Castrol SRF	590 deg. F	518 deg. F
AP Racing Super 600	572 deg. F	311 deg. F
AP Racing 550	550 deg. F	284 deg. F
Sierra SRP Z-10	570 deg. F	n/a

Note: The dry boiling point is the temperature that new fluid will boil. As the fluid is used, it begins to attract moisture, and will boil at a lower temperature. The wet boiling point is the temperature the fluid will boil when it contains about 1–2% moisture.

cars or street rods (even though I know I shouldn't), but as far as racing is concerned, forget it!

If the choice is between dealing with a paint stain on a fender or not being able to stop the damn thing as I approach the end of the back straight of Watkins Glen, I'll take the paint stain any day.

Brake fluid is viewed by most racers as a necessary evil. It's slimy to the touch, it can stain a painted surface, and it always seems troublesome to take the time to bleed the stuff. As a result, many racers completely ignore their fluid with the exception of performing an occasional bleed or top-off. The car's brake fluid is probably the most important fluid in the entire car, since without it, that expensive billboard on wheels won't be able to stop, which can spell the end of the race, the end of a promising season, or even the end of the driver's life.

The point we're driving at is simple: CHANGE YOUR BRAKE FLUID ON A REGULAR BASIS! This isn't a joking matter. Aside from the safety aspect (which should be your primary goal in the first place), contaminated brake fluid will crud up your lines, calipers, proportioning valve and master cylinder, causing long-term operational problems and avoidable replacement of expensive components.

Many a racer has chased a soft or low pedal problem by changing pads, resurfacing rotors, adjusting pedal travel, adjusting balance bars, playing with proportioning valves, changing master cylinders, and more, while the root of the problem was simply tired and water-laden brake fluid that boils quickly. Seriously, get into the habit of draining the old brake fluid, flushing the system and refilling with fresh, quality fluid on a regular basis. That doesn't mean once a year, by the way. Plan to change your brake fluid at least several times during the season. Ideally, the fluid should be changed before or after every race, especially if you've been racing in very humid conditions. Some racing brake fluid manufacturers even note a maximum lifespan for their fluids (Castrol, for example, recommends that their SRF fluid should not be used for more than 18 months).

The outside air doesn't have to "feel" wet for moisture to enter your race car's brake fluid. Air always carries a certain percentage of moisture, and this will wick into your brake system through the microscopic pores in calipers and master cylinders, through seals in the calipers and master cylinder reservoirs, and through improperly tightened fittings. As we mentioned earlier, brake fluid is highly hygroscopic, and it will suck the moisture out of the air, and there's not a damned thing you can do about it. Your only recourse is to routinely change the fluid. Even after having said this, I know the reality is that most racers still won't do this, but I'm telling you that you will gain a benefit from regular brake fluid replacement. You'll have better brakes and more consistent braking performance at every race if you perform a one-bottle flush before each event.

While any DOT-approved brake fluid is probably fine for street applications (by the way, street cars need brake fluid replacement at least once per year, but that's another argument entirely), for racing use, stick with high-grade performance fluids that are designed for racing use. That means the fluid should have good compressibility, exhibit a high boiling point rating, and should have a low rate of moisture absorption. Examples of good fluid choices include Brake Man Brake Fluid, AP 550, AP 600 and Castrol SRF. Any of these fluids are good choices. Castrol SRF is the highest-performing and most stable fluid that I'm currently aware of, but it's expensive (to the tune of about $80/2-pint container). While this is a great brake fluid, it simply isn't needed in anything other than a super-hard-stopping application like Winston Cup or some heavy road race endurance cars. For a sprint race, the extra money doesn't need to be spent. For example, at the time of this writing, Brake Man fluid costs about $7 or so for a 12-ounce bottle (this is great stuff by the way, with excellent compressibility and a low rate of moisture assimilation), and the AP fluids run from about $10 to $17 per 16 oz bottle. Use only racing-quality fluids and change frequently.

Bleeding Brake Fluid

You have several options. You can use a pressure bleeder (not a piece of equipment owned by most do-it-yourselfers, and often inconvenient to use on a trailer-queen street rod due to the chances of scratching paint with master cylinder reservoir hold-downs, and because access may be tight); or you can employ two people, one to pump the pedal and one to open and close the bleeders; or you can use a vacuum pump, not always the most reliable method, as this can induce air into the system. Or you can use one-man bleed valves. One-man bleed valves install in place of the caliper's existing bleed valve. These feature a spring-loaded sealing head. Even when loosened, the valve won't allow air or fluid out until the pedal is depressed. Simply loosen the valve, jump back into the car and depress the pedal. Pedal pressure will overcome the spring-loaded seal, allowing fluid and air to escape. Once the air has been bled, tighten the valve(s). **Note:** in order to avoid making a mess, first attach a clear vinyl hose to the bleed valve, and stick the other end of the hose into a clear bottle or jar. Make sure that the container is stable so that it doesn't fall over during the process. As you bleed, stop once in a while to check (and refill) the master cylinder reservoirs. Don't allow the master cylinder to run dry.

Also, only use a dedicated line wrench to open and close the bleeders to prevent damaging the hex on the bleeders. Several sources offer these one-person bleeders. For example, Earl's offers its "Solo Bleed" valves.

If you prefer to bleed the traditional manual way, you'll need two people.

Fill the overflow bottle with about two inches of clean brake fluid (enough to keep the end of the hose submerged) and insert the end of the tubing below the level of the brake fluid. Attach the other end to the bleed screw.

With the bleed screws tight, the person in the vehicle should pump the brake pedal about six times or so, at a slow to moderate speed, then hold pressure. The person under the vehicle then opens a bleed valve, allowing fluid (and air) to escape. While the brake pedal is still held down, tighten the bleed valve. The brake pedal can then be released. Never release the brake pedal until the bleed valve is closed. Otherwise you'll suck air back into the system (the submerged hose end should prevent that, but don't take chances, as there can be air bubbles in the remote container). Since the hose is clear, the person working the bleed valve can monitor, watching for air bubbles traveling through the hose. Once no more traces of air can be seen, tighten the bleed screw and move to the next caliper.

In terms of order, always start with the caliper (or drum cylinder) furthest away from the master cylinder (in most cases, this will be the right rear).

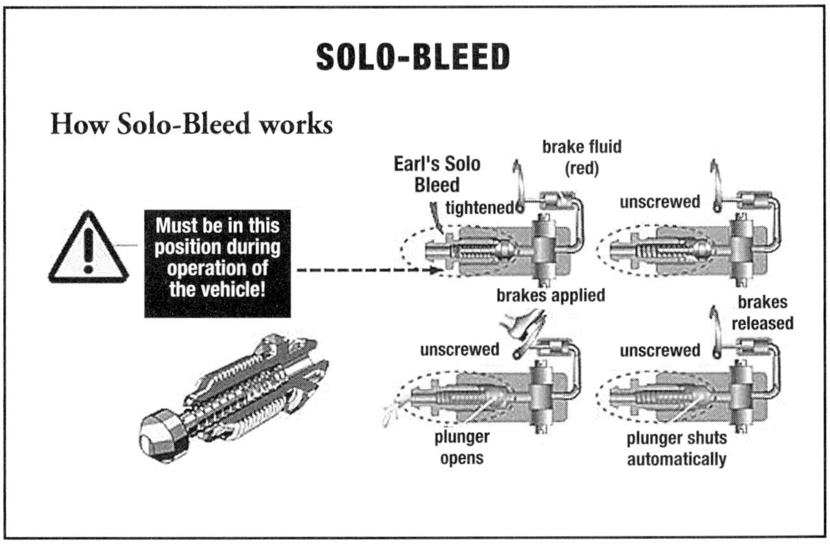

Earl's Solo-Bleed valve. Even with the valve screw loosened, the spring-loaded plunger remains seated to seal the system. As you apply pedal pressure, fluid pressure overcomes this spring, allowing fluid and trapped air to escape. (Courtesy Earl's)

Next will be the left rear, then the right front, and finally the left front. **Note:** Some newer vehicles with anti-lock brakes require specific bleeding procedures, so if you have an owner's manual, pay attention to it.

If you're starting with a new/replacement master cylinder, you'll first need to "bench bleed" the master cylinder before installing it onto the vehicle. Secure the master cylinder in a bench vise (avoid damaging the cylinder housing). The object is to eliminate air inside the cylinder. Fill the reservoirs with fresh brake fluid of the type you plan to use in the brake system. Attach fittings and short tubes to the outlet ports and bend the tubes to submerge the opposite ends into the reservoirs. Using a dowel rod or blunt Phillips screwdriver against the cylinder's pushrod location, slowly stroke the cylinder (all the way in and all the way out) until you don't see any more air bubbles exiting the submerged tubes. A plastic fitting and tube kit (master cylinder bench bleed kit) is readily available at any decent auto parts store). Once the master cylinder has been bled, install it. With your system lines in place and ready to attach, remove the bleed fittings and tubes and attach the lines to the master cylinder. You'll make a small mess, so place a towel under the master cylinder to avoid spilling brake fluid onto the frame, inner fender, etc.

Chapter 13

–AN HOSE AND –AN HARDWARE

Dash sizes, or AN sizes, were first used in military applications, on aircraft, ships, etc., for hydraulic, fuel and coolant plumbing. The tough, braided/reinforced hose, coupled with thread-on connections that were reliable in combat conditions, led early racers to buy military surplus hose, hose ends and fittings in the years following WWII and Korea. The racing community quickly made this type of plumbing popular, resulting in common use in today's performance and racing markets. The term AN (a common abbreviation for Army/Navy specification) sticks with us to this day. That's why the reinforced hose/hose end assemblies we see today are referred to as AN assemblies.

So we refer to these hoses, hose ends and fittings sizes with the word (or symbol) dash, or by the term AN, or by the term –AN. It all means the same thing. For example, "I'm plumbing my fuel system with dash 6 hose; or "I used dash 8 AN hose for my carburetor feed." When in written form, the dash symbol (–) or the word dash might be used.

–AN (ARMY/NAVY) hose sizing is based on single and double-digit identification numbers. Common sizes for performance automotive applications include –3, –4, –6, –8, –10, –12, –16 and –20 (the larger the number, the larger the hose diameter).

–3 and –4 sizes are typically used for brake line applications, small oil lines, some small fuel line applications, pressure gauges and vacuum lines.
–6 size is typically applicable for fuel and oil plumbing
–8 size is typically used for fuel, coolant and oil plumbing
–10 size is typically used for oil, fuel or heater hose plumbing
–12 size is typically used for coolant, large fuel delivery or dry-sump oil
–16 size is typically used for coolant, dry sump oil or large fuel delivery
–20 size is typically used for coolant (radiator hose)

What do these –AN numbers really represent? Actually, there is a logical reason for these numbers, which otherwise might seem like made-up codes. The dash number refers to the hose inside diameter, in denominations of 1/16" of an inch. For example, a –10 size translates to 10/16", or 5/8" inside diameter. This is an easy way to understand dash sizes. Just think in terms of 1/16" increments. A –6 means that the inside diameter is 6/16" (or 3/8"). A –8 size is 8/16" (or 1/2") inside diameter. Just remember that the format is based on 1/16".

Not to make things confusing, but those 1/16" increments are "nominal" numbers that indicate the O.D. of the hose end's internal metal tube (this tube slips into the hose). In reality, most AN hose makers actually make their hoses a bit on the tight-tolerance side, to the tune of about 1/32" smaller than the theoretical nominal diameter size, which aids in hose sealing onto the hose end's tube. So, a –6 hose, which theoretically should have an inside diameter of 6/16" (3/8" or 12/32"), actually has an inside diameter of 11/32". Nevertheless, using the 1/16" theory as your guide will help you to easily visualize what the inside diameter will be. For

–AN HOSE SIZING CHART

Hose Size	Nominal ID	Actual ID	Max Pressure (psi)
3	3/16"	1/8"	1,000–1,500
4	1/4"	7/32"	1,000–1,500
6	3/8"	11/32"	1,000–1,500
8	1/2"	7/16"	1,000–1,500
10	5/8"	9/16"	750–1,250
12	3/4"	11/16"	500–1,000
16	1"	7/8"	750
20	1 1/4"	1 1/8"	750

Note: The above sizes are based on Russell hose. Outside diameters may vary among hose manufacturers and among hose materials. Operating pressure ratings will vary depending on the specific type of hose construction.

–AN THREAD SIZE

The thread size of the AN hose end and its mating fitting are as follows:

AN Size	Thread Size
3	3/8-24 UNJ
4	7/16-20 UNJ
6	9/16-18 UNJ
8	3/4-16 UNJ
10	7/8-14 UNJ
12	1 1/16-12 UNJ
16	1 5/16-12 UNJ
20	1 5/8-12 UNJ

example, a –6 hose will provide about a 3/8" I.D. and a –8 hose will provide about a 1/2" I.D. If you've decided that your fuel line should feature a 1/2" inside diameter, you know that a –8 AN size will be the correct choice.

HOSE ENDS

Hose ends attach directly to the hose and provide a high-pressure seal. The hose end features a two-piece arrangement consisting of a socket (this slips over the end of the hose and features a female threaded opening) and a male/female component. The male/female piece features a male threaded nipple that threads into the hose end socket. The female (exposed) end is then ready to thread onto a male fitting at the desired installation location. These components are called "hose ends" because they comprise the end of the hose.

Note: While this is the most common configuration, not all assembled hose ends feature female threads. Some makers offer special male threaded ends that fit into female fittings.

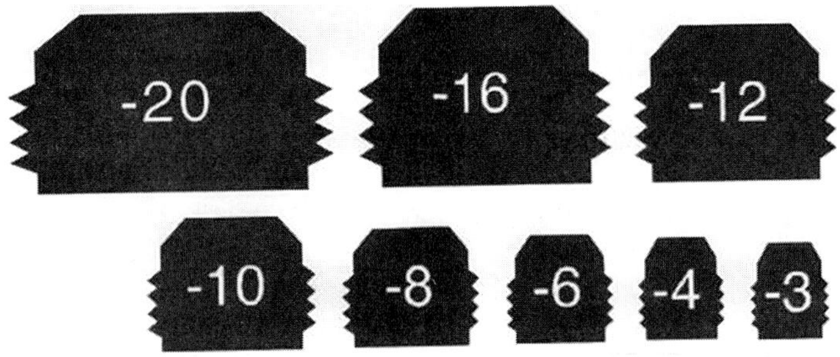

Shown here are examples of AN thread sizes, for comparison.

The "standard" of the AN hose world, at least for automotive performance applications, is a red and blue anodized finish. While aluminum fittings are usually solid blue (sometimes red), two-piece hose ends are generally blue and red. Color shades, as well as finish gloss, will vary among makers. So, if it's important to you that everything looks matched, it's best to stick with one particular manufacturer's hose ends and fittings. Remember: anodizing is not a "coating." It's actually a controlled oxidation process that is used to protect the finish of an alloy, to allow the use of a color dye, and/or to harden the alloy surface. A dull, non-gloss appearance is actually quite normal. If a higher gloss is desired, manufacturers can polish the aluminum beforehand, and they can even use a gloss clear anodizing. The point is to look at the various makers' offerings, and if appearance is critical, decide what color and finish you prefer. Then stick with that one manufacturer's hardware. Otherwise, if you mix it up, you'll have different shades of blues and reds and gloss and flat finishes under the hood. If it's a race car, function is most important. But if it's a show car or street rod, it'll look best if all pieces match. These samples from XRP feature a stunningly beautiful appearance, with rich color and gloss finish.

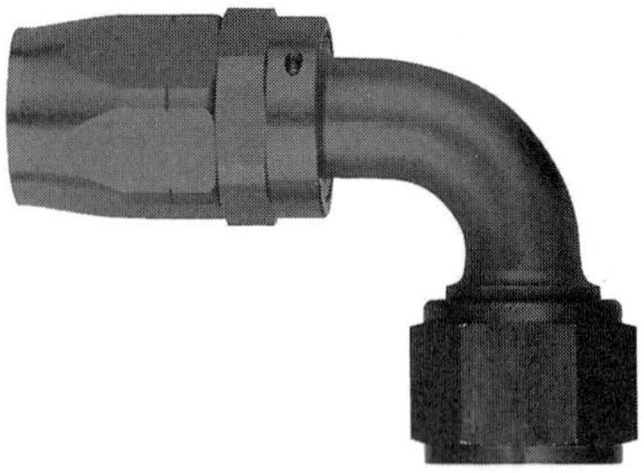

Here's a 90-degree hose end in a standard profile (note the tubular neck). In this example, the red end (upper left in this picture) attaches and seals to the hose, in practical terms becoming an integral part of the hose. The blue end (lower right in this picture) features a female AN thread and 37-degree internal seat, which will attach onto a male AN fitting. (Courtesy Aeroquip)

The inside of this hose end's collar shows the wide-spaced "coarse" threads on the hose side of the collar. These threads, or ridges, help to retain the hose.

Here's a –16 straight hose end with the collar removed. Hose ends feature a nipple that inserts into the I.D. of the hose.

The male threads on hose ends (to retain the hose end collar) are very fine. Even with a new hose end straight out of the box, make sure that the threads are clean and free of any contaminants. The anodizing helps to harden the aluminum surface, but you can still gall the threads if you don't pay attention.

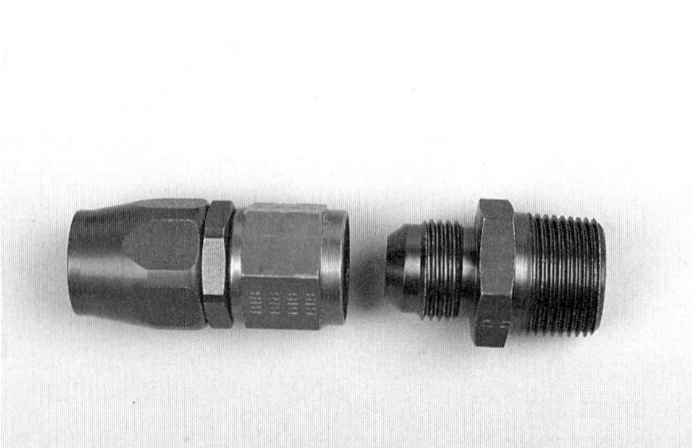

This simply illustrates how a –AN hose end attaches to a –AN adapter. The hose end pictured here features a female threaded port with a 37-degree female (concave) sealing seat. The adapter shown here features a male threaded –AN tip with a 37-degree male (convex) sealing seat. The two angled seats mate, creating the seal. All –AN connections obtain their seal by the mating of these angled seats (these angled seats are sometimes called cones). No additional sealing material is required, so there's no need to add thread tape or compound to the –AN threads. –AN seals at the cone seating. NPT (tapered pipe thread) always requires thread tape or compound. Straight –AN thread that does not feature an angled seat requires an O-ring for sealing.

Super Stock hose end from Earl's. The female thread accepts dash size fittings, but the hose end is a barbed tube. This design offers easy assembly, but is designed to be used only with Earl's Super Stock hose, and will not work with stainless braided hose. This is a popular and economical option for OE replacement hose assemblies. Available in all popular sizes and shapes.

This is really cool. This is an example of Earl's Tube-Mate Auto Fit hose end. This type of hose end allows connection of flexible hose to low pressure tubing. Simply connect the red collar to the hose, flare the tube and join the assembly together. Tubing should be flared, beaded or grooved for secure installation. (Courtesy Earl's)

Hose ends not only allow hose connection and sealing, but because hose ends are available in various shapes, the selection of the hose end shape will allow you to obtain installed angles necessary for clearance and preferred routing. Hose ends are available in straight, 30-degree, 45-degree, 90-degree, 120-degree, 150-degree and even 180-degree shapes. With regard to angled hose ends, these are also offered in two basic "profiles," including standard profile and low profile. Standard profile hose ends feature a tube neck, while low-profile hose ends feature a forged or machined neck that shortens the hose end's distance from the fitting connection to the bend. Low-profile versions are good choices where installation space is at a premium.

Note: Depending on the size of the AN hose end,

This is the plumbing most folks are accustomed to see on a race or street custom engine. Stainless braided hose does its job in providing high pressure connections and in protecting the hose from damage.

This is the cheaper way to go, if appearance is your only concern. Inexpensive plastic braid sleeve covers can be slipped over standard rubber hoses. Some folks think that this provides a pro or racey look. My opinion is that this approach looks cheesey, and it does absolutely nothing to increase hose performance or durability.

Nylon braid-sheathed hose with textile inner braid and synthetic rubber liner. Kevlar-braided hoses, like Earl's Ultra Flex 650 hose, features a high-temperature smooth-bore Teflon lining and a convoluted outer wall for superior flexibility. (Courtesy Earl's)

assemblies with pipe thread connections are available in a full range of NPT sizes, from 1/8" NPT all the way up to1/2" NPT (with 3/4" NPT also available from some makers).

Generally, hose ends that feature female dash-size connections are designed to accept the same hose dash size at each end (for instance, a hose end may be labeled "–6 female to –6 hose," which means that a –6 hose will connect to the tube/sleeve end, and the female threaded connection will accept a male-thread –6 fitting). However, some makers offer sizing differentials. For example, a –6 female to a –8 hose hose-end might be available to suit a specialty need. However, the most common approach for hose ends is a design that accommodates the same dash size at each end.

HOSES

We're all accustomed to seeing braided stainless steel hose with blue and red hose ends on custom cars and race cars. However, great strides have been made by the various hose makers, resulting in a wide array of choices with various benefits. There's certainly nothing wrong with stainless braided hose. It works great and looks great. The only downside of stainless braided hose is its weight and its abrasive outer surface. While the durable stainless braid does a wonderful job in protecting the hose itself, this surface can scratch, or even file-through an adjacent surface if the hose rubs against anything. Where weight savings are needed, and/or abrasion damage needs to be avoided, the industry now offers a selection of optional hose designs that are lighter and that won't abrade nearby surfaces, but still provide the strong reinforcement and protection for the hose itself.

One type isn't better than another…it's simply a matter of selecting the type that best suits a particular application or preference.

Synthetic rubber hose that features an interior braided fabric sheath serve well for fuel, oil or water. These AN dash-sized hoses provide a clean look and are offered in colors. (Courtesy Earl's)

As an alternative to stainless braided hose, lighter-weight high-performance hose is offered with Kevlar outer woven braid or nylon braided sheathing, and special elastomeric hose that features no outer braided cover. Synthetic rubber hose featuring interior braided fabric sheath (for strength) is available with maximum pressure ratings in the 250 psi range, and can be

Stainless steel braid cover sleeves offer abrasion protection to plain-jane hoses. Along with the use of worm-drive clamp-equipped collars, the "look" of AN plumbing can be achieved. Well, that's the theory. Personally, I hate using this stuff. It looks phony, as though it was added purely for appearance, but missed the mark. I know that many people use this stuff to cover everything in sight, including hoses and wiring, but to me it's just tasteless overkill. But, it's available if you want it. As far as I'm concerned, if you're not going to use the real deal (AN plumbing system), the engine bay will look cleaner by leaving your hoses bare and clean. Sometimes, more (shiny stuff) isn't necessarily better. (Courtesy Earl's).

This oiling system features crimp type hose ends, with a combination of Kevlar braided hose and clear-sleeved braided stainless hose.

Shown here are two straight –6 AN adapters, each combined with NPT threads. The top is –6 to 1/2" NPT, while the lower adapter is –6 to 3/8" NPT. Note that all NPT threads must be sealed with Teflon sealing tape or compound.

assembled using hose nipples and beaded tubing. Hoses with a tough, abrasion-resistant outer nylon sheathing bonded to a textile inner braid and synthetic rubber liner can be assembled using traditional AN hose ends or crimp hose ends, with maximum pressure ratings in the 1400 psi range. Ultra-light Kevlar-outer-sheated hoses are available that weigh about 50–60% lighter than stainless steel braided hoses, with maximum pressure ratings in the 1100-1200 psi range, but may require only the use of dedicated crimp hose ends. A wide range of choices are available today, not only in terms of weight, flexibility and materials, but in colors as well.

FITTINGS/ADAPTERS

The fitting (also called an adapter) allows the connection of the hose end to the component. In some cases, the fitting will be provided as part of the component (for example, an engine oil cooler might already feature welded-on –10 fittings, so a –10 hose end will thread directly onto the cooler). In other cases, you'll need a separate fitting to finish the connection. Fittings are available with AN threads (and 37 degree flare seat) on one end, and an appropriate thread on the opposite end to fit the component. For example, let's say that you need to plumb a –10 hose assembly to your water pump-to-heater routing. The water pump may feature a 1/2" NPT female threaded hole. In order to complete the connection, you'll need an AN fitting that features a 1/2" male NPT at one end and a –10 male at the opposite end to attach to your hose end. Fittings are available in a wide variety of both thread size combinations to accommodate any reasonably practical connection between various AN hose ends and NPT threads. Fittings are also available in straight thread (as opposed to tapered pipe thread), for locations that feature straight-thread holes. If the AN fitting features straight thread for the component connection, a sealing O-ring will likely be required.

Fittings, like hose ends, are also available in various shapes, including straight, 45-degree and 90-degree, to further accommodate plumbing routing and clearance issues. With the combination of hose flex and hose end and fitting shapes, it is possible to achieve extremely neat and tidy plumbing routing that would please even the most demanding appearance-conscious builder.

Metric adapters are also available, to accommodate all domestic and import OE vehicle applications for the use of AN 37-degree plumbing connections. The AN threads are the same as always, but the opposite end, instead of featuring NPT or AN straight thread, features metric thread size instead.

Types of Adapters

Note: As far as hose end and fitting materials are concerned, both steel and aluminum are available. For purposes of weight saving and appearance, most people favor aluminum (as you already know, hose hardware makers commonly offer aluminum AN hose ends and fittings in blue and red anodized colors). However, for brake system connections, steel hose ends and fittings are often preferred to avoid the however-unlikely potential for fatigue cracks.

Aluminum NPT to –AN Adapters—Fitting adapters that allow connection between pipe thread locations and AN hose ends.

Special Purpose Adapters—Unique adapters for special applications.

Bulkhead Adapters—are designed to allow

plumbing through wall surfaces, such as firewall, rear bulkhead, etc.

Caps & Plugs—these include caps and plugs for AN 37-degree, AN straight thread and NPT applications.

–AN Tube Nuts/Sleeves—allow connection of hard line tubing using AN thread.

Aluminum Weld Fittings—These aluminum weld fittings, often referred to as bung fittings, permit creating a threaded connection port on an aluminum surface…for instance, when connecting plumbing to an aluminum oil pan or valve cover.

–AN Carburetor, Fuel Injection, Power Steering—Permit AN plumbing to specific types of carburetors, f.i. units and power steering pumps.

Steel Adapters—In addition to the wide array of

This example shows an –AN adapter with a male –AN (37 degree seat) at the right side as shown here, where the hose end will attach; and a straight thread male at the opposite end. Notice the sealing O-ring at the straight thread side. No thread sealant is required for straight threads. If a cone seat is featured, the mating of the 37-degree seats will achieve the seal. If the –AN straight threads feature no seating cone, an O-ring is required for sealing.

Fuel log installed on a Honda engine. Using this –AN fitting allows plumbing with a –8 hose assembly, and avoids hassling with an unreliable slip-on hose and nipple.

Straight male AN to NPT adapter. A wide variety of pipe thread and AN sizes are available to suit any requirement.

Female NPT to male NPT reducer. This allows the use of a smaller NPT male fitting at a larger female NPT hole. For example, if the female NPT hole in an intake manifold featured 1/2" NPT threads at a water jacket port, but you want to connect a 3/8" NPT fitting, this pipe bushing reducer would allow that change.

Dash AN male to straight thread male. This example features a –6 AN male and a 9/16" x 24 male thread with a crush washer, suitable for fuel plumbing or other custom application.

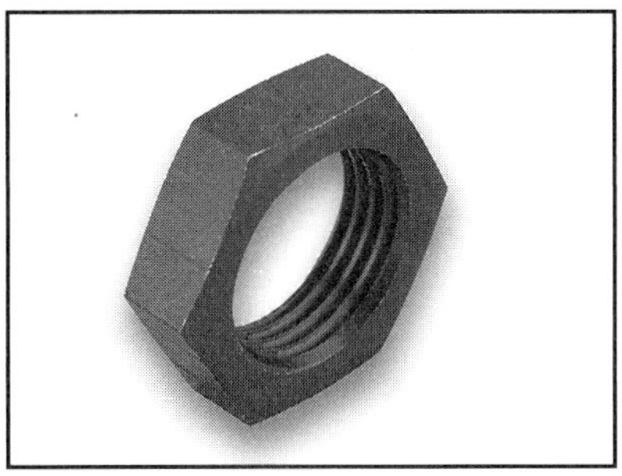

Bulkhead nut. This allows you to secure the bulkhead adapter to a wall. Drill an appropriate-size hole in the wall, pass the long end of the bulkhead adapter through the hole, and secure the adapter with this nut. Always order the bulkhead nut based on the size of the bulkhead adapter (–3, –4, –6, –8, –10, –12 or –16). Always match nut and adapter material. If the adapter is aluminum, use an aluminum nut. If the adapter is steel, use a steel nut.

Extended straight male AN to straight thread. The example shown here features a male –8 AN and a 7/8" x 20 male thread, with a sealing crush washer. This application would suit certain carburetor feed applications.

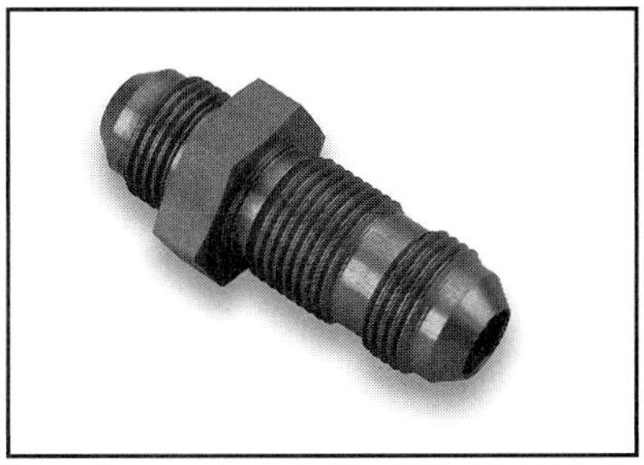

Straight bulkhead adapter. Same size AN male at each end. The longer section passes through a hole in a bulkhead wall. The exposed straight threads accept a nut which secures the adapter to the bulkhead. Separate hoses can then be connected to each end of the adapter, while the adapter remains in a fixed position.

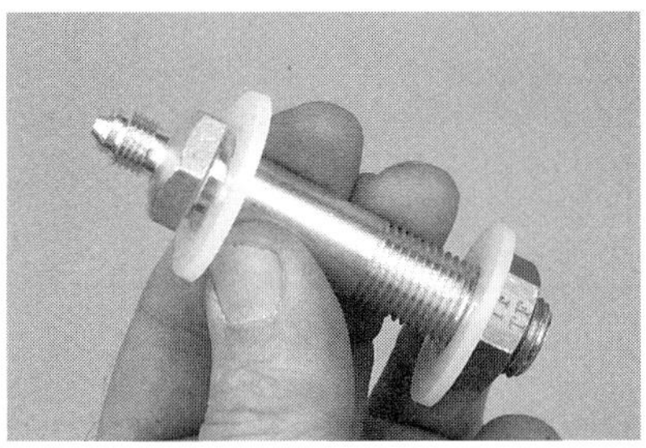

This special bulkhead adapter is designed to mount through a frame rail, allowing connection of a brake line (from caliper hose to hard line), eliminating the need to run a hard line under/over the frame, for a neat installation. One end features a –4 AN male flare and the other end features a female NPT thread.

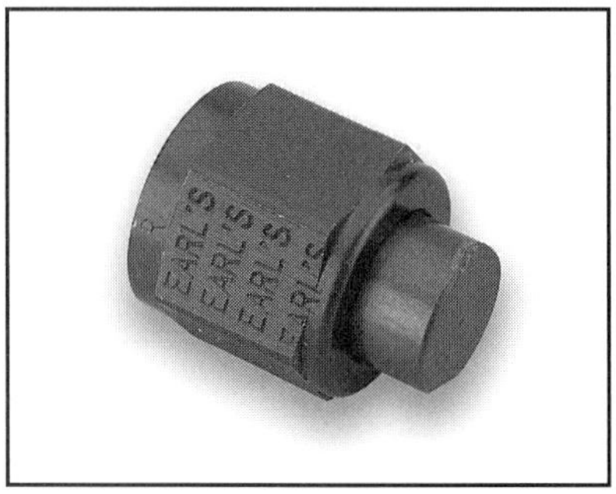

–AN cap. Features a female AN (37 degree sealing cone). Designed to cap-off a male AN fitting. Common uses include capping-off a fitting when hoses are disconnected, to prevent fluid leaks or to protect fluid passages from contaminants (during engine storage, repairs, etc.). All dash AN sizes available.

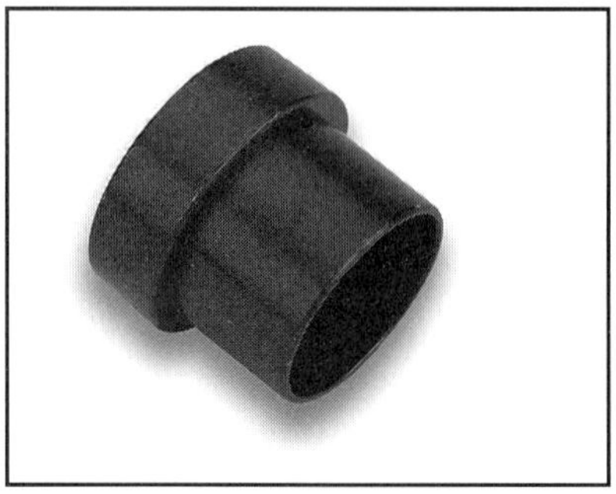

Tube sleeve. Once the tube nut is slid onto the tube, the tube sleeve slips over the tube. Once the tube end is flared, the tube sleeve, captured by the tube nut, slides up to the flare, allowing the tube nut to be threaded onto an AN male. The inside diameter of the exposed sleeve features a 37-degree flare seat that mates to the male AN fitting.

AN female O-ring seal weld fitting. The step-down shoulder fits into a drilled hole and the fitting is then welded in place. The female threads are AN straight threads, requiring an O-ring to seal.

NPT internal plug. Features male NPT thread and a female hex drive. For capping any NPT port. Available in all NPT sizes. Remember: always apply Teflon tape or sealing compound during assembly of any NPT thread.

Tube nut. This pulls the tube sleeve up against the tube flare as the nut is tightened onto a male AN fitting. The nut simply secures the 37-degree seats of the male fitting and tube nut together.

–6 AN male to 9/16"x24 male with sealing crush washer. For single inlet Holley carb.

–6 AN male to 1/2" x 20 male. Typical application for power steering pump or fuel pump. 5/16" tube inverted flare. O-ring included.

–6 AN male to 5/8" x 20 male. Typical for Carter/Edelbrock carburetor. Crush washer included.

–6 AN male to 1" x 20 male, with sealing crush washer. Typical for 75-up Rochester.

–8 AN male to 7/8" x 20 male, extended. Requires included crush washer. Typical for Holley dual feed carb.

–6 AN male to 10mm x 1 male, with crush washer. Typical for Weber carburetor.

–6 AN male to 12mm x 1.25 male with crush washer. Typical for Dellorto, Solex and Mikuni carburetors.

–6 AN male to 14mm x 1.5 male. Typical for some power steering and fuel injection applications. O-ring seal.

–6 AN male to 16mm x 1.5 female. For some fuel injection systems. O-ring seal.

aluminum hose ends and adapters, steel components are also available for high pressure hydraulic and other severe duty applications where steel is preferred.

Bare Aluminum Adapters

For those who wish to have their AN adapters custom-anodized in other colors, bare, un-anodized aluminum adapters are available from select manufacturers (not all makers offer un-anodized items). If you want un-anodized hose ends, you need to check with the various makers. Depending on existing stock or backorder situations, you might be able to place a custom order, or you might get lucky and be able to snatch what you need from existing stock. There are several anodizing shops in the country that do excellent work. While the "standard" colors that all anodizing shops carry usually include black, blue and red, there are other shops that offer a dozen or more colors, and yet other shops that can attempt to match the color you want.

If you purchased already anodized hose ends and adapters, but want to change color, this can be done, but you need to be aware of a few things. If the item has already been anodized, and since anodizing is essentially an oxidation process, the surface is now etched. When the original color is stripped off (chemically), the resulting surface will be rather dull. If you like a "flat" look, you're in business. If you want a gloss surface, you have two choices: each piece must be polished before anodizing, or you can request a clear gloss finish over the color (an option not all anodizing shops offer). From a standpoint of appearance only, be aware that one of the big variables lies with the composition of the aluminum stock. Depending on the hardness and alloy makeup, some aluminum pieces will provide a glossier finish than others. Also, the same color anodizing will appear lighter or darker when applied to different grades of aluminum. If you want all of the custom-anodized pieces to match, be sure that you obtain all of the pieces from the same manufacturer!

Here's an un-anodized aluminum AN to NPT adapter. If you can obtain these in their "bare" form, the results from custom-color anodizing will be outstanding. Otherwise, if the part must be stripped, it may require polishing before anodizing; or a clear-coat may be applied over the anodized color by the anodizing shop.

Seals/Crush Washers for Straight Thread

With NPT threads, simply apply Teflon sealing tape or compound. However, when using an AN adapter that features straight thread at one end, you need a sealing washer. Depending on the applications, you'll need either an O-ring or a crush washer. O-rings are available in both Buna N synthetic rubber and in Viton. Buna is compatible with most fuels, hydraulic fluids and lubricants up to about 275 degrees F. Viton is preferred for synthetic lubricants, oxygen-bearing fuels and additives and non-ether-based brake fluids. Viton also offers better scuff resistance, longer service life and offers a higher operating temperature range of about 400 degrees F.

Crush washers do exactly that...they crush under compression, providing the needed seal between two flat surfaces, where there is no groove relief to accept an O-ring. Crush washers are commonly available in aluminum and copper. Banjo fittings, for example (as used on brake line connections to calipers) require two crush washers...one between the banjo fitting and the caliper, and one between the banjo fitting and banjo bolt head.

Crush washers MUST match the inside diameter of the adapter. Whether you're dealing with aluminum or copper, it is highly recommended to never reuse old crush washers. Always replace with new, since once they're crushed, they don't "spring back" to their original thickness, and the "fingerprint" will have mirrored the parent surface of the original installation, and might not seal against a different surface.

Crush washers are generally available in hole diameters including 3/8", 1/2", 10mm, 12mm, 7/16" and 9/16".

Another and better approach to sealing banjo assemblies are aluminum washers with built-in rubber O-rings. This approach offers the best of both worlds—the strength and crush of aluminum, plus an O-ring that expands to provide a superior seal. This type of sealing washer is available in two basic formats (one is of American design and the

O-rings are available in either BUNA N or VITON.

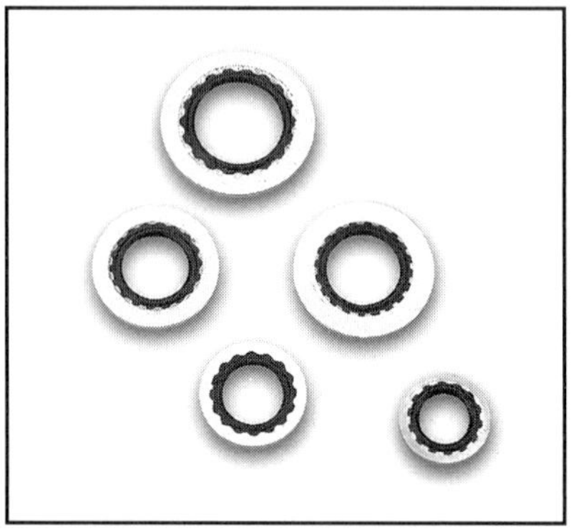

Stat-O-Seals. These feature synthetic rubber O-rings captured inside aluminum washer shells. These buggers work great.

other was born in England). The American design, called the Stat-O-Seal, features a synthetic rubber O-ring captured (mechanically locked) inside the I.D. of an aluminum washer. This is a great choice to seal anything that normally would use an aluminum or copper crush washer.

The English-origin version is called the Dowty seal, which is similar to the Stat-O-Seal, but the O.D. is smaller and the washer is thicker. Either works well, but the Stat-O-Seal is more readily available.

As an example of the superior sealing offered by a Stat-O-Seal, I recently reconfigured the front brakes of a 1973 Duster (fitted with a non-original 426 Hemi) to disc brakes. The owner wanted an "OE" look, so I chose single-piston cast iron calipers from a well-known caliper remanufacturer. I initially installed copper crush washers at the banjo fittings, but could not get them to seal (constant wetness). I tried different-thickness copper crush washers, and even tried aluminum crush washers, but to no avail. Further inspection showed that the sealing seats on the calipers (where the crush washer was position between the caliper seat and the banjo fitting) were never re-faced after the caliper bodies were blasted during cleanup. The sealing seat surface was irregular (not milled). I exchanged the calipers for replacements, but with the same pitiful results. Finally, I replaced the crush washers with Stat-O-Seals. Guess what? Instant perfect seal. No leaks, no more cussing and no more re-bleeding.

While we're on the subject of sealing fluid, here's something many people aren't aware of. If you have a damaged AN 37-degree sealing surface (hose end of adapter) that may have been caused by galling when assembled dirty, etc., you can make a quick and easy field-fix without the need to replace the hose end or adapter. Aluminum conical seals are available in all AN sizes. This is simply a small, shallow conical "cup" that matches male and female seat angles. Simply push the conical seal over the male cone and reassemble the hose end to the adapter in the normal manner.

–AN HOSE ASSEMBLIES

Note: –3 and –4 sizes are relatively small in diameter. Especially for brake hose applications, these are very difficult to assemble by hand, and often require crimped hose ends. For brake lines, or high-pressure hydraulic clutch hoses, don't try to make your own. Buy brake hoses or clutch hoses already-assembled in the length you need.

The following instructions apply only to stainless steel braided hose and appropriate hose ends. Crimp-type hose ends require special crimping dies, available from the hose maker. Dedicated hydraulic crimping machines are also available, and come in handy, especially if you plan to assemble a bunch. Slip-on hoses and barbed nipples are self-explanatory. Simply "slip" the hose over the barbed nipple. Be aware that this isn't as easy as it sounds, since the fit will be very tight. Eat your Wheaties before tackling slip-ons.

Cutting Stainless Braided Hose

When you buy a length of –AN stainless steel braided hose and plan to cut pieces to desired length and make your own assemblies, you first need to know how to properly cut the hose.

Yes, you can use a hacksaw, but frankly, that method stinks. It's difficult to obtain a square cut, you need to use a special hardened blade, and chances are very high that you'll end up with frayed wire ends that you'll then need to snip off with wire cutters (by the way, snipping this hard stainless wire braid isn't as easy as it may seem). If you secure the hose in a vise to make your cut, you'll have a tendency to smash the hose out of round, which will only aggravate the problem.

The best method is to use an abrasive "chop saw." First, wrap the area to be cut with electrical tape or a quality bodyshop masking tape (wrap the hose tightly). Mark your intended cut on the tape, carefully cinch the hose in the saw's vise without distorting it, and let the abrasive wheel slice through. Don't apply a bunch of pressure on the saw arm. Light to moderate pressure works best.

Once the cut is made, remove the tape and thoroughly rinse the hose out using compressed air to remove all wire and rubber particles. This is important! Make sure the hose is clean inside. Naturally, you must always wear eye protection when cutting hose, when trimming stainless braid or when blowing hose clean.

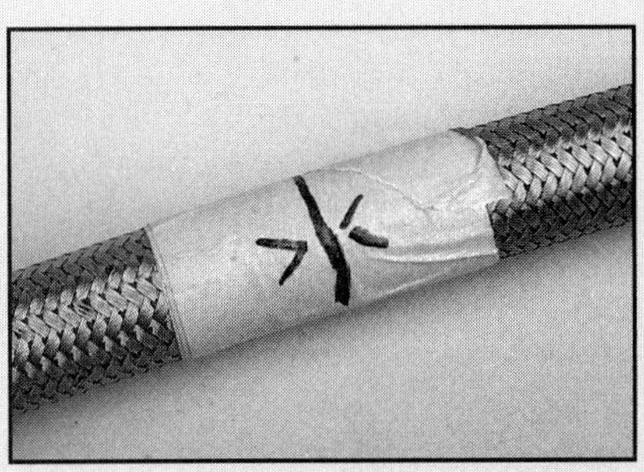

1. Mark the hose at the cut location. Body masking tape works well not only as a background for marking, but to help retain the stainless steel braid ends once the hose is cut.

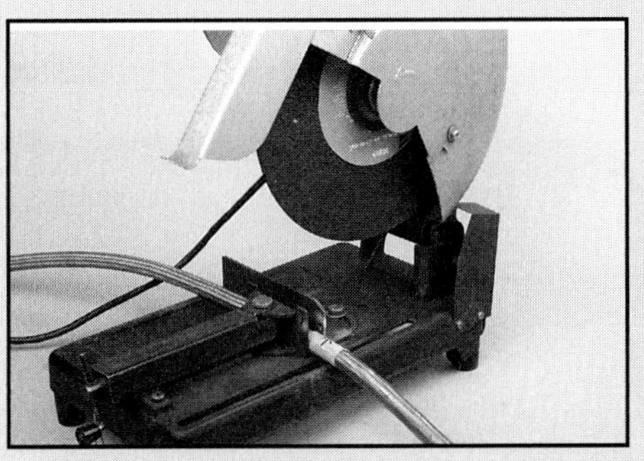

2. An electric chop saw works great for cutting stainless steel braided hose. Position the hose on the saw, making sure that the saw clamp is adjusted for a 90-degree cut. Avoid cutting the hose at an angle.

3. Allow the saw wheel to gain full speed before contacting it to the hose. Maintain steady, moderate pressure. Don't push so hard as to deform the hose.

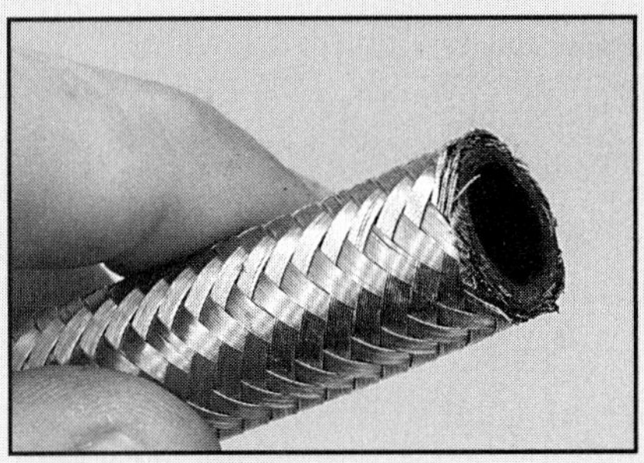

4. This cut is the result of using the abrasive-wheel chop saw. I pulled the tape off after the cut was made. This clean cut required no trimming of frayed braid wire.

5. This is what happens when someone tries to cut stainless steel braided hose with a dull hacksaw, or with "snips" or by cutting the hose without the cut area being wrapped with tape. Don't attempt to install a hose like this to a hose end. If you snip the frayed ends with metal snips, you'll end up making more of a mess. It's best to start over.

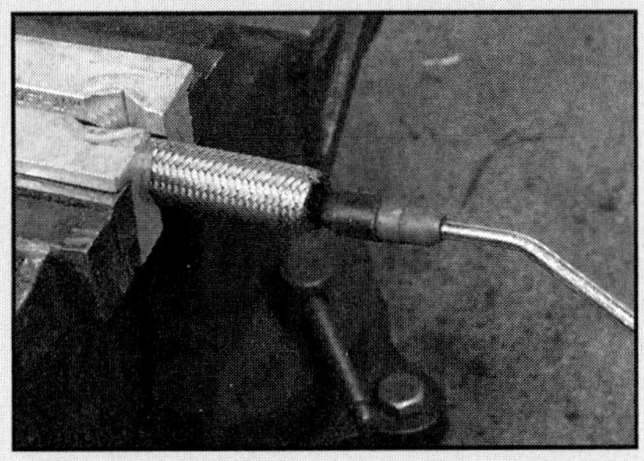

6. This hose was cut on the abrasive wheel chop saw. Once the cut is made, be sure to blow the hose out with compressed air to remove bits of rubber and braid. We lightly secured this hose in a vise only for the photo.

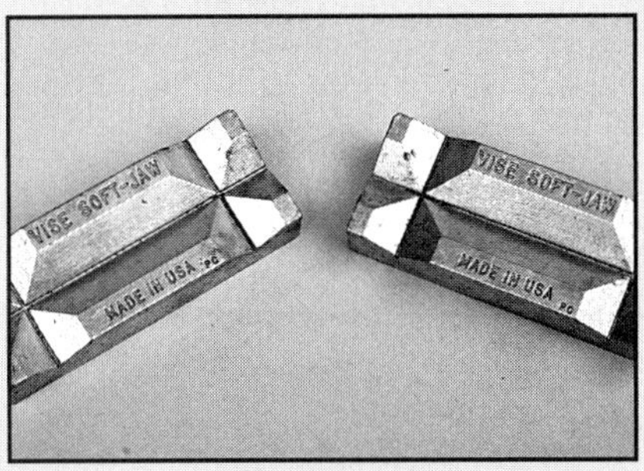

7. If you plan to assemble –AN hose ends, buy a pair of these soft aluminum vise jaws. You can secure aluminum hose ends or adapters in these jaws without damaging the material. If you keep the jaws clean, you can also eliminate surface scratches on the anodized pieces. The jaws feature female V-cuts that accept the hex of the hose end or adapter, in either horizontal or vertical planes, as you wish.

8. The backside of the soft jaws feature a step (this locates the jaw onto the top surface of the vise jaw) and a magnet, which holds the aluminum jaw to the vise.

9. A pair of soft jaws installed on a vise. These jaws will come in handy for plenty of non-related future fab work as well, whenever you need to secure a piece of aluminum round bar stock, etc.

10. Here we insert a stainless steel braided hose into a hose end collar. There are many ways to go about this. Often, I simply screw the hose end collar onto the hose by hand. We're simply showing the collar in a vise for illustration. However you approach this, be careful of the braids at the cut end.

11. Push/twist the hose into the hose end socket roughly 1/8" to 3/16" short of the threaded area.

12. Check to make sure that the hose is fully seated in the hose end socket. Don't push the hose into the threaded area If the hose (and its stainless steel braid) enters the threads, they can jam between the male and female threads when the hose end is assembled.

13. Don't push the hose too far. Make sure that the stainless steel wires won't enter the fine threads.

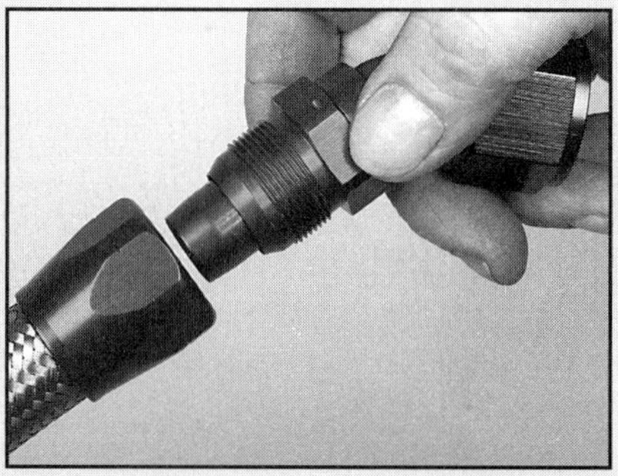

14. The section of the hose end that features the metal sleeve threads into the hose end's socket. Apply a bit of lube to the sleeve and to the threads to aid assembly and to prevent galling. A light lubricant such as WD-40 or lithium grease works fine.

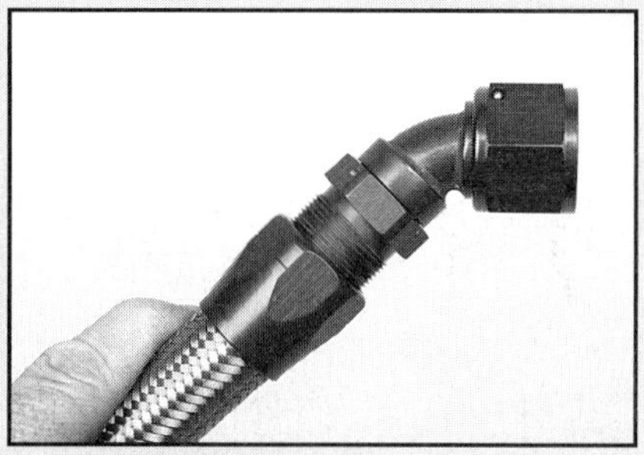

15. A 45-degree hose end shown here before being fully tightened. The hose end's socket (attached to the hose) can be secured in an aluminum vise jaw, while the hex on the hose end's threaded collar is tightened. Note that some hose ends are designed to rotate, while others are designed to retain a fixed position once tightened. If the hose end is not designed for rotation after assembled, make sure that the angle aligns with your intended fitting once installed.

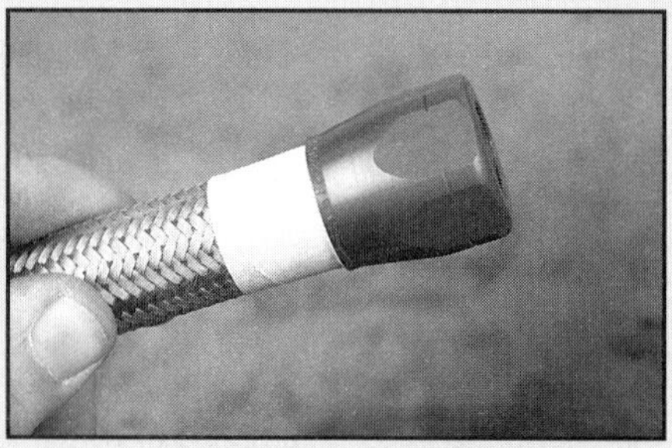

16. Once the hose end collar is installed, I like to place a piece of tape on the hose, flush to the base of the collar. This provides a visual reference if the hose begins to pull out of the collar during the rest of the hose end assembly.

17. Here the hose end collar is secured in aluminum soft jaws on the vise. There's no need to tighten the living daylights out of the jaws. Just snug the jaws to prevent the collar from moving.

18. Apply a bit of lube to the hose end's nipple and threads.

19. Insert the hose end's tube into the collar, being careful to center the nipple to the hose I.D. If you try to insert the nipple off-center, you can force the nipple edge into the hose rubber, damaging the hose. Take your time. You'll be able to feel the nipple entering the hose.

20. Continue to insert the nipple until you can engage the threads. Make sure the threads are not crossed. Tighten as far as you can with your fingers, verifying that thread engagement feels/looks good.

21. If you care about the finish of your hose end, use a clean aluminum –AN wrench to tighten the hose end assembly. Here we're using a –10 wrench, which fits our –10 hose end.

22. Continue to tighten the socket/tube into the collar, using moderate pressure. Do not overtighten! Remember: these are aluminum parts. Also, as you tighten, the nipple seals into the hose I.D., creating a leak-proof connection. Note: it's a good idea to apply a bit of a push of the hose towards the collar during tightening, just to make sure the hose isn't pushed out of the collar. Again, observe your reference tape.

23. You'll know the assembly is fully tightened when it stops or offers too much resistance. This narrow nut is almost touching the collar.

24. Blow with compressed air again. Once a hose is fully assembled, I like to run hot soapy water through it, followed by compressed air, followed by air-drying. Do everything you can to make sure the hose is clean inside. I know plenty of guys who gripe about clogged carburetor jets or stuck needles and seats, when it's their own fault because they didn't take the time to inspect and clean new hose assemblies before installing them.

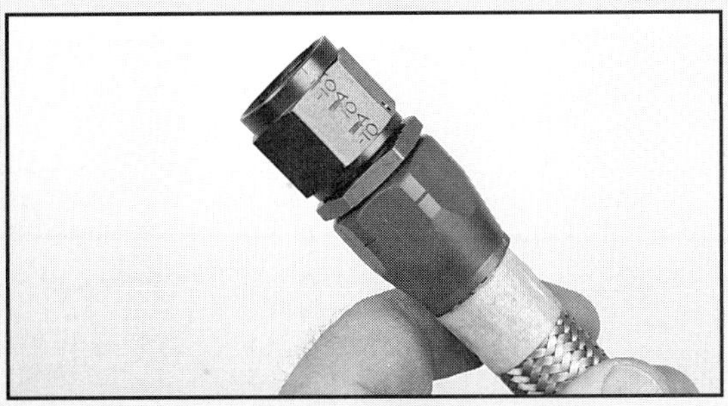

25. Hose end fully installed onto a braided hose. Note that the reference tape only moved about 0.020", which is fine. Using the tape helps. If the hose walked out noticeably, the hose end socket/nipple must be removed, the hose repositioned and reassembled.

1. These are the hose end installation tools from Koul Tool. Two kits are available. The kit on the left handles sizes –4, –6 and –8, while the kit on the right handles sizes –10, –12 and –16.

3. Assemble the two clamshell halves together.

2. Simply place the hose end collar into one half of the clamshell tool, with the hose entry port facing the tool's funnel end.

4. Lube the funnel port, and insert the braided hose, "screwing" the hose into the tool in a clockwise movement. You'll feel the hose stop against the collar threads.

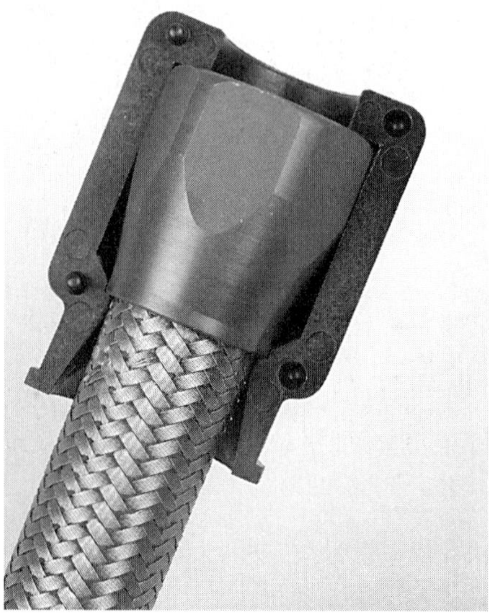

5. With one half of the clamshell removed, you can see the hose installed into the collar. This is a very "slick" (pun intended) way to install a braided hose into a collar. Note: because some maker's hose end collars vary in length, the Koul Tool kits include plastic spacers, and instructions regarding the need for a spacer (and spacer size), depending on the specific brand of hose end being assembled. If a spacer is needed, it is positioned between the collar's threaded port (the hex side) and the inside of the tool. If a certain hose end maker's collar is on the short side, as it is pushed back into the tool during hose insertion, the hose entry port might move back behind the funnel port, possibly allowing the hose braid to expand. The object is to place the hose end collar's hose port directly at the base of the tool's funnel. This isn't difficult. The tool instructions clearly tell you when a spacer is needed, so there's no need for you to figure anything out. Here we used an Aeroquip –10 hose end collar, where no spacer was required.

6. As long as the hose and collar are in the tool, you can take advantage of this and insert the hose end socket/nipple. Here, we removed the hose and tool from the vise and flipped it 180 degrees to expose the collar's threaded port. Remember to lube the nipple and threads.

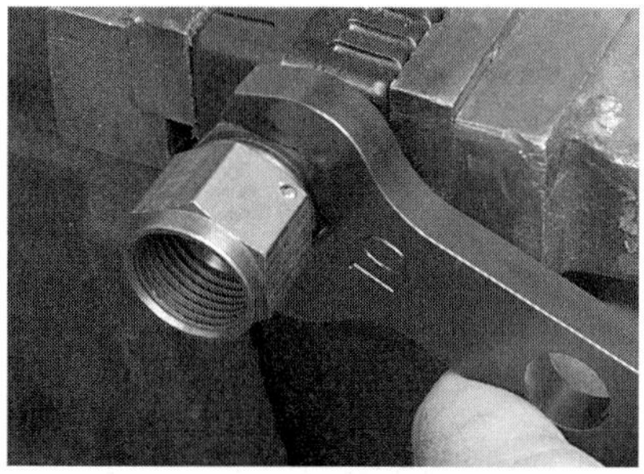

7. Here we tighten the hose end assembly using an aluminum –AN wrench.

Handy Assembly Tool

While I'm on the subject, here's a very cool specialty tool that eases the pain of hose assembly. I just came across this tool recently, and I like it. It's called the Koul Tool (aptly named). While you don't absolutely need this tool, if you want to save your fingertips from needle punctures from frayed stainless steel wire, I'd highly recommend it.

The tool is essentially a guide that allows you to feed the hose onto the hose end's collar, while encapsulating the sharp tips of the wire braid.

The tool is offered in kit form, to cover all popular hose sizes. Kit P/N 468 covers sizes –4, –6 and –8; while Kit P/N 1016 covers hose sizes –10, –12 and –16.

Here's how it works: choose the correct size tool for the hose end to be installed. The tool is made of two pieces that clamshell together. Place the hose end collar in one side of the tool, enclose it with the mating side, secure the tool in a vise, lube the funnel entrance, and install the hose using a twisting motion. The tool's funnel entrance guides the hose neatly into the collar with no muss or fuss.

In fact, while the hose end cap is still in the clamshell, with the hose inserted, you can leave the tool in the vise and install the remainder of the hose end, threading it into the collar while the collar is held stationary.

The kits include several adapters, since the various hose end and fitting makers often produce their own unique lengths. The kit instructions advise you regarding the need for adapter spacers. For instance, the kit's #1 adapter is required for Aeroquip –16, while –16 Fragola, Goodridge, Earl's

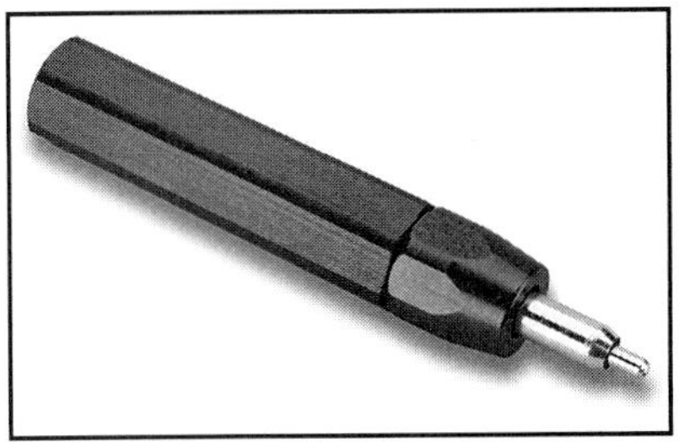

This is a braid spreader, designed as an aid for assembly of –3 and –4 braided hose with Teflon liners. The stainless braid must be separated from the Teflon liner in order to insert the hose end's "olive" into the hose, which can be very tricky. This tool is inserted into the cut end of the hose and pushed, spreading the braid. The olive is then placed on the tool mandrel, pushed over the hose liner and under the braid until the liner bottoms against the shoulder of the olive. Of course, you could take the easy way out and just buy –3 or –4 hose assemblies in the length you need instead of messing around.

AN wrenches are clearly marked for the intended AN size, plus they're usually color-coded per size to make it easy to identify them. Notice the small radiused cutouts at the corners on this wrench, which prevents corner-edge contact. If a bit of dirt has accumulated in the corners, these cutouts help keep grit away from the hose end.

While aluminum AN wrenches are designed for this dedicated task, bear in mind that not all hose end and fitting hex dimensions are created equal, as some hose end and fitting manufacturers' hex dimensions may differ. However, these wrenches will fit the vast majority of applications as intended. The fit of this –12 wrench on this XRP –12 hose end is perfect.

Notice where a stainless steel braided hose rests on this engine's left valve cover. Since it's a race application, the owner might not be concerned about the braided hose scratching the aluminum cover. However, with prolonged contact under vibrational conditions, the braid could eventually wear through the cover. Installation of a support bracket (to keep the hose from making contact) or shielding that small section of hose with a protective rubber sleeve would be advised.

and XRP hose ends require the kit's #2 adapter. The hose end must fit inside the tool tightly, with no end-play. The adapters simply serve as spacers to prevent the hose end from walking inside the tool.

PROTECTING HOSE ENDS AND FITTINGS

Whenever you're dealing with aluminum hose ends and fittings, be aware of two precautions: aluminum is softer than steel, so your steel tools can gouge or burr these items. Also, if you're concerned about appearance (street rod, custom application, etc.), you certainly don't want to burnish off the attractive anodized finish. Instead of taping a wrench or trying to jam a piece of cloth between the fitting and a wrench, make the investment and buy a selection of aluminum –AN wrenches. The aluminum wrenches, unless they're dirty and gritty, won't damage your pretty hose ends or fittings.

Also, be aware that an anodized component's finish can be damaged by aggressive solvents. Anodizing isn't really a coating. Rather, it's a chemical process similar to oxidation that hardens the surface and, where required, provides a die-colored appearance. You can damage the coloration by using some brake cleaner solvents or thinners. However, some anodized items feature a protective clear anodizing treatment over the color-dyed treatment that helps to protect the appearance. Regardless, don't take chances. Use only a mild cleaner, and never use any cleaner that contains abrasives (such as buffing compound).

Slip-On Barbed Nipple AN

If you opt to use AN hose ends with barbed nipples and the proper reinforced hose designed for these hose ends, in theory the task is easy. Simply insert the barbed nipple tube into the hose until the end of the hose seats into the shallow stop-collar. However, this is anything but easy, due to the tight interference fit of the nipple in the hose. Lube the nipple with WD40 or lithium grease, and push the hose onto the nipple. You'll need your strength for this. I've found that it helps to first dip the end of the hose into very hot water (to slightly soften and expand the hose). Once the hose is fully inserted, it's not coming off.

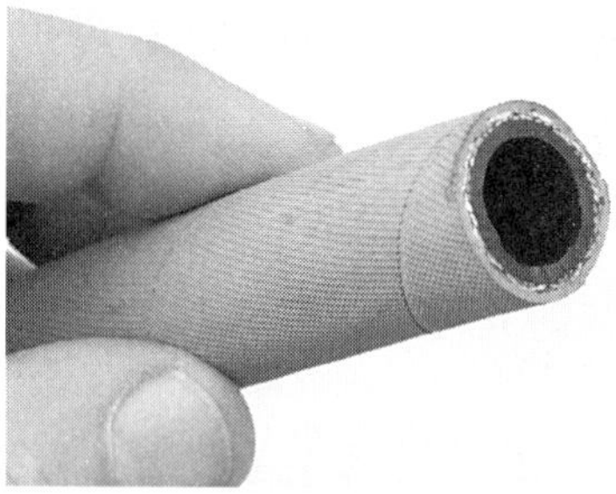

Reinforced hose designed for use with barbed nipple hose ends. This is tough stuff. You'll need a very sharp razor to cut this.

Shown here is a 90-degree barbed-nipple AN hose end. The two barbs provide an extremely secure retainment to the hose.

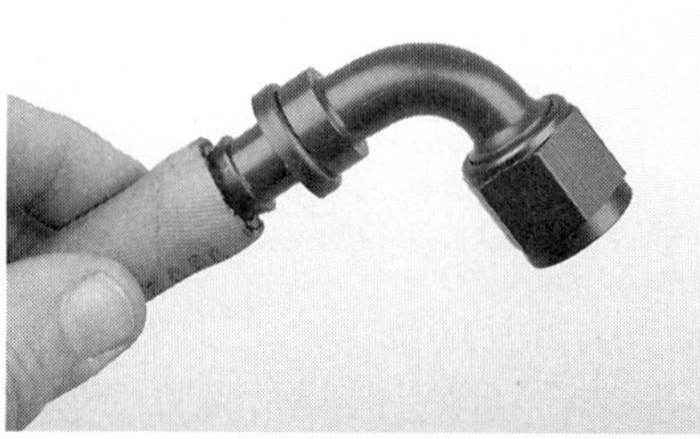

Here the hose is pushed onto the nipple about halfway. Keep working with it until the end of the hose seats fully into the shallow stop-collar.

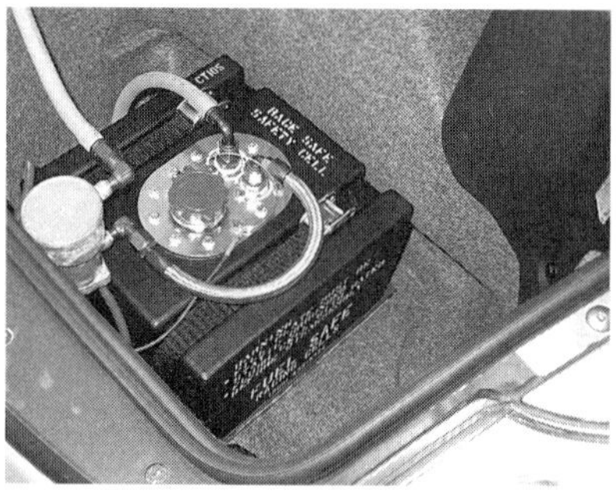

Here's a small fuel cell that we installed in the front boot of a Porsche 911 that was prepped to run in the Mojave desert. This provided extra emergency fuel in the field. We plumbed the feed line (tank to pump) using a –8 stainless braided hose assembly, and the output line (from the pump to a refueling hose) using –8 kevlar reinforced hose connected to a –8 AN hose end with barbed nipple. We could have used slip-ons everywhere, but we were simply dressing up the appearance by mixing the two types of hose.

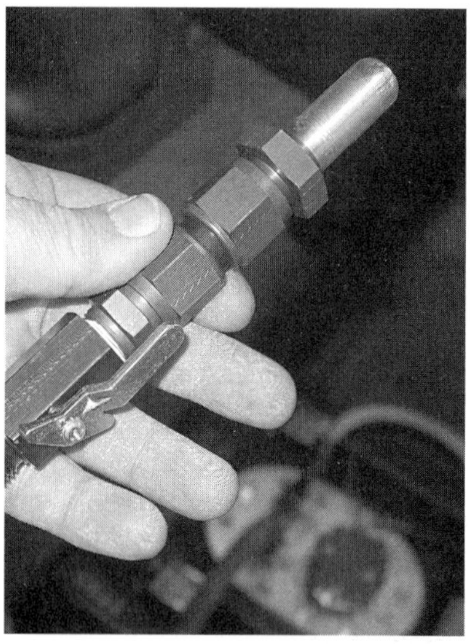

Here's the refueling hose assembly that runs from the fuel cell. Note the slip-on hose at the lower left of this photo. We conjured a setup to allow in-field refueling by transferring the extra fuel from the fuel cell to the car's original fuel tank with this little –8 petcock valve. The outlet nipple started off as a –8 straight bulkhead adapter that I turned-down on the lathe.

QUICK DISCONNECTS

Quick-connect/disconnect fittings are available that allow you to quickly (with no tools) disconnect or connect fluid hose assemblies, and with no fluid loss. Jiffy-Tite (probably the most popular of this genre) offers these self-sealing connectors that are suitable for fuel, oil and water applications. The coupler features a spring-loaded valve that shuts off when the coupler is disconnected, and opens when connected. This is perfect for race cars, where engines are serviced or changed on a regular basis, since this quick-connect feature saves time. Simply slide the coupler collar back to release the connection. You can pop the couplers loose and yank an engine without spilling fluids. Even though these couplers are easy to disconnect and connect by hand, once coupled, the connection is secure, with no worries about accidental disconnection or leakage.

These self-sealing fittings are available in a variety of thread styles including AN, NPT pipe thread, reusable hose ends and barb type hose ends. Every imaginable configuration is also available (male/male, male/female, female/female, etc.).

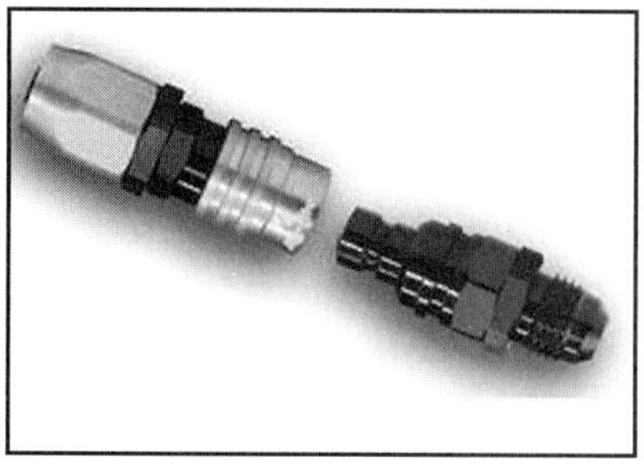

Straight hose end (left) with self-sealing quick-disconnect coupler. This connects to the coupler fitting (right), which in turn features a male –AN flare. (Courtesy Jiffy-Tite)

Quick-connect fittings are also available with NPT thread.

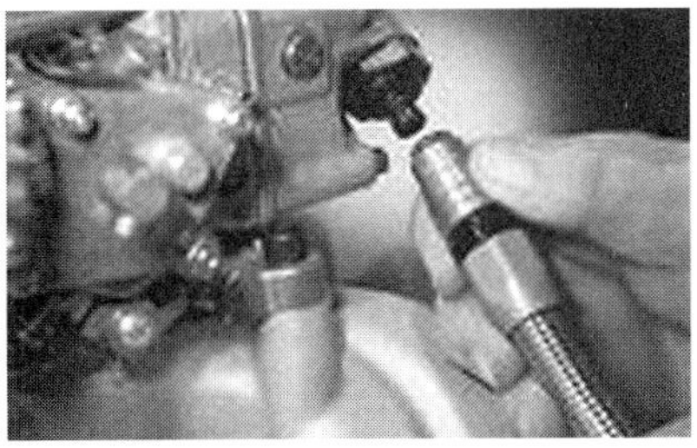

Special quick-connect carburetor adapters are available (installed at the carb bowl as seen here) that accept the quick-connect hose end. This allows you to pull the spring-loaded coupler back, and disconnect the fuel line from the carb with no fuel loss, since the hose end seals itself when uncoupled. (Courtesy Jiffy-Tite)

This dry sump oil system connects to its remote oil reservoir using a Jiffy-Tite quick-connect hose end coupler. This particular engine is on a display stand, but in a race car, this allows you to quickly disconnect the oil hoses without spilling oil. As soon as the coupler is disconnected, a spring-loaded valve shuts and seals.

This quick-connect coupler features a male –AN (where it mates to the 90-degree hose end).

TUBE PLUMBING WITH –AN TUBE NUTS AND TUBE SLEEVES

If hard lines are preferred as opposed to flexible hose, but you want to be able to connect using AN 37-degree seating/sealing, you'll use "tube nuts and tube flares" on metal tubing. The hard line (metal tube) must be flared, using a 37-degree flaring tool. A tube sleeve slips on immediately behind the flare. This tube sleeve features a concave seat that mates to the back side of the tube flare. Behind the tube sleeve is a tube nut. This draws the tube sleeve and the flare on the tube to the male 37-degree AN fitting or adapter.

When making tube assemblies, first slip the tube nut over the unflared tube, followed by the tube sleeve. The tube nut threaded port must face the end of the tube that will be flared. The tube sleeve's smaller O.D. shoulder faces the tube nut, with the larger-O.D. sleeve shoulder facing the end of the tube that will be flared.

AN Dash Size	Equivalent
3	3/16"
4	1/4"
6	3/8"
8	1/2"
10	5/8"
12	3/4"
16	1"

Once the tube nut and sleeve are in position on the tube, then the tube is flared. Don't forget to install the tube nut and sleeve onto the tube before flaring (this is a common oops that many people make).

As far as sizing is concerned, the AN dash size refers to the outside diameter of the tubing.

Hard tubing, equipped with the appropriate size tube nut and sleeve, will allow tubing connection to –3, -4, –6, –8, –10, –12 or –16 male AN fittings/adapters. This hard tubing plumbing may be used for any plumbing application where flexibility is not required. Examples are plumbing fuel and nitrous oxide from the engine-mounted distribution block to the nitrous oxide injectors; or plumbing a fuel line from an engine-mounted fuel pump to the carburetor. If any movement is anticipated between the two points, use a flexible hose plumbing system instead. For example, never run a hard line from a chassis-mounted location to the engine, as engine rock will require plumbing flex.

In some cases, hard tubes are chosen for appearance, where a street rodder has decided that a clean hard line plumbing will best complete the visual theme of the build. Hard tube assemblies are covered in the next chapter.

TUBE NUT/TUBE SLEEVE

Tube nuts are used with tube sleeves on metal tubing which has had the end flared with a 37° flaring tool. It mates to 37° flare end fittings.

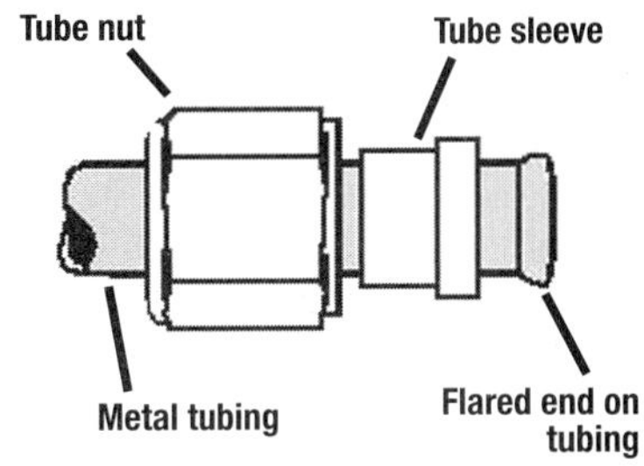

(Courtesy Russell Performance)

Hard line plumbing isn't relegated to small diameter lines only. Notice the curved –10 hard tube, tube nut and sleeve that attaches to the water jacket feed.

Closeup of the spark plug wire tubing arrangement on this custom engine. This was simply an exercise in "doing things a little differently." In this particular build, I just wanted to do something that you don't see everyday.

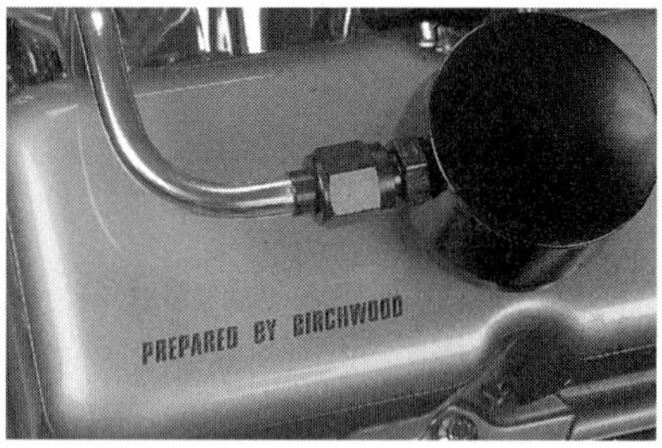

During the build of this custom small-block Chevy, I even hard-line plumbed the PCV valve using 3/8" tubing, and a –6 tube nut & sleeve, mating to a male –6 adapter at the PCV valve.

This hard-tube plumbing setup on a Honda engine shows what can be done with a simple hand-held tubing bender and a little bit of patience. See the next chapter for details on how to make your own hard tube plumbing lines.

Straight tubing runs from the fuel log to the carb inlets, with tube nuts and sleeves at each end.

The hard fuel line is rigidly mounted from the mechanical fuel pump to the carb. Here the hard line, equipped with a tube nut and tube sleeve, attaches to a –6 90-degree adapter that features a male –6 AN flare at one end (for the tube nut attachment) and a female –6 AN at the opposite end, which threads onto a male –6 fitting at the fuel pump.

This custom street rod engine takes full advantage of metal tubing for all of its plumbing.

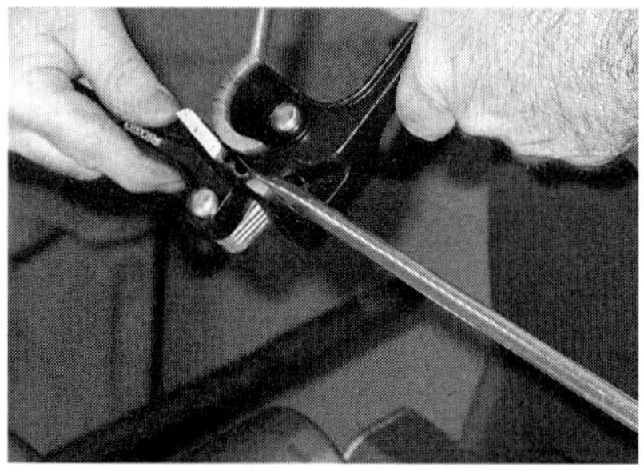

Hand-held tubing bender tools are readily available. This one from Rigid Tools is inexpensive and works fairly well. Again, see Chapter 14 for step by step instructions.

This nitrous oxide plumbing setup features 37-degree cone "jets" that mate to the tube flare and tube sleeve.

Chapter 14

HOW TO MAKE HARD LINE TUBING ASSEMBLIES

The plug wire tubes on this custom small-block Chevy illustrate how clean and neat tubing can be. All of these bends were done with a two-piece hand bender.

TUBE BENDING

For automotive applications, the types of tubing you may be faced with include steel, aluminum or stainless steel (for some vintage applications, possible copper as well). Bending tubes with your hands may achieve the desired function, but appearance will suffer. Also, you may not be able to achieve the desired band radius you need without kinking or collapsing the tube. So, you'll need a tubing bender. While the pro outfits may own elaborate bending setups or computer-aided design (CNC) banding machines, you and I are force to use hand-operated bending tools. These tools will work just fine, for both simple bends and even (to a degree) complex bend combinations. All it takes is patience and a bit of practice.

The two type of common hand bending tools are the one-piece type that features multiple-groove heads (for accommodating 3/16", 1/4", 5/16" and 3/8" tubes). These feature a fixed stopper arm that contacts the roof of the tube. With the tube resting in its groove, hold the exposed portion of the tube with one hand, while you rotate the tool (by its grip handle). The stopper arm allows you to push against the tube as you roll the tube around the radiused groove until you achieve the desired bend angle. The other type features a hinged body with two handles. The head features the grooves (one for each tube diameter size), while the separately-hinged rolling head (also featuring the same tube grooves) compression-forms the bend as its handle is drawn down. A built-in stopper arm also provides a resting contact for the opposite end of the tube. These benders usually feature an incremented head, so you can monitor the bend degree. The common tools of this type usually include a zero mark, a 45-degree mark and a 90-degree mark.

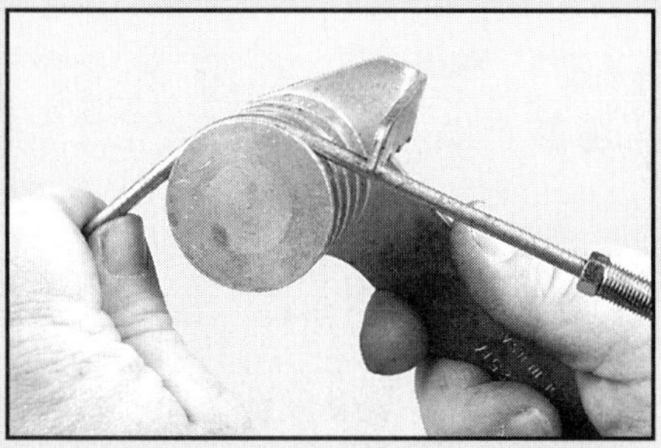

1. The one-piece fixed bender provides die grooves and a stopper arm. While you hold the handle with one hand, you can bend the tube with the other hand. Bends as severe as 180 degrees are possible with this type.

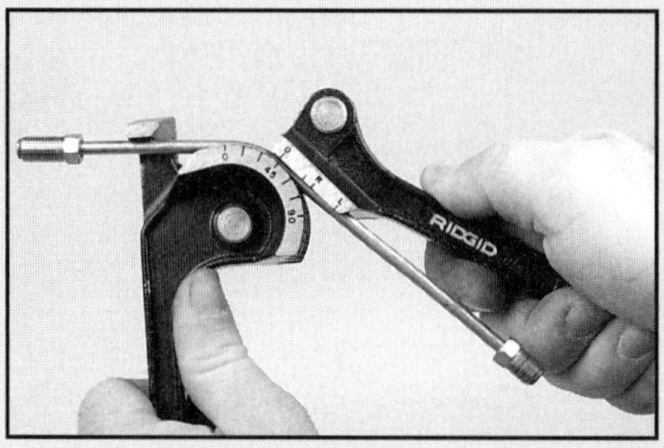

2. The two-piece hinged bender features a stopper arm (seen in the upper left of this photo), while the hinged roller arm draws the tube along the groove, allowing you to monitor the bend angle. Here we've bent a 3/16" tube to an angle of 45 degrees. These tools do an excellent job, and are inexpensive.

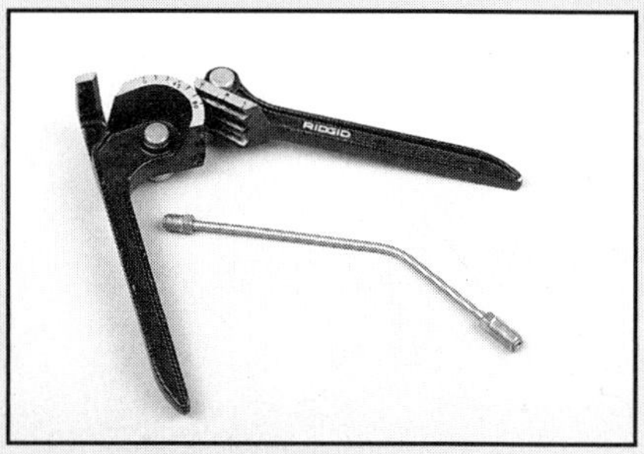

3. A perfect 45-degree bend.

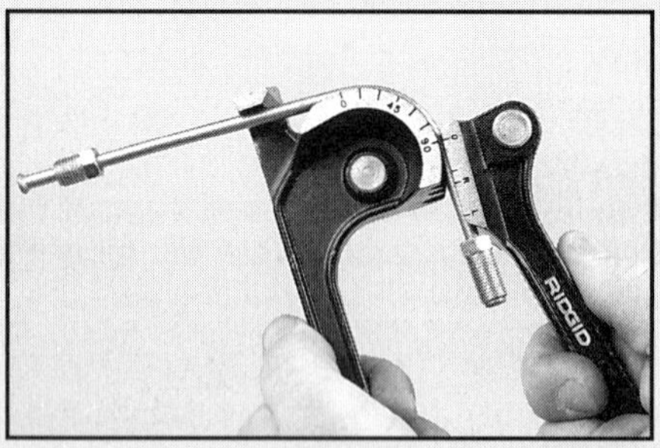

4. Here, we perform a 90-degree bend. While you can achieve correct angles with the one-piece bender, much more attention and constant measuring is needed. Also, the two-piece tool provides matching repeatability, especially important if you're trying to create a number of identical tubes.

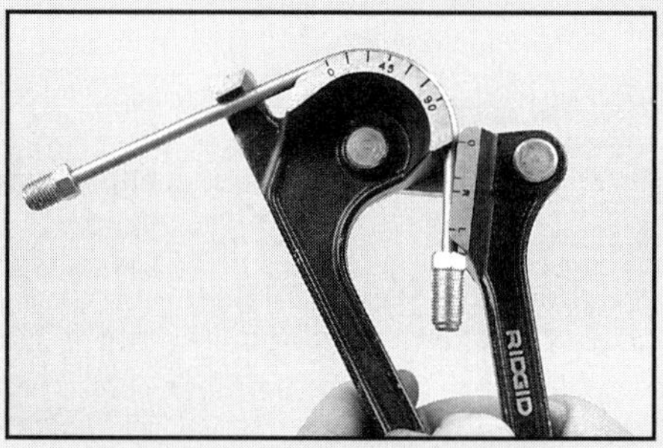

5. Though the scale does not read beyond 90 degrees, a greater degree of bend is possible with the two-piece tool, limited only by the final contact of the two tool handles. This is about 120-degree bend.

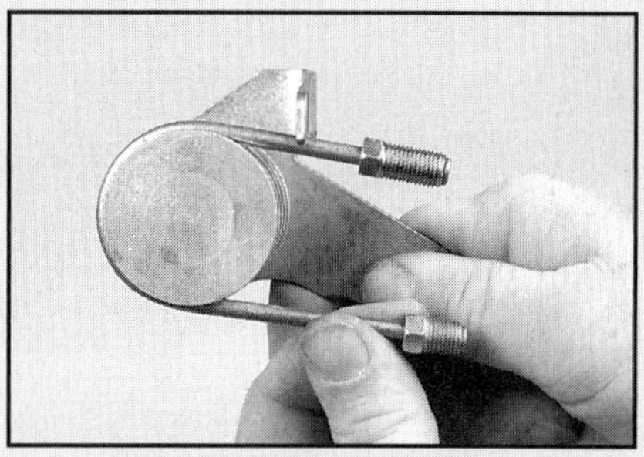

6. The one-piece bender can achieve as much as a 180-degree bend, although you'll need to use care when extracting the tube, since it will be somewhat captured inside the die groove.

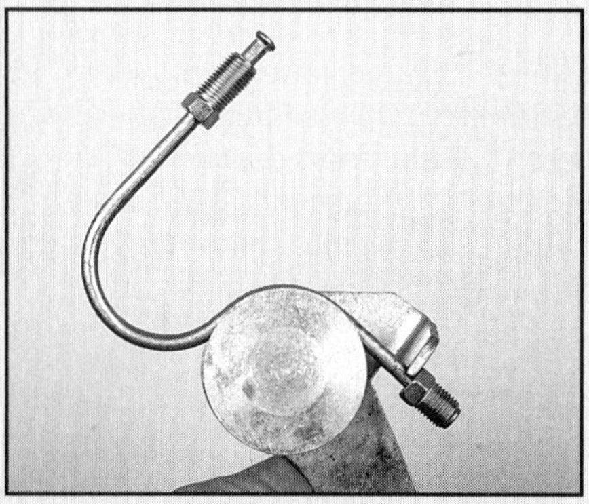

7. Once you get the hang of single bends, you can move on to multiple bends. Here we started with a 180-degree bend, followed up with a 90-degree bend. Performing even multiple bends in the same plane isn't that difficult. However, when you start making bends on different planes, you'll need a lot more patience.

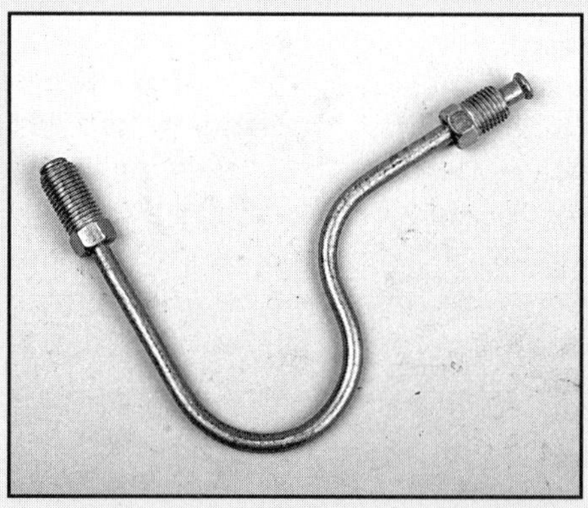

8. While plumbing a frame for brake lines, one location called for a simple 90-degree bend. However, because of an obstruction in the intended path, we formed this drop-down loop. This is the same piece shown earlier, using a combination of 180 and 90-degree bends.

9. Here I used 3/8" aluminum tubing to plumb a custom engine's PCV valve line. I bent the tube, slipped on a tube nut and sleeve, then formed a single flare, mating to a –6 AN male adapter.

10. On this custom small-block Chevy intended for a street rod, we formed 3/8" aluminum tubing as "holders" for 7mm spark plug wires.

11. Our plug wire tubes terminate into the floor of a dual-quad air cleaner, attached using –6 AN tube nuts and sleeves on the tubes and –6 AN to 3/8" NPT adapters at the air cleaner floor. All of these bends were done with a two-piece hand bender.

12. This 1953 Chevy's inline 6-cylinder engine's tri-power carb setup is plumbed with polished copper tubing. Copper is very soft, and easy to bend, but it loves to kink, so a tube bender tool is a must.

13. These 3/16" tubes for a nitrous oxide system started out as eight straight pieces. We used a two-piece hand-held bender for each. The incremented scale on the bender made it much easier to match bends for a uniform appearance.

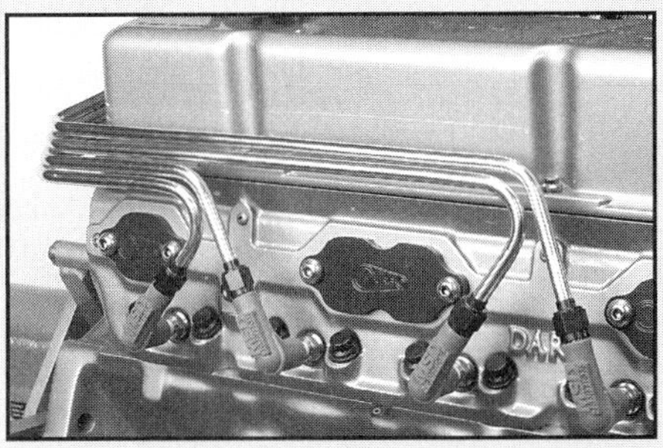

14. This spark plug tube setup flows beautifully. However, this wasn't something that we could rush through. In all honesty, this tube job took us about 11 hours, but we felt it was worth the effort.

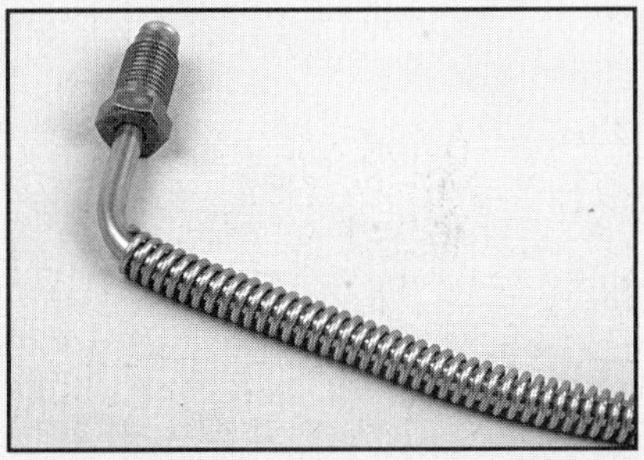

15. These spiral wraps found on some brake lines simply offer abrasion protection for the tube. They are basically "stone guards" to protect the lines. Unfortunately, they also promote rust and corrosion on all-steel assemblies, since they trap moisture and salt. This spiral-shrouded line from Fine Lines uses stainless steel tubing and stainless spiral wire, so problem solved.

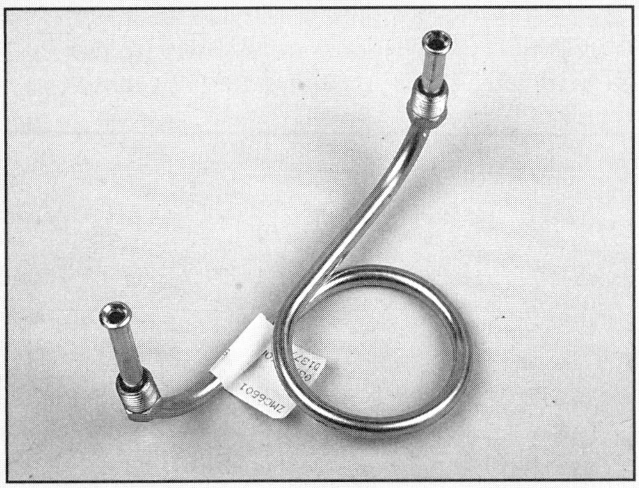

16. One or multiple loops are commonly found in master cylinder plumbing lines. Although there are many theories on the need for these loops, the real reason is to make the master cylinder serviceable (giving you some deflection room during master cylinder removal and installation), and to compensate for potential firewall flex under braking. That's it. The loops have nothing to do with brake fluid cooling, holding residual fluid, or any other old wives' tale. By the way, the pros use a special mandrel to wrap the tube around to create these loops. You can accomplish a loop (or multiple loops) by finding a rod that features an O.D. that matches your desired loop I.D., then carefully bending the tube around the rod. If you plan to do this, use some scrap tube and practice first. Grip the tube as close as possible to the forming rod as you wrap.

FLARING

Flaring creates the seat against which the tube tip will seal against its connection. If the connection (the female threaded hole) features a male angled snout at its base, the tube must have a female, or inverted flare tip. If the connection (female threaded hole) features a female, or concave angled seat, the tube must feature a male, or convex tip.

When flaring, it's important to not only create the proper seat angle, but to create a seat that is concentric (not offset).

In order to obtain a finished tube, you have four choices: purchase a mild steel tube of the correct length, already flared and equipped with fittings. If bends are needed, you perform the bending; or buy an already-finished stainless steel tube assembly that is ready-to-install (that duplicates your original tube in terms of length, bends and flares; or you can purchase a length of straight tubing and the required fittings and make the assembly on your own.

If you decide to make your own, choose mild steel tubing. Don't try to flare stainless steel yourself, since this material is harder to manipulate and is much more prone to splitting during the flaring process. If you want stainless, buy it in ready-to-install form. If the tubing fabricators don't have what you need in stock, you can send them your original tube for duplication in stainless.

The common (and cheap) flaring tools available at hardware stores and auto parts stores often produce poor quality flares, especially in unskilled hands.

Using commonly available hand-flaring tools, it's relatively easy to create a single flare

Common tubing sizes include 3/16" (the most common for brake lines), 1/4" (for some brake lines), 3/8" (for many fuel and power steering lines) and 1/2" (for larger-volume fuel delivery where needed).

Types of Flares

- Single flare (45 degree is common; 37-degree flares are designed to mate to –AN flare seats)
- Double flare
- Convex "bubble" flare (this flare features a 45-degree under-head and a convex bubble, or "ball" flare at the tip face
- Convex ISO flare (this flare features a flat under-head surface and a slightly convex flare at the tip face)

Cutting the Tube

Never use a hacksaw or chop saw to cut a tube. You'll need a square cut (in order to leave the same amount of material around the perimeter for the flare). Always use a tubing cutter. This is a tool that features an adjustable jaw (to accommodate various tube diameters, and to increase the depth of the cut as the job is performed). Position the tube onto the cutter's double rollers (placing the tube in the center of the rollers registers the tube) and adjust the cutter by rotating the tool's knob (or T-handle as the case may be) until the cutter makes contact. Begin to gradually increase cutter pressure (by turning the knob clockwise) as you hand-rotate the tube. Do this bit by bit, making sure that you're achieving a cut around the entire circumference of the tube. Continue to rotate the tube, apply more cut pressure, rotate the tube, apply more cut pressure, etc. until the tube separates.

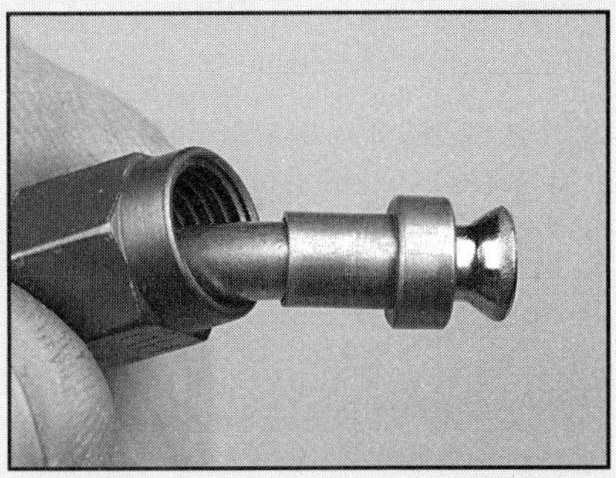

1. A single flare features a concave tube tip and angled underhead. This flare is performed in a single step, with no rolling of the tube edge. It's a simple bell. A 37-degree single flare works with various applications, such as use with –AN tube sleeves and nuts.

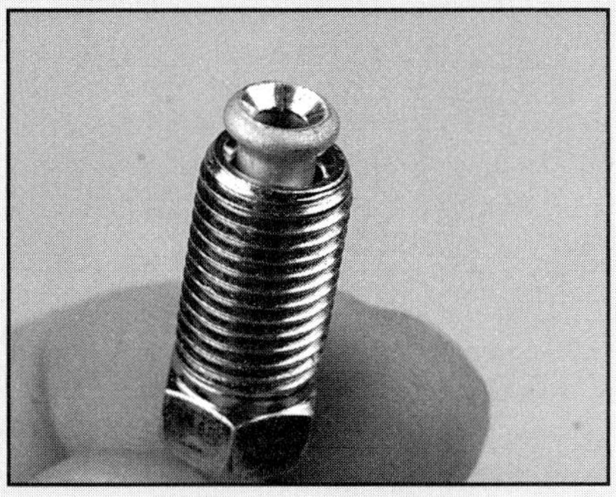

2. A double flare features a concave tube tip and a slightly angled underhead. This view also shows the chamfered concave seat area at the tube tip.

3. British bubble flare. Note the convex bubble tip and the angled under-head. The fitting tip features a chamfered seat that accepts the angled under-head.

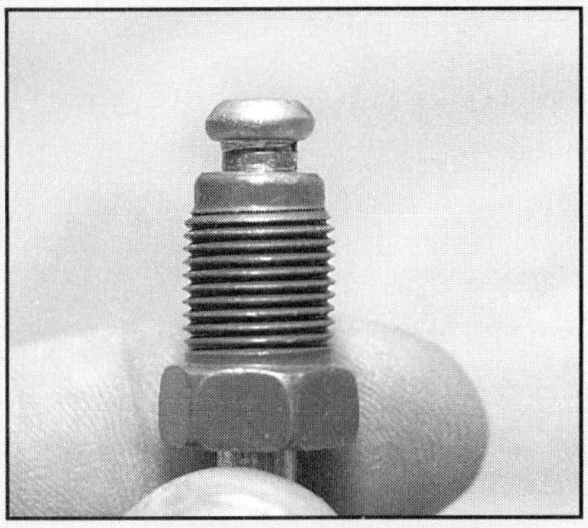

4. Convex ISO flare. This features a bubble flare with a flat underhead.

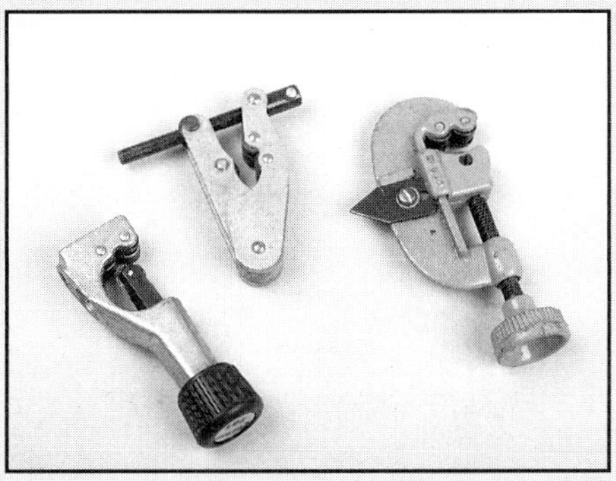

5. Tubing cutters of various designs are available, but they all accomplish the same task. The cutter on the left is good for tight spaces. The cutter on the right is the most robust and durable. The T-handle cutter in the center works, but it's a cheapie. I've detailed the cutting job in the double-flare steps here.

6. Here's a really good single flaring tool (this one is from Rigid). It's only designed to do single flares, but it does a great job at that task.

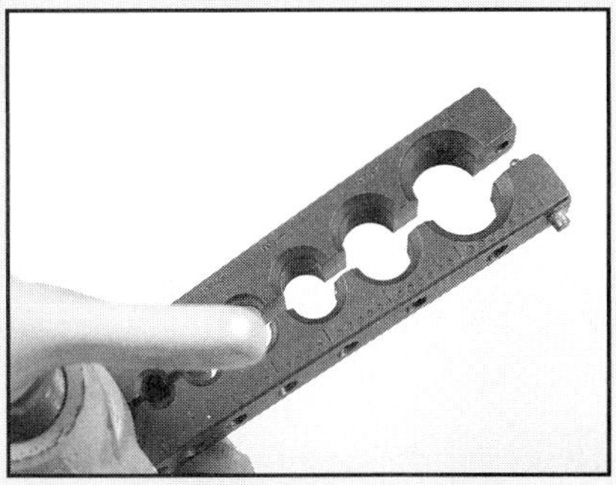

7. The flaring bar hinges apart to allow insertion of the tube. Once the tool head is aligned with the appropriate-diameter tube hole and tightened, the jaws secure the tube in place.

8. This single-flaring tool features a chamfered driver that isn't stationary. The driver actually rotates as the T-handle is turned and runs eccentrically, sweeping around the tube edge to create a perfectly round seat.

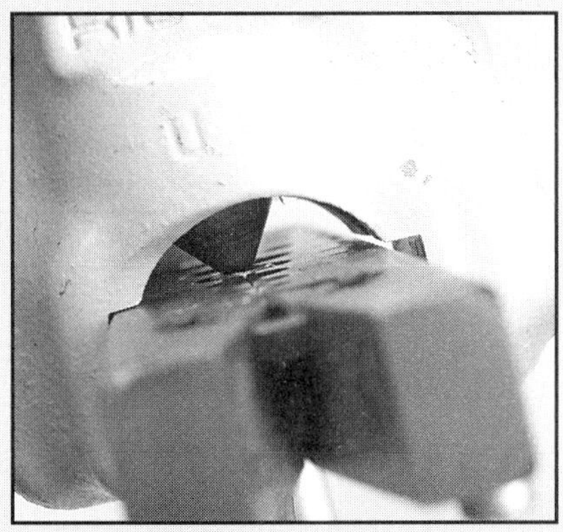

9. Once the seat is created, this tool's handle breaks free to prevent overtightening (overtightening could cause the tube to split).

10. A finished single flare on a 5/16" aluminum fuel line tube.

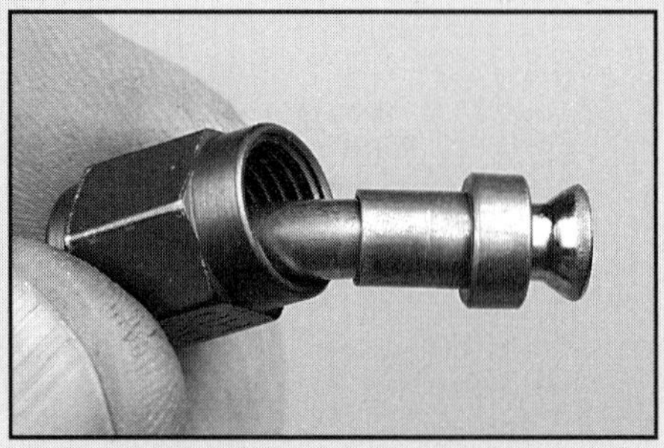

11. Here's a 3/16" tube with a single flare, fitted with a –3 AN tube sleeve and tube nut.

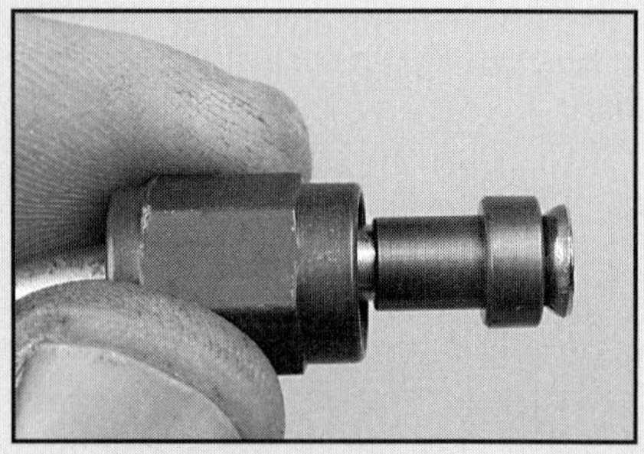

12. The single flare underhead chamfer is captured by the tube sleeve's convex chamfer.

13. The concave single flare of this tube seats against the 37-degree convex seat on this nitrous oxide injector fitting.

Creating a Double Flare

These instructions are based on the use of a commonly-available double-flaring tool kit and the use of thinwall steel tubing, aluminum or 0.040"-wall soft copper tubing.

Cut the tubing end squarely using a tubing cutter. Be sure to deburr both the O.D. and I.D. of the cut. When deburring the I.D., you should create a slight chamfer. Be sure to eliminate the sharp outer edge.

If the tube currently features any bends, be sure to slip the appropriate-size threaded fitting onto the tube before flaring. If you forget to do this, you'll be forced to cut the flare off and start over.

Insert the tube into the appropriate-sized hole in the flaring bar (3/16" tube into the hole marked 3/16", etc.). Position the tube in the flaring bar so that the cut end extends above the flaring bar to match the thickness of the die adapter's head thickness (you can lay the die upside down with its head resting flat on top of the flaring bar, next to the tube, then adjust the exposed height of the tube to match the thickness of the die head).

Tighten the wing nuts on the flaring bar, starting with the wing nut closest to the tube. Tighten both wingnuts securely to prevent the tube from being able to move.

Insert the die stem into the exposed tube.

Position the yoke tool over the flaring bar (so that the yoke's jaws contact the underside of the flaring bar and the yoke's swivel flaring driver tip contacts the center dimple of the die). Tighten the yoke's driver, using the yoke's T-handle, until the pointed swivel rests in the die's dimple.

Turn the T-handle clockwise (tightening) until the flat underside of the die rests flush on the flaring bar surface. This forms the tube end into a "bell" shape.

Next, loosen and remove the yoke tool. Remove the die adapter from the tube. Reinstall the yoke tool, engaging the yoke's driver against the tube (the point of the driver must contact the center of the tube hole). Continue to tighten the yoke, driving the pointed swivel driver against the tube's bell. This will force the convex face of the tube's bell to fold inward, creating a double-wall at the flare.

Remove the yoke and remove the tube from the flaring bar. Inspect the flare for symmetry (verify that the flare is round and consistent, with no off-center or crooked angles).

If the tube is still straight, be sure to slip the threaded fitting onto the tube, and run the fitting up against the new flare before bending the tube. Always be aware of your fittings. You can't slip a fitting over a flare or past tight bends, so always think ahead with regard to the fittings.

Once your tube is finished (flared at both ends, fittings in place and bent to shape), always blow the tube clean to remove any possible contaminants before placing the tube into service.

Note: Using this type of double-flaring tool is not always easy. It is very common to create sloppy, off-center flares at first. Most people require practice to become proficient, so my advice is to practice on a bunch of scrap tubes first until you get the hang it this procedure. Personally, I despise these cheap flaring tools. Unfortunately, this is the type most folks are forced to use, since this is the only type that's readily available and affordable. A professional shop that performs tube flaring on a regular basis can justify spending hundreds of dollars on a hydraulic flaring tool, but the hobbyist isn't likely to spend that kind of dough if he only needs to flare a couple of tubes per year.

The best advice is to avoid performing your own flares altogether. Go to the auto parts store, and buy lengths of finished tubing (pick the diameter needed, already flared and equipped with the appropriate size threaded fittings for your application). If you need to make long runs, you can join two lengths together using female/female union couplers. That way, you're only required to make your bends, without the hassle of trying to achieve perfect double flares.

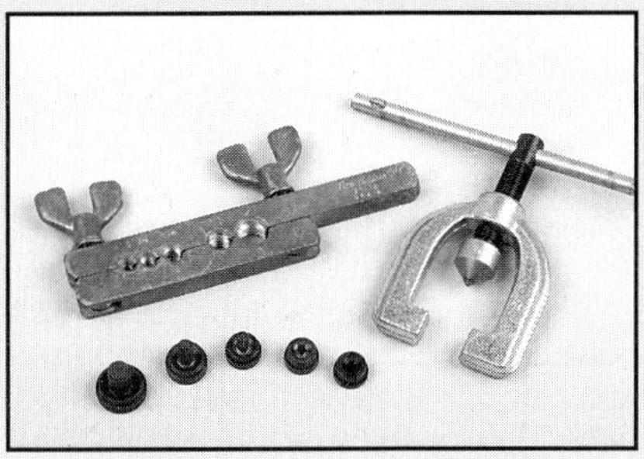

1. Typical double flaring tool set. These things are finicky to use, but unless you pony up for a pro tool, this is what you'll deal with.

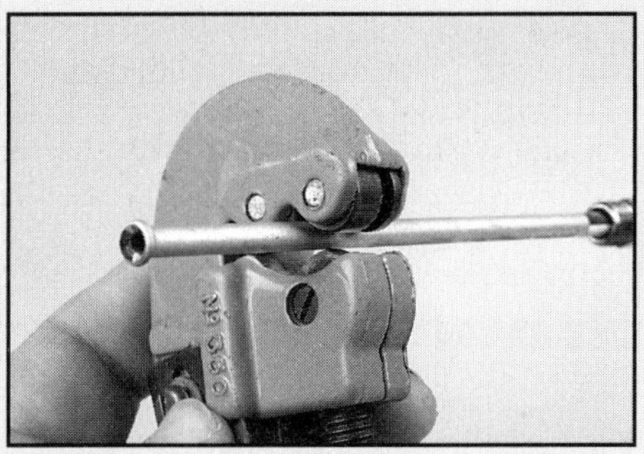

2. Position the tube in the cutter, with the tube resting against the double rollers (seen here above this tube). The cutter (seen here at the bottom of the tube) drives into the tube as the tool's handle is tightened.

3. Continually rotate the tube as the cutter applies pressure. There's no need to rush this. Simply tighten the handle a bit (not too severely) and rotate the tube to apply the cut around the tube circumference.

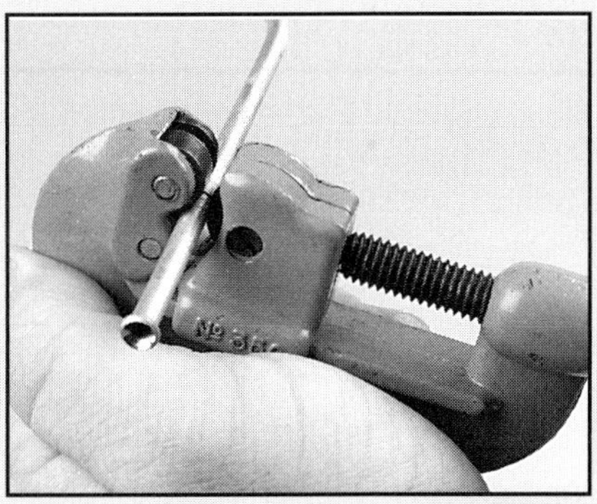

4. In gradual succession, continue to rotate the tube as you apply more cutting pressure. Continue this until the tube is cut completely through.

5. This tube has just been cut. Notice that the material has been compressed into the I.D. This must be cleaned out by deburring.

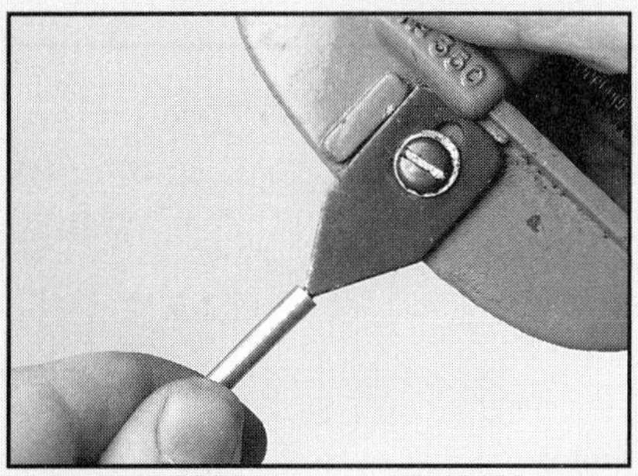

6. Many tubing cutters feature a built-in tube deburring knife. Insert the tip of the tool into the tube and rotate. Be sure to make full turns. Don't concentrate in one area, since this might result in egg-shaping the hole.

Creating A Double Flare (cont.)

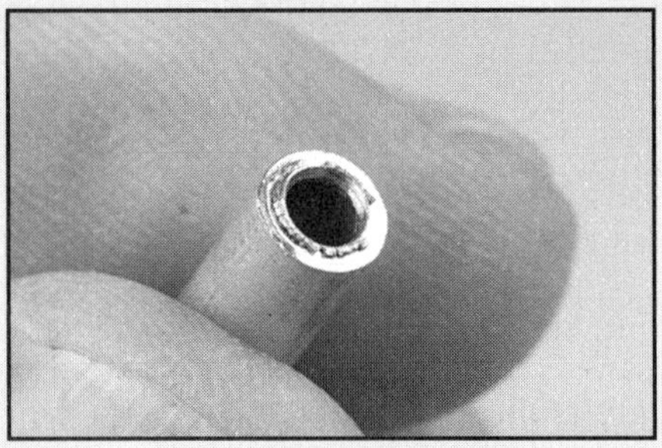

7. You can see how the hole has been opened after only one sweep with the deburring tool. Continue until you remove the entire burr. You want to create a slight chamfer as well. The object is to remove any restriction, and to remove all internal and external sharp edges.

8. This is a very nifty way to deburr the tube I.D. A chamfered pilot drill inserts into the hole, and its chamfer then quickly removed the burr and provides a smooth female chamfer in one quick pass.

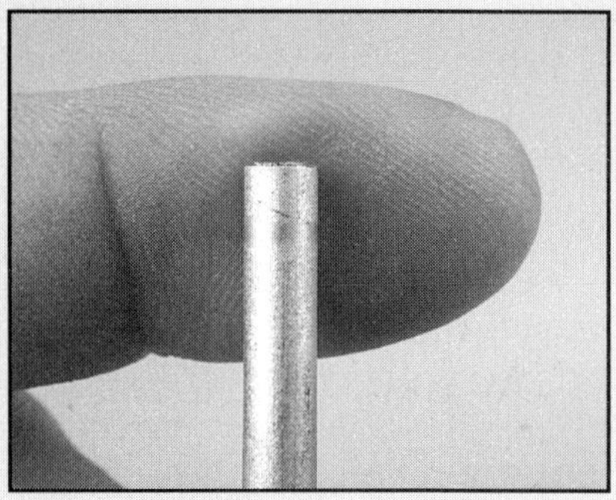

9. Make sure that the tube cut is square. Some cheap tubing cutters have a tendency to cut the tube at an angle. If the end isn't square, you won't be able to achieve a usable flare.

10. Use a fine metal file to remove the sharp edge around the O.D.

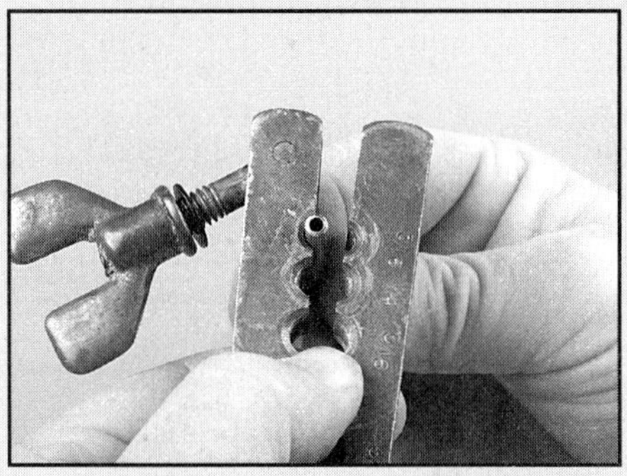

11. Separate the flaring bar's jaws and insert the tube into the appropriate-diameter hole. Remember to slip your threaded fitting onto the tube before flaring!

12. Adjust the exposed tube tip height to match the flaring die that will be used for this diameter tube. The tube should be exposed above the flaring bar surface as seen here, matching the height of the die's larger-diameter shoulder.

13. With the flaring bar securely tightened (make sure it's tight, to prevent the tube from being pushed out of the flaring bar during flare forming), insert the die stem into the I.D. of the tube.

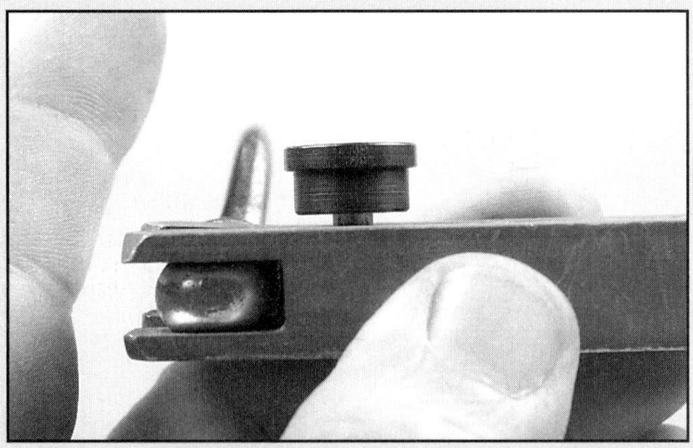

14. Here, the die is inserted into the tube and is ready to be driven into the flaring bar.

15. The yoke is positioned on the flaring bar and the chamfered pilot driver tip is engaged into the die's centering dimple.

16. Using the yoke's T-handle, tighten until the die is flush with the flaring bar surface.

17. The first step of the double flaring process is finished. With the die removed, you can see the bubble flare that has been created (by the way, if what you needed was a British bubble flare, you're done at this point).

18. To continue creating a double flare, reinstall the yoke, without the die. Engage the driver point into the tube I.D.

19. Carefully tighten, forcing the driver into the tube. This will fold the convex bubble down, forcing the metal to fold over, creating a concave seat. The tricky part is to avoid overtightening, which will smash the metal and split the edge.

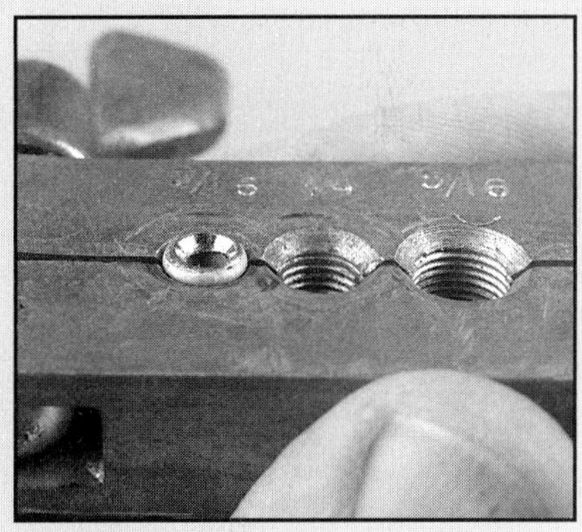

20. With the yoke removed, we see a finished double flare.

21. The finished product. The small scars from the flaring bar's jaws won't hurt anything, but you can carefully buff them down if you wish.

22. Hopefully, you remembered to install the fitting before you flared the tube. The fitting should mate to the underhead uniformly around the circumference.

LINE FITTINGS

Thread diameter and pitch will range widely according to requirements for connection. Line fittings all feature straight thread, since all sealing occurs at the tube-to-connection seats. Typical inch fitting thread sizes include 3/8" x 24, 7/16" x 24, 1/2" x 20 and 9/16" x 18. Common metric fitting sizes include 12mm–14mm, with thread pitches ranging from 1.0, 1.25 and 1.5.

TUBE BULGES

Tubing ends designed to accept a slip-on hose may feature one of three "bulge" designs (a radiused bulge protrusion around the perimeter of the tube). An end bulge may be created at the very tip of the tube. This simply serves as a larger-than-tube outer diameter area to provide an interference fit for the hose. Once the hose is installed onto the tube, a hose clamp is installed and tightened on the smooth tube area inboard of the bulge.

A "hose bulge" is created slightly inboard of the tube tip (leaving a short section of tube unchanged before the bulge), serving the same purpose...to provide a tight sealing fit for the hose. The "stopper bulge" is created far inboard from the tube tip...maybe 1 1/2" or so, serving as a stopping point for the end of the hose. If a stopper bulge is featured, the hose is not intended to slip over that bulge. Rather, the installation depth of the hose is stopped by this bulge.

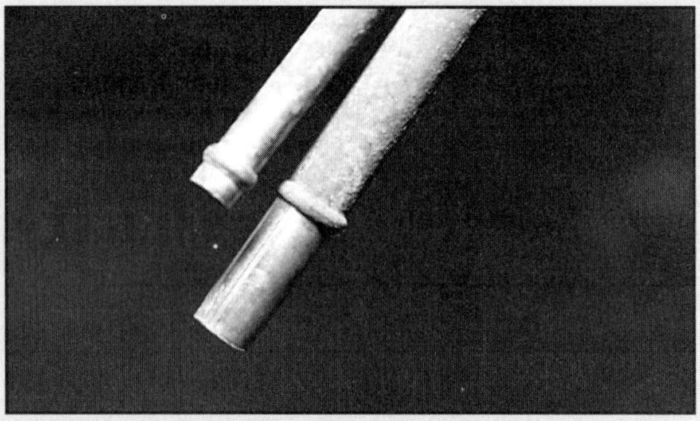

The extended-shoulder hose bulge (upper) serves to retain the hose. Slip the hose on about an inch and a half beyond the bulge and install a clamp. The stopper bulge (lower tube here) serves to stop the hose. Install the hose until the hose contacts this bulge, then install a clamp at about the halfway point between the bulge and tube tip.

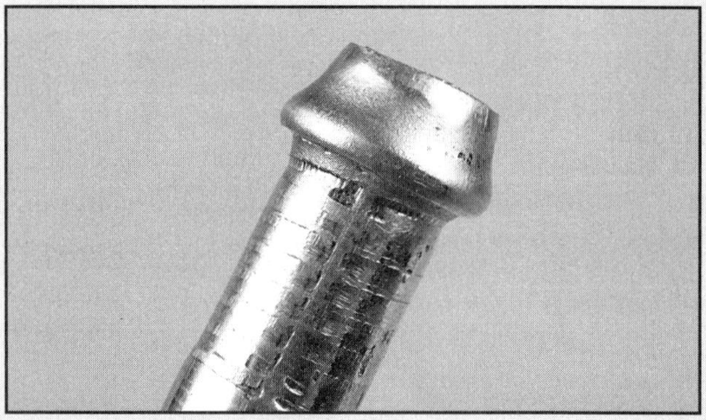

The end-bulge on this tube is designed to offer better hose retainment, increasing the hose interference fit.

SOURCES

FASTENER SOURCES

A&A Manufacturing
19033 174th Ave.
Spring Lake, MI 49456
616-846-1730
www.aa-mfg.com

AEBS
8270 Miramar Rd.
San Diego, CA 92126
858-693-3200
www.aebsracing.com

A1 Technologies
7022 Alondra Blvd.
Paramount, CA 90723-3926
562-408-1808
www.a1technologies.com

AbboTec
101 Canyon Ct.
Weatherford, TX 76087
817-441-1570

ARP
1863 Eastman Ave.
Ventura, CA 93003
800-826-3045
www.arp-bolts.com

Bolt Depot.Com
866-337-9888
www.boltdepot.com

Carrillo Industries
990 Calle Amanecer
San Clemente, CA 92673-6211
949-498-1800
www.carrilloind.com

Chrome Hardware Supply
3838 E. Grove St.
Phoenix, AR 85040
888-629-2476
www.chromehardwaresupply.com

Competition Cams
3406 Democrat Rd.
Memphis, TN 38118
800-999-0853
www.compcams.com

Crane Cams
530 Fentress Blvd.
Daytona Beach, FL 32114
386-252-1151
www.cranecams.com

Crower Cams & Equipment
6180 Business Center Ct.
San Diego, CA 92154-5604
619-661-6477
www.crower.com

Dorman Products
R&B Colmar
3400 E. Walnut St.
Colmar, PA 18915

Emhart Technologies
Automotive Division
P.O. Box 868
Mt. Clemens, MI 48046
586-949-0440

Fastener Specialties
1005 Callowhill Rd.
Perkasie, PA 18944
215-822-7201

The Flight Shop
P.O. Box 602
Brigham City, UT 84302
435-723-3469
www.theflightshop.com

Hartwell Dzus
950 S. Richfield Rd.
Placentia, CA 92870
800-441-0327
dzusracing@hartwell-dzus.com

Manley Performance
1960 Swarthmore Ave.
Lakewood, NJ 08701
800-526-1362
www.manleyperformance.com

Maryland Metrics
P.O. Box 261
Owings Mills, MD 21215
800-638-1830
www.mdmetric.com

McMaster Carr Supply Co.
200 Aurora Rd.
Aurora, OH 44202
330-995-9555
www.mcmaster.com

Milodon
2250 Agate Ct.
Simi Valley, CA 93065
805-577-5950

Moroso Performance
80 Carter Dr.
Guilford, CT 06437
203-453-6571
www.moroso.com

Mr. Gasket
10601 Memphis Ave. Ut 12
Cleveland, OH 44144
216-688-8300

MS Performance
13928 Balboa Blvd.
Sylmar, CA 91342
818-833-9095
www.msaerospace.com

Pioneer
5184 Pioneer Rd.
Meridian, MS 39301
601-483-5211

Protex Fasteners, Ltd.
Arrow Rd.
Redditch Worcs B98 8PA
United Kingdom
+44 1527 63231
www.protex.com

Specialty Fasteners
1537 W. McKinley St. Ste 18
Azusa, CA 91702-3265
626-969-6789

SPS Technologies
301 Highland Ave.
Jenkintown, PA 19046
215-572-3308

Stage 8 Locking Fasteners
64 Louise St.
San Rafael, CA 94901
800-843-7836
www.stage8.com

THREAD CHEMICAL SOURCES

Fel-Pro
7450 McCormick Blvd.
Skokie, IL 60076-4046
708-674-7700

Loctite Corp.
1001 Trout Brook Crossing
Rocky Hill, CT 06067-3910
203-571-5100

Silver Seal Products
19224 Allen Rd.
Trenton, MI 48183-1050
313-479-2255
800-521-2936

TORQUE WRENCH SUPPLIERS/REPAIR FACILITIES

Angle Repair Service Inc.
175 Angle Dr.
Beckley, WV 25801
304-253-5729
www.anglerepair.com

Central Tools Inc.
456 Wellington Ave.
Cranston, RI 02910
401-467-8211
800-866-5287

Gauge Repair Service
365 Van Ness Way
Torrance, CA 90501
310-212-0912

K-D Tools
805 Estelle Dr.
Lancaster, PA 17604-3767
800-866-5753
717-898-6571

Mac Tools
4635 Hilton Corporate Drive
Columbus, OH 43232
614-755-7117

Snap-On Inc.
2801 80TH ST.
Kenosha, WI 53141-1410
414-656-5530

TORQUE PLATES

B-H-J Products, Inc.
37530 Enterprise Ct.
Newark, CA 94560
510-797-6780
www.bhjinc.com

Goodson Shop Supplies
P.O. Box 847
Winona, MN 55987-0847
800-533-8010
www.goodson.com

Peterson Machine Tool
5425 Antioch Dr.
Shawnee Mission, KS 66202
800-255-6308

Sunnen Products Co.
7910 Manchester Ave.
St. Louis, MO 63143
800-325-3670
www.sunnen.com

TAPS & DIES

McMaster-Carr Supply Co.
330-995-5500
www.mcmaster.com

Grainger Industrial Supply
888-361-8649
www.grainger.com

Maryland Metrics
800-638-1830
www.mdmetric.com

SPHERICAL BEARINGS & RODS ENDS

Alinabal
28 Woodmont Rd.
Milford, CT 06460
800-254-6763
www.alinabal.com

Aurora Bearing Co.
970 S. Lake St.
Aurora, IL 60506-5954
630-859-2030
www.aurorabearing.com

F.K. Bearings
865 W. Queen St.
Southington, CT 06489
800-662-4003
www.fkrodends.com

Fluro Metric Rod Ends
Hoerr Racing Products
4802 Competition Lane
Peoria, IL 61615
866-851-7223
www.fluro.de

National Rod Ends
2425 Waters Dr.
Mendota Heights, MN 55120
800-233-6213
www.nationalrodends.com

QA1 Motorsports
800-721-7761; 952-985-5675
www.qa1.net

Radial Bearing Corp.
21 Taylor St.
Danbury, CT 06810
800-377-1937
www.radialbearing.thomasregister.com

Rod End Supply
800-284-2902
www.rodendsupply.com

RIVETS & RIVET GUNS

Aircraft Spruce & Specialty
225 Airport Circle
Corona, CA 92880
877-477-7823
www.aircraftspruce.com

Emhart Technologies
Automotive Division
P.O. Box 868
Mt. Clemens, MI 48046
586-949-0440
www.emhart.com

Gage Bilt Rivet Tools
14500 Barber
Warren, MI 48088
800-521-1363
www.gagebilt.com

Hanson Rivet & Supply
13241 Weidner St.
Pacoima, CA 91331
800-777-4838
www.hansonrivet.com

Marson Alcoa Fastening Systems
44 Campanelli Pkwy
Stoughten, MA 02072
800-343-3633
www.marsoncorp.com

Specialty Fasteners
1537 W. McKinley St., Suite 18
Azusa, CA 91702-3265
626-969-6789

Speedbear Fasteners
103 Enlow Rd.
Imperial, PA 15126
800-433-3444

SRI (Supplies for Racing & Industry)
801 B Performance Rd.
Mooresville, NC 28115
888-624-6611
www.sri-supplies.com

Textron Fastening Systems
714-545-5511
www.textronfasteningsystems.com

CLECOS

Airparts, Inc.
2400 Merriam Lane
Kansas City, KS 66106
800-800-3229
www.airpartsinc.com

Cokid Supply Co.
4320 Morning Sun Ave. #24
Colorado Springs, CO 80918
www.cokid.com

Eastwood Co.
800-343-9353
www.eastwoodco.com

Pegasus Auto Racing Supplies
2475 S. 179th St.
New Berlin, WI 53146
800-688-6946
www.pegasusautoracing.com

Quickcar Racing Products
170 Business Park Dr.
Lebanon, TN 37090
800-997-7333; 615-453-2120
www.quickcar.net

The Yard Aircraft Supply
725 E. central
Wichita, KS 67202
800-888-8991
www.yardstore.com

CLAMPS

Adel Wiggins Group
(Adel clamps)
5000 Triggs St.
Los Angeles, CA 90022
323-269-9181
www.adelwiggins.com

Billet Specialties
500 Shawmut Ave.
La Grange, IL 60526
708-588-0505
www.billetspecialties.com

Breeze Industrial Products Corp.
(worm-drive and T-bolt clamps)
3582 Tunnelton Rd.
Saltsburg, PA 15681
724-639-3571
www.breezeclamps.com

Clampco
(worm-drive, T-bolt, V-band and spring-loaded band clamps)
1743 Wall Rd.
Wadsworth, OH 44281
330-336-8850
www.clampco.com

Earl's Performance Products
(div. of Holley Performance Products)
1801 Russellville Rd.
Bowling Green, KY 42101
270-782-2900
www.holley.com

Ideal Tridon
3200 Parker Dr.
St. Augustine, FL 32084-0897
904-829-1000
www.idealclamps.com

Murray Corp.
260 Schilling Circle
Hunt Valley, MD 21031
410-771-0380
www.murraycorp.com

Russell
(division of Edelbrock)
2700 California St.
Torrance, CA 90503
310-781-2222
www.edelbrock.com

PERFORMANCE HOSE ASSEMBLIES

A-1 Performance Plumbing
1927-2 Stout Dr.
Warminster, PA 18974
800-669-8442; 215-675-8442
www.a1racing.com

Aeroquip Performance Products
(14615 Lone Oak Rd.
Eden Prairie, MN 55344
419-877-9664

BMRS
(reinforced hose, hose ends, fittings, adapters)
Concord, NC
704-793-4319

Brake Quip
1470 Amherst Rd.
Knoxville, TN 37909
865-251-9193
www.brakequip.com

Classic Tube
(brake and fuel hoses and lines)
800-882-3711
www.classictube.com

Daily Engineering
(dry sump pumps)
42095 Zevo Dr. Unit 7
Tenecula, CA 92590
951-296-2110
www.dailyengineering.com

Earl's Performance Products
(reinforced hose, hose ends, fittings, adapters, coolers)
Holley Performance Products
P.O. Box 10360
Bowling Green, KY 42102
270-782-2900
www.holley.com

Earl's Indy
302 Gasoline Alley
Indianapolis, IN 46222-3967
800-331-4639
www.earlsindy.com

Fine Lines
127 Hartman Rd.
Wadsworth, OH 44281
800-778-8237; 330-335-5000
www.finelinesinc.com

Fragola Performance Systems (reinforced hose, hose ends, fittings, adapters)
888 W. Queen St.
Southington, CT 06489
866-337-2739

Goodridge USA
529 Van Ness
Torrance, CA 90501
310-533-1924

Goodyear Tire & Rubber
1376 Tech Way Dr., Bldg. 72
Akron, OH 44316
330-796-4589

Jiffy-Tite
4437 Walden Ave.
Lancaster, NY 14086
888-605-7788
www.jiffy-tite.com

Russell
P.O. Box 2936
Torrance, CA 90509
310-781-2222
www.russellperformance.com

Seals-It
164 Maple St.
Ellington, CT 06029
860-979-0060
www.sealsit.com

Stainless Steel Brakes
11470 Main Rd.
Clarence, NY 14031
800-448-7722
www.ssbrakes.com

XRP
5630 Imperial Hwy
South Gate, CA 90280
562-861-4765

PLUMBING TOOLS

Aeroquip Performance Products
(14615 Lone Oak Rd.
Eden Prairie, MN 55344
419-877-9664
www.aeroquip.com

Koul Tools
405 Jones Dr.
Lake Havasu City, AZ 86406
928-854-6706
www.koulttools.com

ABOUT THE AUTHOR

Mike Mavrigian has written a broad range of automotive technical articles for the past 26 years involving a variety of industry areas, including engine rebuilding, performance and race engine building, tires, wheels, brakes, wheel alignment, suspension, ride control, steering, drivetrain, engine diagnostics, air compressors, pneumatic tools, metallurgical issues relating to engine building/rebuilding, vibrational stress relief and cryogenic treatment of metals, paint and body, climate control, electric winches, trailer towing/hitch systems, SUV upgrades, sport/compact upgrades, race car preparation, street rod construction and automotive-related threaded fasteners/clamping forces.

In addition to his magazine work, Mike has authored consumer-oriented books, including such titles as *Performance Wheels & Tires* (HPBooks), *Brake Systems* (HPBooks), *Rebuilding the Gen V/VI Chevrolet Big Block Engines* (HPBooks), *Rebuilding the Small-block Chevrolet LT1/LT4 Engine* and several other books dealing with general automotive repair. Mike also writes each issue of *Modern Tire Dealer* magazine's popular annual *Performance Handbook*, and has authored technical training manuals for McGraw-Hill and a host of automotive parts and tool manufacturing clients.

Mike serves as editor of *Precision Engine* magazine, a technical publication focused on professional performance and racing engines. Mike has authored automotive technical articles for a host of other industry publications, including *Modern Tire Dealer*, *Motor Age*, *Autobody Repair News*, *Truck & SUV Performance*, *Aftermarket Business*, *SEMA News*, *Jobber Retailer* and *Parts Professional* magazines. Consumer magazine work includes *Mopar Action*, *Street Rodder* and *Sport Compact Car* magazines.

Corporate Project Vehicle Programs—Mike owns and operates Birchwood Automotive Group, based in Creston, Ohio, which prepares specialty vehicles and engines for a range of automotive industry manufacturers and other clients.

Birchwood's client list includes Porsche Cars North America, *Field & Stream* magazine, *Outdoor Life* magazine, Holley Performance Products, BFGoodrich, Michelin and AERA, to name but a few. In addition to executing project vehicle programs for its corporate clients, Birchwood produces street rods, muscle cars, custom cars and sport-compact performance cars for a range of consumer customers.

Motorsports Programs—The Birchwood operation also maintains an 18-person experienced and skilled motorsports team that specializes in endurance road racing programs. The team has competed in over fourteen 24-hour and 12-hour endurance races at tracks including Nelson Ledges, Garrettsville, OH; Mid-Ohio Sports Car Course, Lexington, OH; Watkins Glen International, Watkins Glen, NY; Summit Point Raceway, Summit Point, VA and Moroso Motorsports Park in West Palm Beach, FL, where the team garnered a class win during a 1999 24-hour race. In 2005, Mike was contracted by Porsche Cars North America to facilitate Porsche's Speed Record Run event at Talladega Superspeedway, involving the firm's Carrera GT cars, driven by pro driver David Donohue and celebrity Jay Leno.

GENERAL MOTORS

Big-Block Chevy Engine Buildups: 1-55788-484-6/HP1484
Big-Block Chevy Performance: 1-55788-216-9/HP1216
Camaro Performance: 1-55788-057-3/HP1057
Camaro Restoration Handbook ('67–'81): 0-89586-375-8/HP1375
Chevelle/El Camino Handbook: 1-55788-428-5/HP1428
How to Customize Your Chevy Silverado/GMC Sierra Truck: 978-1-55788-526-5/HP1526
How to Hot Rod Big-Block Chevys: 0-912656-04-2/HP104
How to Hot Rod Small-Block Chevys: 0-912656-06-9/HP106
How to Rebuild Small-Block Chevy LT-1/LT-4: 1-55788-393-9/HP1393
John Lingenfelter: Modify Small-Block Chevy: 1-55788-238-X/HP1238
LS1/LS6 Small-Block Chevy Performance: 1-55788-407-2/HP1407
Powerglide Transmission Handbook:1-55788-355-6/HP1355
Rebuild Big-Block Chevy Engines: 0-89586-175-5/HP1175
Rebuild Small-Block Chevy Engines: 1-55788-029-8/HP1029
Small-Block Chevy Engine Buildups: 1-55788-400-5/HP1400
Small-Block Chevy Performance: 1-55788-253-3/HP1253
Turbo Hydramatic 350 Handbook: 0-89586-051-1/HP1051

FORD

Ford Windsor Small-Block Performance: 1-55788-323-8/HP1323
How to Customize Your Ford F-150 Truck: 978-1-55788-529-6/HP1529
Mustang Performance (Engines): 1-55788-193-6/HP1193
Mustang Performance 2 (Chassis): 1-55788-202-9/HP1202
Mustang Restoration Handbook ('64–'70): 0-89586-402-9/HP1402
Rebuild Big-Block Ford Engines: 0-89586-070-8/HP1070
Rebuild Ford V-8 Engines: 0-89586-036-8/HP1036
Rebuild Small-Block Ford Engines: 0-912656-89-1/HP189

MOPAR

Big-Block Mopar Performance: 1-55788-302-5/HP1302
How to Hot Rod Small-Block Mopar Engine Revised: 1-55788-405-6/HP1405
How to Maintain & Repair Your Jeep: 1-55788-371-8/HP1371
How to Modify Your Jeep Chassis/Suspension for Offroad: 1-55788-424/HP1424
How to Modify Your Mopar Magnum V8: 1-55788-473-0/HP1473
How to Rebuild and Modify Chrysler 426 Hemi Engines: 978-1-55788-525-8/HP1525
How to Rebuild Your Mopar Magnum V8: 1-55788-431-5/HP1431
Rebuild Big-Block Mopar Engines: 1-55788-190-1/HP1190
Rebuild Small-Block Mopar Engines: 0-89586-128-3/HP1128
The Mopar Six-Pack Engine Handbook: 978-1-55788-528-9/HP1528
Torqueflite A-727 Transmission Handbook: 1-55788-399-8/HP1399

IMPORTS

Baja Bugs & Buggies: 0-89586-186-0/HP1186
Honda/Acura Engine Performance: 1-55788-384-X/HP1384
How to Hot Rod VW Engines: 0-912656-03-4/HP103
Mitsubishi & Diamond Star Performance Tuning: 978-1-55788-496-1/HP496
Porsche 911 Performance: 1-55788-489-7/HP489
Rebuild Air-Cooled VW Engines: 0-89586-225-5/HP1225

HANDBOOKS

Automotive Detailing: 1-55788-288-6/HP1288
Auto Electrical Handbook: 0-89586-238-7/HP1238
Auto Math Handbook: 1-55788-020-4/HP1020
Automotive Paint Handbook: 1-55788-291-6/HP1291
Auto Upholstery & Interiors: 1-55788-265-7/HP1265
Classic Car Restorer's Handbook: 1-55788-194-4/HP1194
Engine Builder's Handbook: 1-55788-245-2/HP1245
Engine Cooling Systems: 978-1-55788-425-1/HP1425
Fiberglass & Other Composite Materials Rev.: 1-55788-498-6/HP1498
High Performance Fasteners & Plumbing: 978-1-55788-523-4/HP1523
Metal Fabricator's Handbook: 0-89586-870-9/HP1870
Paint & Body Handbook: 1-55788-082-4/HP1082
Performance Ignition Systems: 1-55788-306-8/HP1306
Pro Paint & Body: 1-55788-394-7/HP1394
Sheet Metal Handbook: 0-89586-757-5/HP1757
Welder's Handbook Revised: 978-1-55788-513-5/HP1513

INDUCTION

Holley 4150: 0-89586-047-3/HP1047
Holley Carbs, Manifolds & F.I.: 1-55788-052-2/HP1052
Rochester Carburetors: 0-89586-301-4/HP1301
Turbochargers: 0-89586-135-6/HP1135
Street Turbocharging: 1-55788-488-9/HP1488
Weber Carburetors: 0-89586-377-4/HP1377

RACING & CHASSIS

Chassis Engineering: 1-55788-055-7/HP1055
4Wheel & Off-Road's Chassis & Suspension: 1-55788-406-4/HP1406
Dirt Track Chassis & Suspension: 978-1-55788-511-1/HP1511
How to Make Your Car Handle: 0-912656-46-8/HP146
How to Build a Winning Drag Race Chassis & Suspension: 978-1-55788-462-6/HP1462
Racing Engine Builder's Handbook: 1-55788-492-7/HP1492
Stock Car Racing Engine Technology: 978-1-55788-506-7/HP1506
Stock Car Setup Secrets: 1-55788-401-3/HP1401

STREET RODS

How to Build a 1934–'35 Chevy Street Rod: 978-1-55788-514-1/HP1514
How to Build a 1935–'40 Ford Street Rod: 1-55788-493-5/HP1493
Street Rodder magazine's Chassis & Suspension Handbook: 1-55788-346-7/HP1346
Street Rodder's Handbook, Rev.: 1-55788-409-9/HP1409
Street Rodding Tips & Techniques: 978-155788-515-9/HP1515

ORDER YOUR COPY TODAY!
All books can be purchased at your favorite retail or online bookstore such as amazon.com and barnesandnoble.com (use ISBN number), or auto parts store (use HP part number). You can also order direct from HPBooks by calling toll-free at 800-788-6262, ext. 1, or at www.penguin.com.